Special Functions in Queuing Theory

And Related Stochastic Processes

Special Functions in Queuing Theory

And Related Stochastic Processes

H. M. SRIVASTAVA
Department of Mathematics
University of Victoria
Victoria, British Columbia, Canada

B. R. K. KASHYAP
Operations Management Group
Indian Institute of Management
Calcutta, India

1982

ACADEMIC PRESS
A Subsidiary of Harcourt Brace Jovanovich, Publishers

New York London
Paris San Diego San Francisco São Paulo Sydney Tokyo Toronto

ACADEMIC PRESS, INC.
111 Fifth Avenue, New York, New York 10003

United Kingdom Edition published by
ACADEMIC PRESS, INC. (LONDON) LTD.
24/28 Oval Road, London NW1 7DX

Library of Congress Cataloging in Publication Data

Srivastava, H. M.
Special functions in queuing theory.

Bibliography: p.
Includes indexes.
1. Queuing theory. 2. Functions, Special.
3. Stochastic processes. I. Kashyap, B. R. K.
II. Title.
T57.9.S69 519.8'2 81-17683
ISBN 0-12-660650-1 AACR2

(AMS) 1980 Subject Classifications
Primary: 60K25, 33A30
Secondary: 60G20, 33A40

PRINTED IN THE UNITED STATES OF AMERICA

82 83 84 85 9 8 7 6 5 4 3 2 1

To my parents, and to my wife, Rekha Srivastava
H. M. SRIVASTAVA

To the memory of my parents, and to my wife, Raj Kashyap
B. R. K. KASHYAP

Contents

CHAPTER 2 Poisson Queues II

CHAPTER 3 Queues with Variable Parameters

CHAPTER 4 Queues with Poisson Arrivals or Service

CHAPTER 5 Queues with General Arrival and General Service Distributions

CHAPTER 6 Miscellaneous Results

Preface

The aim of this book is to give comprehensive coverage of results in those queuing and related theories that incorporate members of the family of mathematical functions known as the special functions. We attempt to present systematically the most recent developments on the applications of a wide variety of special functions (such as beta, gamma, incomplete beta, incomplete gamma, Bessel, modified Bessel, Legendre, Whittaker, hypergeometric, confluent hypergeometric, generalized hypergeometric, G, and H) to different types of queuing systems and to allied stochastic processes. The book has been made self-contained by the inclusion of necessary background material in queuing theory (Preliminaries I) and in the theory of special functions involved (Preliminaries II). A valuable feature of the book is the extensive and up-to-date bibliography on the subject.

In Chapters 1 and 2 on Poisson queues we describe the uses of special functions in obtaining various results for queuing systems associated with Poisson arrivals and negative-exponential service times. Also considered are Poisson queues with priorities and those with bulk service or bulk arrivals. Chapter 3 deals with studies in queuing theory where special functions have been used and where the interarrival and service times both follow the negative-exponential distribution, the parameters of either or both of these distributions being variable. In Chapter 4 we describe work done by various researchers, including ourselves, on $M/G/1$ and $G/M/1$ queues. Chapter 5 presents a discussion of queuing systems where arrival and service mechanisms are both characterized by general distributions. The last chapter briefly outlines similar applications of special functions in some related fields, including population genetics, collision processes, inventories, and stochastic programming. Then follows a review of the recent work on probability distributions in which various authors have attempted to obtain more and more generalized distributions using special functions.

This book is written primarily for two seemingly diverse groups of research workers and graduate students: first, those queuing and probability theorists and mathematically oriented workers in operations research who are interested in learning how the special functions can provide closed-form solutions in their fields of study; and second, those working in the area of special functions who wish to develop their understanding of the applications of these functions to queues and other phenomena in business and other applications areas, that is, in several important areas beyond the traditional ones of theoretical and quantum mechanics and of mathematical physics.

Several persons have contributed significantly to this book, both directly and indirectly. Contribution of subject matter is duly acknowledged throughout the text and in the bibliography. We are grateful to various authors whom we have freely consulted and who often provided invaluable references and advice serving for the enrichment of the matter presented here. We are especially thankful to Professor B. W. Conolly, of the University of London, and Professor A. M. Mathai, of McGill University; it was in the course of discussions on the subject with the latter, several years ago, that this monograph was conceptualized.

The collaboration of the authors was initiated and a technical report (listed in the bibliography) was prepared during the second author's visit to the University of Victoria in July and August 1974 under a travel grant from the Ford Foundation. This collaboration has led to the present monograph, many chapters of which were written during parts of the academic years 1978–1979 and 1979–1980 while the second author was at the University of Victoria on leave from the Indian Institute of Management—Calcutta. Thanks are due to the Indian Institute of Management—Calcutta for granting leave to the second author (and for inviting the first author to the Institute in January and February 1976 and again in August 1980), and to the Natural Sciences and Engineering Research Council of Canada and the University of Victoria for providing financial assistance and other facilities for this project.

We also express our gratitude to Mrs. Georgina Smith and Miss Elaine Cumming for their painstaking efforts, devotion, and patience in the expert preparation of the typescript.

PRELIMINARIES I

Basic Concepts in Queuing Theory

I.1 Queuing Systems

Queuing (or waiting-line) theory is a branch of applied probability theory and serves to study service systems prone to congestion. The subject grew out of the pioneering work done in this field by Agner Krarup Erlang (1878–1929), a Danish telephone engineer, on the use of probability theory in solving telephone congestion problems [see Brockmeyer *et al.* (1948), Saaty (1957b), Gross and Harris (1974), Conolly (1975), Kleinrock (1975), and White *et al.* (1975)].

A queuing system may be described as one having a service facility at which units of some kind (generically called "customers") arrive for service and, whenever there are more units in the system than the service facility can handle simultaneously, a queue, or waiting line, develops. The waiting units take their turn for service according to a preassigned rule, and after service they leave the system. Thus, the input to the system consists of the customers demanding service, and the output is the serviced customers. A queuing system is characterized by the following:

(a) *The input process.* Let the customers arrive at instants $t_0, t_1, t_2, \ldots$; then the interarrival times are

$$u_r = t_r - t_{r-1} \qquad (r = 1, 2, \ldots). \tag{1}$$

The random variables u_r are, in general, assumed to be statistically independent, and their probability distribution $A(u)$ is called the interarrival time distribution, or simply the arrival distribution, or the input distribution. The customers may come from an infinite source, as in the case of telephone calls and as is assumed in many queuing studies, or from a finite source. They may arrive singly or in groups of fixed or varying sizes. The queuing system may have an upper limit on the number that can be admitted into the system, as in the case of finite waiting space.

(b) *The queue discipline.* This may be described as the rule determining the formation of the queue or queues and the manner in which a customer or customers are selected for service from those waiting. The most common queue discipline is "first come, first served," according to which the units enter service in order of their arrival. Other possibilities are random selection for service, a priority rule, or even the "last come, first served" rule. In the case of priority, we may have two classes, viz., the priority class and the nonpriority class, or we may have several priority classes representing different levels of priority. Further, we may have a preemptive priority discipline according to which a lower priority unit is taken out of service whenever a higher priority unit arrives, the service on the preempted unit resuming only when there are no higher priority units in the system. Contrary to this is the nonpreemptive, or the head-of-the-line, priority rule, which takes priorities into account only at the commencement of service, and once started, the service is continued until completion. It is customary to include under this heading queuing phenomena, like balking and reneging, depicting the behavior of the waiting customers. Customers are said to *balk* when, looking at the size of the queue, and estimating therefrom the time they may have to wait before service, they do not join the queue. After joining the queue, the customers are said to *renege* if they get fed up with waiting and leave the queue before service starts.

(c) *The service mechanism.* The time that elapses while a unit is being served is called its service time. The service times $v_1, v_2, \ldots$ of the successive units are assumed to be independent of each other and of the input distribution, and their probability distribution $B(v)$ is called the service time distribution or briefly service distribution. The specification of service mechanism includes the number of servers. Thus, we have single-server and multiserver systems (sometimes called single-channel and multichannel from telephone parlance). Again, in a bulk service system, service may be provided in batches of fixed or varying sizes.

As is common in probability theory, if we can write

$$dA(u) = a(u)\,du, \qquad dB(v) = b(v)\,dv, \tag{2}$$

then $a(u)$ and $b(v)$ are, respectively, the interarrival time and service time density functions. In this book we shall use $A(x)$ for the interarrival time density function for notational convenience. However, at most places the symbols have been defined as and when they occur, especially where confusion is likely.

I.2 Queue Parameters

For an understanding, and efficient management, of queuing systems, the following three queue parameters are generally investigated.

(1) *The queue length* is the number of units waiting in the queue, or present in the system. In the latter case it is sometimes called the system size. The determination of the probability distribution of this discrete random variable is important for the design of the system, e.g., for providing an extra server or a larger waiting space in the case of large queues.

(2) *The waiting time* is the time a customer has to wait in the queue before service. Sometimes, the service time is also included in the waiting time. Some authors, like Cox and Smith (1961), use the terms queuing time and waiting time according as the service time is excluded or included. This is a continuous variable and its probability distribution is important from the customer's point of view.

(3) *The busy period* starts with the arrival of a customer who finds the system empty and ends when the system next becomes empty. However, the initial busy period may start with any number in the system, and for multiserver systems we may define a busy period during which a given number k of servers are throughout busy [see, for example, the work of Chaudhry and Templeton (1973), described in Section 1.6]. The distribution of the duration of a busy period is important from the server's point of view.

I.3 Notation

We shall use Kendall's (1953) notation of using a triple $\cdot/\cdot/\cdot$ to specify a queuing system, in which the first two constituents are letters representing the form of input and service time distributions, respectively, while the third is a number denoting the number of servers. The following letters are used for signifying distributions.

(i) M (standing for Markov) for negative exponential interarrival or service time distribution:

$$A(u) = 1 - e^{-\lambda u} \qquad (0 \leqq u < \infty) \tag{1}$$

or

$$B(v) = 1 - e^{-\mu v} \qquad (0 \leqq v < \infty), \tag{2}$$

according as M occurs in the first or the second place in the triple. From (1), it follows that the mean interarrival time is

$$\int_0^\infty u\,dA(u) = \frac{1}{\lambda}. \tag{3}$$

Similarly, from (2), the mean service time is $1/\mu$. Alternatively, we may say that λ und μ are, respectively, the mean arrival and service rates per unit time. It can easily be seen, and is commonplace in elementary probability theory, that (1) is equivalent to saying that the number of arrivals in a given duration of time follows the Poisson distribution:

$$\text{Prob}\{n \text{ arrivals in time } t\} = (\lambda t)^n e^{-\lambda t}/n! \qquad (n = 0, 1, 2, \ldots). \tag{4}$$

That is why very often (1) is spoken of as Poisson arrivals. Similarly, (2) implies that the number of services rendered in a given time duration is Poisson distributed.

(ii) D (standing for deterministic) for constant interarrival or service times:

$$A(u) = \begin{cases} 0 & (u < 1/\lambda) \\ 1 & (u \geqq 1/\lambda) \end{cases} \tag{5}$$

or, equivalently, the density function

$$a(u) = \delta(u - 1/\lambda), \tag{6}$$

where $\delta(x)$ is the Dirac delta function. The case is similar for $B(v)$ if D is in the second place.

(iii) E_k for the k-Erlang distribution:

$$a(u) = \{\lambda^k/(k-1)!\}u^{k-1}e^{-\lambda u} \qquad (0 \leqq u < \infty); \tag{7}$$

similarly for $B(v)$. When $k = 1$, it reduces to the exponential distribution. As $k \to \infty$ so that k/λ is finite, we have the constant interarrival time case, as is seen below.

From (7), the Laplace transform of $a(u)$ is

$$\mathscr{L}[a(u)] = [\lambda/(\lambda + s)]^k \to e^{-\Lambda s}, \tag{8}$$

as $k \to \infty$ if we set $k/\lambda = \Lambda$, a finite quantity. But $e^{-\Lambda s}$ is the Laplace transform of the Dirac delta function $\delta(u - \Lambda)$, which completes the proof.

(iv) G for general or arbitrary interarrival or service time distribution. Some authors write GI for general independent input to stress independence.

Thus, for example, we write $M/E_k/1$ to denote a single-server queuing system with Poisson arrivals and k-Erlang service time distribution, and $GI/G/c$ for the most general case of general independent input, general service times, and c servers (channels).

Where desirable, we use the extended notation of Lee (1966), where, for example, we write $M/E_k/1:N$ to denote that N is the maximum allowable number in the system at any time.

I.4 Queue Equations

From the viewpoint of mathematical tractability, the negative exponential interarrival or service time distribution is the simplest and, fortunately, these distributions serve as a good fit in many practical problems. We give below a mathematical model for the simplest queuing system, viz., $M/M/1$. Let

$$(1) \qquad P_n(t) = \text{Prob}\{n \text{ units in system at time } t\}.$$

Now, with Poisson arrivals, we have from I.3(4)

$$(2) \quad \text{Prob}\{\text{one arrival during time interval } (t, t+\Delta t)\} = \lambda\,\Delta t\, e^{-\lambda \Delta t} = \lambda\,\Delta t + o(\Delta t),$$

$$(3) \qquad \text{Prob}\{\text{no arrival during } (t, t+\Delta t)\} = 1 - \lambda\,\Delta t + o(\Delta t),$$

$$(4) \qquad \text{Prob}\{\text{more than one arrival during } (t, t+\Delta t)\} = o(\Delta t),$$

where $o(\Delta t)$ stands for terms negligible in comparison to Δt. Similar equations, with μ in place of λ, hold for the number of services completed during $(t, t+\Delta t)$. Thus, considering the possible transitions during $(t, t+\Delta t)$, we have, using the multiplication and addition theorems of probability,

$$(5) \qquad P_n(t+\Delta t) = P_n(t)\{1 - \lambda\,\Delta t + o(\Delta t)\}\{1 - \mu\,\Delta t + o(\Delta t)\} + P_{n-1}(t)\{\lambda\,\Delta t + o(\Delta t)\} + P_{n+1}(t)\{\mu\,\Delta t + o(\Delta t)\},$$

whence, on taking limits as $\Delta t \to 0$,

$$(6) \quad \frac{d}{dt} P_n(t) = -(\lambda + \mu)P_n(t) + \lambda P_{n-1}(t) + \mu P_{n+1}(t) \qquad (n = 1, 2, \ldots).$$

It is easy to see that for $n = 0$ we have

$$(7) \qquad \frac{d}{dt} P_0(t) = -\lambda P_0(t) + \mu P_1(t).$$

Equations (6) and (7) are a system of differential-difference equations. Applying Laplace transforms of the form

$$(8) \quad \mathscr{L}[P_n(t)] = p_n(s) = \int_0^\infty e^{-st} P_n(t)\, dt \qquad [\text{Re}(s) \geqq 0;\ n = 0, 1, 2, \ldots],$$

the system can be reduced to pure difference equations, whereas using the generating function

$$P(z, t) = \sum_{n=0}^{\infty} z^n P_n(t), \tag{9}$$

it is reduced to a differential equation. Very often, we use both the transforms (8) and (9), and thus reduce the differential-difference equations (6) and (7) to an algebraic equation.

Most of the initial work on queuing theory was done under a steady-state assumption, under which the probabilities P_n are independent of time, and therefore the derivatives on the left-hand side of (6) and (7) vanish, thus giving rise to difference equations that are easier to solve. Also, most of the work was initially done for Poisson queues, i.e., queues with Poisson arrival and negative exponential service times. The negative exponential distribution has the important Markovian or "forgetfulness" property that if, for instance, the service times have a negative exponential distribution, then at any time the *residual* service time also has the same distribution. That is why in (6) and (7) we did not have to take into account the time since service on the unit being served began, which is called the elapsed service time. This is not the case with other distributions. However, for mathematical convenience, most of the other distributions have also been approached through the negative exponential.

I.5 Non-Poisson Queues

We give below some well-known methods of queuing theory by which general interarrival or service time distributions are approached.

(1) *Imbedded Markov chain technique.* Suppose we are considering an $M/G/1$ system. If we study the system only at the epochs when customers depart after service, then between any two such consecutive epochs the only possible transitions are customer arrivals for which the Markovian property is available. Such epochs or time points are called regeneration points and the process considered at these epochs is called a Markov chain imbedded in the total process, which is non-Markovian. For a $GI/M/1$ system, the arrival epochs are the regeneration points. This method has been used by Kendall (1951, 1953) and by many authors since.

(2) *Supplementary variable technique.* According to this method, we study the system in continuous time, but enhance the description of the system by including supplementary variables denoting the time since the last regeneration point. For example, for an $M/G/1$ queuing process we define $P_n(x, t)\,dx$ as the probability that at time t the number in the system is n and the

elapsed service time of the unit in service lies between x and $x + dx$. Similarly, the $GI/M/1$ system may be dealt with by the inclusion of elapsed time since the last arrival.

(3) *Phase technique.* This method consists of dividing the interarrival time or the service time into a number of phases so that the time occupied by each phase has the negative exponential distribution. Erlang used this method by regarding service to be completed in k fictitious phases, each of which has the same negative exponential distribution, thus obtaining what is called a k-Erlang distribution and is the k-fold convolution of the exponential distribution with itself. For arrival distributions, we can in the same way postulate an arrival timing channel consisting of a number of phases. Authors like Gaver (1954) and Luchak (1956) considered the generalization, in which the number of phases is itself a random variable. More recently, Neuts (1975a,b, 1978) has exploited the concept of phases to study more complex systems through the use of Markov chains.

The subject of queuing theory has vast applications, ranging from day-to-day department store, airport, and hospital queues to industrial queues like inventories, assembly lines, and computer time-sharing systems. In some cases we even have nonphysical queues like the names of the candidates, say, for a job, which constitute a queue on paper. From the rates of papers on the subject, one can still call it a growing subject 75 years after Erlang.

PRELIMINARIES II

Applied Special Functions

II.1 Gamma, Beta, and Related Functions

The Gamma Function

The familiar factorial function $n!$ is well defined when n is a positive integer [or even zero, since the relation $n! = n(n-1)!$ implies, when $n = 1$, that $0! = 1$]. In an attempt to give a meaning to $x!$ when x is any positive number, Euler in 1729 undertook the problem of interpolating $n!$ between the positive integral values of n. He was thus led to what is now well known as the *gamma function*, which is encountered fairly frequently in the study of special functions. Unlike the exponential, circular, and hyperbolic functions, the gamma function $\Gamma(z)$ does not satisfy any differential equation with *rational* coefficients.

We shall begin by presenting three *equivalent* definitions of the gamma function, which are due essentially to Euler, Gauss, and Weierstrass:

Definition 1 (Euler)

(1)
$$\Gamma(z) = \int_0^\infty t^{z-1} e^{-t}\, dt, \qquad \operatorname{Re}(z) > 0.$$

Definition 2 (Gauss)

(2)
$$\Gamma(z) = \lim_{n\to\infty} \left\{ \frac{n!\, n^z}{z(z+1)\cdots(z+n)} \right\}, \qquad z \neq 0, -1, -2, \ldots.$$

Definition 3 (Weierstrass)

(3)
$$\frac{1}{\Gamma(z)} = z e^{\gamma z} \prod_{n=1}^{\infty} \left[\left(1 + \frac{z}{n}\right) \exp\left(-\frac{z}{n}\right) \right],$$

where γ is the Euler–Mascheroni constant defined by

(4) $$\gamma = \lim_{n\to\infty}\left\{1 + \frac{1}{2} + \frac{1}{3} + \cdots + \frac{1}{n} - \ln n\right\} \cong 0.5772156649\ldots.$$

By integration by parts, Definition 1 yields the recurrence relation for $\Gamma(z)$:

(5) $$\Gamma(z+1) = z\Gamma(z),$$

which enables us to use Definition 1 to define $\Gamma(z)$ on the entire z-plane (except when z is zero or a negative integer) as follows:

(6) $$\Gamma(z) = \begin{cases} \displaystyle\int_0^\infty t^{z-1}e^{-t}\,dt, & \mathrm{Re}(z) > 0, \\ \dfrac{\Gamma(z+1)}{z}, & \mathrm{Re}(z) < 0, \quad z \neq -1, -2, -3, \ldots. \end{cases}$$

The recurrence relation (5) can be proved for all $z \neq 0, -1, -2, \ldots$ by using Definitions 2 and 3. Since the function $\Gamma(1/w)$ has simple poles at $w = -1, -\frac{1}{2}, -\frac{1}{3}, \ldots$, which implies that $w = 0$ is an accumulation point of poles of $\Gamma(1/w)$, the gamma function $\Gamma(z)$ has an essential singularity at infinity. Furthermore, from Definition 3 it follows immediately that $1/\Gamma(z)$ has no poles, and therefore $\Gamma(z)$ is never zero.

Definition 3 in conjunction with the well-known product formula

(7) $$\sin \pi z = \pi z \prod_{n=1}^{\infty}\left(1 - \frac{z^2}{n^2}\right)$$

also yields the useful result

(8) $$\Gamma(z)\Gamma(1-z) = \frac{\pi}{\sin \pi z}, \qquad z \neq 0, \pm 1, \pm 2, \ldots,$$

which incidentally provides an immediate analytic continuation of $\Gamma(z)$ from the right to the left half of the z-plane.

Some special values of $\Gamma(x)$, when x is real, are worthy of note. Indeed, from Definition 1 we note that

(9) $$\Gamma(1) = \int_0^\infty e^{-t}\,dt = -e^{-t}\big|_0^\infty = 1,$$

and that $\Gamma(x) > 0$ for all x in the open interval $(0, \infty)$. Thus (7) with $z = \frac{1}{2}$ immediately yields

(10) $$\Gamma(\tfrac{1}{2}) = \sqrt{\pi},$$

which, in view of Definition 1, implies

(11) $$\int_0^\infty \frac{e^{-t}}{\sqrt{t}}\,dt = \sqrt{\pi},$$

or, equivalently,

$$\int_0^\infty \exp(-\tau^2)\,d\tau = \frac{\sqrt{\pi}}{2}. \tag{12}$$

By repeated applications of the recurrence relation (5), it follows that

$$\Gamma(n+1) = n!, \qquad \Gamma(n+\tfrac{1}{2}) = \frac{(2n)!\sqrt{\pi}}{2^{2n}n!}, \tag{13}$$

$$\Gamma(-n+\tfrac{1}{2}) = (-1)^n\sqrt{\pi}\,\frac{2^{2n}n!}{(2n)!}, \qquad n = 0, 1, 2, \ldots,$$

where the last two results also require the use of (10). Evidently, the formulas listed under (13) enable us to compute $\Gamma(x)$ when x is a positive integer and when x is half an odd integer, positive or negative.

Pochhammer's Symbol and the Factorial Function

Throughout this work we shall find it convenient to employ the Pochhammer symbol $(\lambda)_n$ defined by

$$(\lambda)_n = \begin{cases} 1 & \text{if} \quad n = 0, \quad \lambda \neq 0 \\ \lambda(\lambda+1)\cdots(\lambda+n-1) & \text{if} \quad n = 1, 2, 3, \ldots. \end{cases} \tag{14}$$

Since $(1)_n = n!$, $(\lambda)_n$ may be looked upon as a generalization of the elementary factorial; hence the symbol $(\lambda)_n$ is also referred to as the factorial function.

In terms of gamma functions, we have

$$(\lambda)_n = \frac{\Gamma(\lambda+n)}{\Gamma(\lambda)}, \qquad \lambda \neq 0, -1, -2, \ldots, \tag{15}$$

which can easily be verified. Furthermore, the binomial coefficient may now be expressed as

$$\binom{\lambda}{n} = \frac{\lambda(\lambda-1)\cdots(\lambda-n+1)}{n!} = \frac{(-1)^n(-\lambda)_n}{n!}, \tag{16}$$

or, equivalently,

$$\binom{\lambda}{n} = \frac{\Gamma(\lambda+1)}{n!\,\Gamma(\lambda-n+1)}. \tag{17}$$

It follows from (16) and (17) that

$$\frac{\Gamma(\lambda+1)}{\Gamma(\lambda-n+1)} = (-1)^n(-\lambda)_n,$$

which, for $\lambda = \alpha - 1$, yields

(18) $$\frac{\Gamma(\alpha - n)}{\Gamma(\alpha)} = \frac{(-1)^n}{(1 - \alpha)_n}, \qquad \alpha \neq 0, \pm 1, \pm 2, \ldots .$$

Equations (15) and (18) suggest the definition

(19) $$(\lambda)_{-n} = \frac{(-1)^n}{(1 - \lambda)_n}, \qquad n = 1, 2, 3, \ldots; \quad \lambda \neq 0, \pm 1, \pm 2, \ldots .$$

Equation (15) also yields

(20) $$(\lambda)_{m+n} = (\lambda)_m (\lambda + m)_n,$$

which, in conjunction with (19), gives

(21) $$(\lambda)_{n-k} = \frac{(-1)^k (\lambda)_n}{(1 - \lambda - n)_k}, \qquad 0 \leqq k \leqq n.$$

For $\lambda = 1$, we have

(22) $$(n - k)! = \frac{(-1)^k n!}{(-n)_k}, \qquad 0 \leqq k \leqq n,$$

which may alternatively be written in the form

(23) $$(-n)_k = \begin{cases} \dfrac{(-1)^k n!}{(n - k)!}, & 0 \leqq k \leqq n \\ 0, & k > n. \end{cases}$$

Legendre's Duplication Formula

In view of the definition in (14) it is not difficult to show that

(24) $$(\lambda)_{2n} = 2^{2n} \left(\frac{\lambda}{2}\right)_n \left(\frac{\lambda + 1}{2}\right)_n, \qquad n = 0, 1, 2, \ldots,$$

which follows also from Legendre's duplication formula for the gamma function, viz.,

(25) $$\sqrt{\pi}\, \Gamma(2z) = 2^{2z-1} \Gamma(z) \Gamma(z + \tfrac{1}{2}), \qquad z \neq 0, -\tfrac{1}{2}, -1, -\tfrac{3}{2}, \ldots .$$

Gauss's Multiplication Theorem

For every positive integer m, we have

(26) $$(\lambda)_{mn} = m^{mn} \prod_{j=1}^{m} \left(\frac{\lambda + j - 1}{m}\right)_n, \qquad n = 0, 1, 2, \ldots,$$

which obviously reduces to (24) when $m = 2$. Starting from (26) with $\lambda = mz$, it can be proved that

$$\Gamma(mz) = (2\pi)^{(1-m)/2} m^{mz-(1/2)} \prod_{j=1}^{m} \Gamma\left(z + \frac{j-1}{m}\right), \tag{27}$$

$$z \neq 0, -1/m, -2/m, \ldots; \qquad m = 1, 2, 3, \ldots,$$

which is known in the literature as Gauss's multiplication theorem for the gamma function.

Stirling's Formula for $n!$ and Its Generalizations

For a large positive integer n, it naturally becomes tedious to compute $n!$. An easy way of computing an approximate value of $n!$ was found by Stirling in 1730 and modified subsequently by De Moivre, who showed that

$$n! \sim (n/e)^n \sqrt{2\pi n}, \qquad n \to \infty, \tag{28}$$

or, more generally,

$$\Gamma(x+1) \sim (x/e)^x \sqrt{2\pi x}, \qquad x \to \infty, \quad x \text{ real}, \tag{29}$$

where e is the base of the natural logarithm.

For a complex number z we have the following asymptotic expansion:

$$\Gamma(z) = z^z e^{-z} \sqrt{\frac{2\pi}{z}} \left[1 + \frac{1}{12z} + \frac{1}{288z^2} - \frac{139}{51{,}840z^3} - \frac{571}{2{,}488{,}320z^4} + O(z^{-5})\right], \tag{30}$$

$$|z| \to \infty, \qquad |\arg(z)| \leqq \pi - \varepsilon, \quad \varepsilon > 0.$$

Some useful consequences of (30) include the asymptotic expansions

$$\ln \Gamma(z + \alpha) = (z + \alpha - \tfrac{1}{2}) \ln z - z + \tfrac{1}{2} \ln(2\pi) + O(z^{-1}) \tag{31}$$

and

$$\frac{\Gamma(z+\alpha)}{\Gamma(z+\beta)} = z^{\alpha-\beta}\left[1 + \frac{(\alpha-\beta)(\alpha+\beta-1)}{2z} + O(z^{-2})\right], \tag{32}$$

$$|z| \to \infty, \qquad |\arg(z)| \leqq \pi - \varepsilon, \qquad |\arg(z+\alpha)| \leqq \pi - \varepsilon, \quad \varepsilon > 0,$$

where α and β are bounded complex numbers.

Yet another interesting consequence of (30) is the following asymptotic expansion of $|\Gamma(x+iy)|$ where x and y are real numbers:

$$|\Gamma(x+iy)| \sim \sqrt{2\pi}\, |y|^{x-(1/2)} e^{-\pi|y|/2}, \qquad |y| \to \infty, \tag{33}$$

x being finite.

The Logarithmic Derivative of $\Gamma(z)$

The psi function $\psi(z)$ defined by

$$\psi(z) = \frac{d}{dz} \ln \Gamma(z) = \frac{\Gamma'(z)}{\Gamma(z)} \qquad \text{or} \qquad \ln \Gamma(z) = \int_1^z \psi(t)\, dt, \tag{34}$$

satisfies the following properties:

$$\psi(z) = \lim_{n \to \infty} \left[\ln n - \sum_{k=0}^{n} (z+k)^{-1} \right]; \tag{35}$$

$$\begin{aligned} \psi(z) &= -\gamma - \frac{1}{z} + \sum_{n=1}^{\infty} \frac{z}{n(z+n)} \\ &= -\gamma + (z-1) \sum_{n=0}^{\infty} [(n+1)(n+z)]^{-1}, \end{aligned} \tag{36}$$

γ being the Euler–Mascheroni constant defined by (4).

These results clearly imply that $\psi(z)$ is meromorphic (that is, analytic everywhere in the bounded complex plane except for poles) with simple poles at $z = 0, -1, -2, \ldots$. Also,

$$\psi(1) = -\gamma, \tag{37}$$

which follows at once from (36).

The following additional properties of $\psi(z)$ can be deduced from results for $\Gamma(z)$:

$$\psi(z+n) = \psi(z) + \sum_{k=1}^{n} (z+k-1)^{-1}, \qquad n = 1, 2, 3, \ldots; \tag{38}$$

$$\psi(z) - \psi(-z) = -\pi \cot \pi z - z^{-1}; \tag{39}$$

$$\psi(z) - \psi(1-z) = -\pi \cot \pi z; \tag{40}$$

$$\psi(\tfrac{1}{2} + z) - \psi(\tfrac{1}{2} - z) = \pi \tan \pi z; \tag{41}$$

$$\psi(mz) = \ln m + \frac{1}{m} \sum_{j=0}^{m-1} \psi\left(z + \frac{j}{m}\right), \qquad m = 1, 2, 3, \ldots. \tag{42}$$

The Beta Function

The *beta function* $B(\alpha, \beta)$ is a function of two complex variables α and β, defined by the *Eulerian integral of the first kind*

$$B(\alpha, \beta) = \int_0^1 t^{\alpha-1}(1-t)^{\beta-1}\, dt, \qquad \operatorname{Re}(\alpha) > 0, \quad \operatorname{Re}(\beta) > 0, \tag{43}$$

or, equivalently,

$$(44)\quad B(\alpha,\beta)=2\int_0^{\pi/2}(\sin\theta)^{2\alpha-1}(\cos\theta)^{2\beta-1}\,d\theta,\qquad \mathrm{Re}(\alpha)>0,\quad \mathrm{Re}(\beta)>0,$$

which follows from (43) on setting $t=\sin^2\theta$. Clearly, we have

$$(45)\qquad B(\alpha,\beta)=B(\beta,\alpha).$$

Putting $t=u/(1+u)$ in (43), we obtain the following representation of $B(\alpha,\beta)$ as an infinite integral:

$$(46)\qquad B(\alpha,\beta)=\int_0^\infty \frac{u^{\alpha-1}}{(1+u)^{\alpha+\beta}}\,du,\qquad \mathrm{Re}(\alpha)>0,\quad \mathrm{Re}(\beta)>0.$$

The beta function is closely related to the gamma function; in fact, we have

$$(47)\qquad B(\alpha,\beta)=\frac{\Gamma(\alpha)\Gamma(\beta)}{\Gamma(\alpha+\beta)},\qquad \alpha,\beta\neq 0,-1,-2,\ldots,$$

which not only confirms the symmetry property (45), but also continues the beta function analytically for all complex values of α and β, except when α, $\beta=0,-1,-2,\ldots$. Thus we may write, by analogy with (6), that

$$(48)\quad B(\alpha,\beta)=\begin{cases}\displaystyle\int_0^1 t^{\alpha-1}(1-t)^{\beta-1}\,dt, & \mathrm{Re}(\alpha)>0,\quad \mathrm{Re}(\beta)>0\\[2ex] \displaystyle\frac{\Gamma(\alpha)\Gamma(\beta)}{\Gamma(\alpha+\beta)}, & \mathrm{Re}(\alpha)<0,\quad \mathrm{Re}(\beta)<0,\\ & \alpha,\beta\neq -1,-2,-3,\ldots.\end{cases}$$

Next we combine the relation (47) with (44) and (46), and we obtain the useful integral formulas

$$(49)\qquad \int_0^{\pi/2}\sin^\mu\theta\cos^\nu\theta\,d\theta=\frac{\Gamma[\frac{1}{2}(\mu+1)]\,\Gamma[\frac{1}{2}(\nu+1)]}{2\Gamma[\frac{1}{2}(\mu+\nu+2)]},$$

$$\mathrm{Re}(\mu)>-1,\quad \mathrm{Re}(\nu)>-1,$$

and

$$(50)\qquad \int_0^\infty \frac{u^{\lambda-1}}{(1+u)^\mu}\,du=\frac{\Gamma(\lambda)\,\Gamma(\mu-\lambda)}{\Gamma(\mu)},\qquad 0<\mathrm{Re}(\lambda)<\mathrm{Re}(\mu),$$

respectively.

It may be remarked in passing that the integral (49) provides a generalization of Wallis's formula of elementary calculus, and that (50) with $\mu=1$ yields the infinite integral

$$(51)\qquad \int_0^\infty \frac{u^{\lambda-1}}{1+u}\,du=\Gamma(\lambda)\Gamma(1-\lambda)=\frac{\pi}{\sin\pi\lambda},\qquad 0<\mathrm{Re}(\lambda)<1,$$

which is usually evaluated in the literature by contour integration. [See, for example, Copson (1935, p. 139, Example 1).]

The Incomplete Gamma Functions

The *incomplete gamma function* $\gamma(z, \alpha)$ and its *complement* $\Gamma(z, \alpha)$ (also known as *Prym's function*) are defined by

$$\gamma(z, \alpha) = \int_0^\alpha t^{z-1} e^{-t}\, dt, \qquad \mathrm{Re}(z) > 0, \quad |\arg(\alpha)| < \pi, \tag{52}$$

$$\Gamma(z, \alpha) = \int_\alpha^\infty t^{z-1} e^{-t}\, dt, \qquad |\arg(\alpha)| < \pi, \tag{53}$$

so that

$$\gamma(z, \alpha) + \Gamma(z, \alpha) = \Gamma(z). \tag{54}$$

For fixed α, $\Gamma(z, \alpha)$ is an entire (integral) function of z, while $\gamma(z, \alpha)$ is a meromorphic function of z, with simple poles at the points $z = 0, -1, -2, \ldots$. In fact, we have

$$\gamma(z, \alpha) = \sum_{n=0}^{\infty} \frac{(-1)^n}{n!} \frac{\alpha^{z+n}}{z+n} = z^{-1} \alpha^z \, {}_1F_1\left[\begin{matrix} z; \\ z+1; \end{matrix} \; -\alpha\right] \tag{55}$$

$$= z^{-1} \alpha^z e^{-\alpha} \, {}_1F_1\left[\begin{matrix} 1; \\ z+1; \end{matrix} \; \alpha\right], \qquad z \neq 0, -1, -2, \ldots,$$

where ${}_1F_1$ denotes the *confluent hypergeometric function*. (See Sections II.2 and II.3 for details.)

The following recursion formulas are worthy of note:

$$\gamma(z+1, \alpha) = z\gamma(z, \alpha) - \alpha^z e^{-\alpha}, \tag{56}$$

$$\Gamma(z+1, \alpha) = z\Gamma(z, \alpha) + \alpha^z e^{-\alpha}. \tag{57}$$

The Incomplete Beta Functions

The *incomplete beta function* $B_x(\alpha, \beta)$ defined by

$$B_x(\alpha, \beta) = \int_0^x t^{\alpha-1}(1-t)^{\beta-1}\, dt \tag{58}$$

and the associated function

$$I_x(\alpha, \beta) = B_x(\alpha, \beta)/B(\alpha, \beta) \tag{59}$$

occur frequently in this book. In terms of the hypergeometric function ${}_2F_1$ (cf. Section II.2), we have

$$(60) \qquad B_x(\alpha, \beta) = \sum_{n=0}^{\infty} \frac{(1-\beta)_n x^{\alpha+n}}{n!(\alpha+n)} = \alpha^{-1} x^{\alpha} {}_2F_1\left[\begin{matrix}\alpha, 1-\beta; \\ \alpha+1;\end{matrix} x\right].$$

For the associated function $I_x(\alpha, \beta)$, we note here the following properties which are easily verifiable by using the definition in (59):

$$(61) \qquad I_x(\alpha, \beta) = 1 - I_{1-x}(\beta, \alpha);$$

$$(62) \qquad I_x(k, n-k+1) = \sum_{j=k}^{n} \binom{n}{j} x^j (1-x)^{n-j}, \qquad 1 \leqq k \leqq n;$$

$$(63) \qquad I_x(\alpha, \beta) = x I_x(\alpha-1, \beta) + (1-x) I_x(\alpha, \beta-1);$$

$$(64) \qquad (\alpha+\beta-\alpha x) I_x(\alpha, \beta) = \alpha(1-x) I_x(\alpha+1, \beta-1) + \beta I_x(\alpha, \beta+1);$$

$$(65) \qquad (\alpha+\beta) I_x(\alpha, \beta) = \alpha I_x(\alpha+1, \beta) + \beta I_x(\alpha, \beta+1).$$

The Error Functions

The *error function* erf(z), also known as the *probability integral* $\Phi(z)$, is defined by

$$(66) \qquad \operatorname{erf}(z) = \frac{2}{\sqrt{\pi}} \int_0^z \exp(-t^2)\, dt = \Phi(z),$$

and its *complement* by

$$(67) \qquad \operatorname{erfc}(z) = 1 - \operatorname{erf}(z) = \frac{2}{\sqrt{\pi}} \int_z^{\infty} \exp(-t^2)\, dt.$$

Clearly,

$$(68) \qquad \operatorname{erf}(0) = 0, \qquad \operatorname{erfc}(0) = 1,$$

and, in view of the well-known result (12), we also have

$$(69) \qquad \operatorname{erf}(\infty) = 1, \qquad \operatorname{erfc}(\infty) = 0.$$

The alternative notations

$$(70) \qquad \operatorname{Erf}(z) = \frac{\sqrt{\pi}}{2} \operatorname{erf}(z), \qquad \operatorname{Erfc}(z) = \frac{\sqrt{\pi}}{2} \operatorname{erfc}(z)$$

are sometimes used for the error functions.

For finite values of $|z|$, the error function erf(z) can be expressed as an infinite series by expanding $\exp(-t^2)$ in (66) in powers of t^2 and integrating term by term. We thus obtain

(71) $$\operatorname{erf}(z) = \frac{2z}{\sqrt{\pi}}\,{}_1F_1[\tfrac{1}{2};\tfrac{3}{2};-z^2], \qquad |z| < \infty,$$

in terms of the confluent hypergeometric function. (See Sections II.2 and II.3.)

In terms of the incomplete gamma functions, it is easily verified that

(72) $$\operatorname{erf}(z) = \frac{1}{\sqrt{\pi}}\gamma(\tfrac{1}{2}, z^2), \qquad \operatorname{erfc}(z) = \frac{1}{\sqrt{\pi}}\Gamma(\tfrac{1}{2}, z^2).$$

Remark Many authors use the notations Erf(z) and $\widehat{\mathrm{E}}$rf(z) for the error functions erf(z) and Erf(z), respectively, defined by (66) and (70), and Erfc(z) and $\widehat{\mathrm{E}}$rfc(z) for their complements; see, for example, 4.9(21) and 4.9(28).

II.2 The Gaussian Hypergeometric Function

The Hypergeometric Equation

The second-order linear differential equation

(1) $$z(1-z)\frac{d^2w}{dz^2} + [c - (a+b+1)z]\frac{dw}{dz} - abw = 0,$$

or, equivalently,

(2) $$\{\delta(\delta + c - 1) - z(\delta + a)(\delta + b)\}w = 0, \qquad \delta = z(d/dz),$$

in which a, b, and c are real or complex parameters, is called the *hypergeometric equation.* Its only singularities are at

$$z = 0, 1, \infty;$$

each singularity is easily seen to be *regular*.

The hypergeometric equation (1) or (2) is the most celebrated equation of the Fuchsian class. Its importance stems, in part, from a well-known theorem that *every homogeneous linear differential equation of the second order, whose singularities (including the point at infinity) are regular and at most three in number, can be transformed into the hypergeometric equation.*

Series solutions of the hypergeometric equation (1) valid in the neighborhoods of $z = 0, 1$, or ∞ can be developed by direct application of the classical method of Frobenius. Thus, if c is not an integer, the general solution of (1) valid in a neighborhood of the origin is found to be

(3) $$w = A\,{}_2F_1(a, b; c; z) + Bz^{1-c}\,{}_2F_1(a - c + 1, b - c + 1; 2 - c; z),$$

where A and B are arbitrary constants, and (for convenience)

$$(4)\qquad {}_2F_1(a,b;c;z) = 1 + \frac{ab}{1\cdot c}z + \frac{a(a+1)b(b+1)}{1\cdot 2\cdot c(c+1)}z^2 + \cdots$$

$$= \sum_{n=0}^{\infty} \frac{(a)_n(b)_n}{(c)_n}\frac{z^n}{n!}, \qquad c \neq 0, -1, -2, \ldots,$$

in terms of the Pochhammer symbol defined by II.1(15).

Gauss's Hypergeometric Series

The infinite series in (4) evidently reduces to the elementary geometric series

$$(5)\qquad \sum_{n=0}^{\infty} z^n = 1 + z + z^2 + \cdots + z^n + \cdots$$

in its special cases when

$$(6)\qquad \text{(i)}\quad a = c \quad \text{and} \quad b = 1, \qquad \text{(ii)}\quad a = 1 \quad \text{and} \quad b = c.$$

Hence it is called the *hypergeometric series*, or, more precisely, Gauss's hypergeometric series after the famous German mathematician Carl Friedrich Gauss (1777–1855), who in 1812 introduced this series into analysis and gave the F notation for it. (See also Section II.4, where the general notation ${}_pF_q$ will be introduced, p and q being any nonnegative integers.)

By d'Alembert's ratio test, it is easily seen that the hypergeometric series in (4) converges absolutely *within* the unit circle, that is, when $|z| < 1$, provided that the denominator parameter c is neither zero nor a negative integer. Note, however, that if either or both of the numerator parameters a and b in (4) is zero or a negative integer, the hypergeometric series terminates, and the question of convergence does not enter the discussion.

Further tests show that the hypergeometric series in (4), when $|z| = 1$ (that is, on the unit circle), is

(i) absolutely convergent if $\mathrm{Re}(c - a - b) > 0$;
(ii) conditionally convergent if $-1 < \mathrm{Re}(c - a - b) \leqq 0,\ z \neq 1$;
(iii) divergent if $\mathrm{Re}(c - a - b) \leqq -1$.

As a matter of fact, in case (i) we are led to the well-known *Gauss's summation theorem*:

$$(7)\qquad {}_2F_1(a,b;c;1) = \frac{\Gamma(c)\Gamma(c-a-b)}{\Gamma(c-a)\Gamma(c-b)},$$

$$\mathrm{Re}(c - a - b) > 0, \quad c \neq 0, -1, -2, \ldots .$$

An obvious special case of (7) occurs when the numerator parameter a or b is a nonpositive integer $-n$, say. We thus have the summation formula

(8) $F(-n, b; c; 1) = (c-b)_n/(c)_n, \quad n = 0, 1, 2, \ldots, \quad c \neq 0, -1, -2, \ldots,$

which incidentally is equivalent to *Vandermonde's convolution theorem*:

(9) $$\sum_{k=0}^{n} \binom{\lambda}{k}\binom{\mu}{n-k} = \binom{\lambda+\mu}{n}, \quad n \geqq 0,$$

λ and μ being any complex numbers.

For a number of summation theorems for the hypergeometric series (4) when z takes on other special values, see Bailey (1935, pp. 9–11), Erdélyi *et al.* (1953, Vol. I, pp. 104–105), Slater (1966, p. 243), and Luke (1975, pp. 271–273).

The Hypergeometric Function and Its Analytic Continuation

We have seen that the hypergeometric series in (4) converges absolutely when $|z| < 1$ and thus defines a function

$$_2F_1(a, b; c; z),$$

which is analytic when $|z| < 1$, provided that c is neither zero nor a negative integer. This function is correspondingly called the *hypergeometric function* or Gauss's hypergeometric function.

The hypergeometric function $_2F_1(a, b; c; z)$ can, in fact, be continued analytically beyond the unit circle in a number of ways. If, for convenience, we use this same notation for the analytically continued function, one way is to employ *Euler's integral representation*:

(10) $$_2F_1(a, b; c; z) = \frac{\Gamma(c)}{\Gamma(a)\Gamma(c-a)} \int_0^1 t^{a-1}(1-t)^{c-a-1}(1-zt)^{-b}\, dt,$$

$$\mathrm{Re}(c) > \mathrm{Re}(a) > 0, \qquad |\arg(1-z)| < \pi,$$

or, equivalently,

(11) $$_2F_1(a, b; c; z) = \frac{\Gamma(c)}{\Gamma(b)\Gamma(c-b)} \int_0^1 t^{b-1}(1-t)^{c-b-1}(1-zt)^{-a}\, dt,$$

$$\mathrm{Re}(c) > \mathrm{Re}(b) > 0, \qquad |\arg(1-z)| < \pi,$$

since, by the series definition (4),

$$_2F_1(a, b; c; z) \equiv {}_2F_1(b, a; c; z).$$

An analytic continuation without such parametric constraints as in (10) and (11) can be provided by employing the following contour integral representation of the *Mellin–Barnes type*:

$$\frac{\Gamma(a)\Gamma(b)}{\Gamma(c)}\,{}_2F_1(a,b;c;z)=\frac{1}{2\pi i}\int_{-i\infty}^{i\infty}\frac{\Gamma(a+\zeta)\Gamma(b+\zeta)\Gamma(-\zeta)}{\Gamma(c+\zeta)}(-z)^{\zeta}\,d\zeta, \tag{12}$$

$$a,b\neq 0,-1,-2,\ldots,\qquad |\arg(-z)|<\pi,$$

where the path of integration is the imaginary axis (in the complex ζ-plane) which is indented, if necessary, to ensure that the poles of $\Gamma(a+\zeta)\Gamma(b+\zeta)$, viz.,

$$\zeta=-a-n,\qquad \zeta=-b-n,\quad n=0,1,2,\ldots, \tag{13}$$

lie to the left of the path and the poles of $\Gamma(-\zeta)$, viz.,

$$\zeta=0,1,2,\ldots, \tag{14}$$

lie to its right.

The integrals in (10)–(12) define analytic functions of z that are single valued in the domain $|\arg(-z)|<\pi$, that is, in the whole z plane with the exception of the points on the nonnegative real axis. Since the Gaussian function ${}_2F_1(a,b;c;z)$ is defined by the hypergeometric series (4) everywhere within the unit circle $|z|=1$ (including, for example, at points on the nonnegative real axis from $z=0$ to $z=1-\varepsilon$, ε being an arbitrarily small positive number), we can use one or the other of these integrals to effect the analytic continuation of the hypergeometric function to the whole z plane cut along the real axis from $z=1$ to $z=\infty$. This analytic continuation of ${}_2F_1(a,b;c;z)$ is conveniently denoted by the same symbol ${}_2F_1(a,b;c;z)$, which will henceforth represent a branch (the *principal branch*) of the analytic function generated by the Gaussian hypergeometric series in (4).

By evaluating the contour integral in (12) using the sum of the residues of the integrand at the poles given by (13), it can be shown that

$$\begin{aligned}\frac{1}{\Gamma(c)}\,{}_2F_1(a,b;c;z)&\\ &=\frac{\Gamma(b-a)}{\Gamma(b)\Gamma(c-a)}(-z)^{-a}\,{}_2F_1(a,1-c+a;1-b+a;z^{-1})\\ &\quad+\frac{\Gamma(a-b)}{\Gamma(a)\Gamma(c-b)}(-z)^{-b}\,{}_2F_1(b,1-c+b;1-a+b;z^{-1}),\end{aligned} \tag{15}$$

$$a-b\neq 0,\pm 1,\pm 2,\ldots,\qquad |\arg(-z)|<\pi.$$

This is an important result for it may be used to provide the analytic continuation of ${}_2F_1(a,b;c;z)$ into the domain $|z|>1$, that is, outside the circle of convergence of the hypergeometric series (4). It also shows that the function so defined possesses a branch point at infinity and that, asymptotically,

$$(16)\quad {}_2F_1(a,b;c;z) \sim A(-z)^{-a} + B(-z)^{-b}, \qquad |z| \to \infty, \quad |\arg(-z)| < \pi,$$

where, for convenience,

$$(17)\qquad A = \frac{\Gamma(c)\Gamma(b-a)}{\Gamma(b)\Gamma(c-a)}, \qquad B = \frac{\Gamma(c)\Gamma(a-b)}{\Gamma(a)\Gamma(c-a)}.$$

Linear, Quadratic, and Cubic Transformations

The linear transformations of the hypergeometric function, known as *Euler's transformations*, may be recalled here as follows:

$$(18)\qquad {}_2F_1(a,b;c;z) = (1-z)^{-a}\,{}_2F_1\left(a, c-b; c; \frac{z}{z-1}\right),$$

$$c \neq 0, -1, -2, \ldots, \qquad |\arg(1-z)| < \pi;$$

$$(19)\qquad {}_2F_1(a,b;c;z) = (1-z)^{-b}\,{}_2F_1\left(c-a, b; c; \frac{z}{z-1}\right),$$

$$c \neq 0, -1, -2, \ldots, \qquad |\arg(1-z)| < \pi;$$

$$(20)\qquad {}_2F_1(a,b;c;z) = (1-z)^{c-a-b}\,{}_2F_1(c-a, c-b; c; z),$$

$$c \neq 0, -1, -2, \ldots, \qquad |\arg(1-z)| < \pi.$$

The transformations (18) and (19) can be proved, for example, by setting $t = 1 - s$ in Euler's integral representations (10) and (11), respectively, and then appealing to the principle of analytic continuation. The last transformation (20) follows if we perform the transformations (18) and (19) successively.

For an extensive list of quadratic and cubic transformations of the Gaussian hypergeometric function, see Erdélyi *et al.* (1953, Vol. I, pp. 110–114) and Luke (1969, Vol. I, p. 92 et seq.).

Hypergeometric Representations of Elementary Functions

$$(21)\qquad (1-z)^{-a} = \sum_{n=0}^{\infty} \frac{(a)_n z^n}{n!} = {}_2F_1(a,b;b;z);$$

$$(22)\qquad \ln(1+z) = z\,{}_2F_1(1,1;2;-z);$$

$$(23)\qquad \ln\left(\frac{1+z}{1-z}\right) = 2z\,{}_2F_1(\tfrac{1}{2}, 1; \tfrac{3}{2}; z^2);$$

(24) $$\sin^{-1} z = z\,{}_2F_1(\tfrac{1}{2}, \tfrac{1}{2}; \tfrac{3}{2}; z^2);$$

(25) $$\tan^{-1} z = z\,{}_2F_1(\tfrac{1}{2}, 1; \tfrac{3}{2}; -z^2);$$

(26) $$\{\tfrac{1}{2}(1 + \sqrt{1 - z})\}^{1-2a} = {}_2F_1(a, a - \tfrac{1}{2}; 2a; z)$$
$$= \sqrt{1 - z}\,{}_2F_1(a, a + \tfrac{1}{2}; 2a; z);$$

(27) $$(1 + \sqrt{z})^{-2a} + (1 - \sqrt{z})^{-2a} = 2\,{}_2F_1(a, a + \tfrac{1}{2}; \tfrac{1}{2}; z).$$

Hypergeometric Representations of Other Functions

Legendre functions of the first and second kinds

(28) $$P_\nu^\mu(z) = \frac{1}{\Gamma(1-\mu)} \left(\frac{z+1}{z-1}\right)^{\mu/2} {}_2F_1\left(-\nu, \nu + 1; 1 - \mu; \frac{1-z}{2}\right);$$

(29) $$Q_\nu^\mu(z) = \frac{\sqrt{\pi}\, e^{i\mu\pi}}{2^{\nu+1}} \frac{\Gamma(\mu + \nu + 1)}{\Gamma(\nu + \frac{3}{2})} \frac{(z^2 - 1)^{\mu/2}}{z^{\mu+\nu+1}}$$
$$\cdot\, {}_2F_1[\tfrac{1}{2}(\mu + \nu + 1), \tfrac{1}{2}(\mu + \nu) + 1; \nu + \tfrac{3}{2}; z^{-2}].$$

The incomplete Beta function [cf. II.1(60)]

(30) $$B_z(\alpha, \beta) = \alpha^{-1} z^\alpha\, {}_2F_1(\alpha, 1 - \beta; \alpha + 1; z).$$

Complete elliptic integrals of the first and second kinds

(31) $$\mathbf{K}(k) = \int_0^{\pi/2} \frac{d\theta}{\sqrt{1 - k^2 \sin^2\theta}} = \frac{1}{2}\pi\, {}_2F_1(\tfrac{1}{2}, \tfrac{1}{2}; 1; k^2);$$

(32) $$\mathbf{E}(k) = \int_0^{\pi/2} \sqrt{1 - k^2 \sin^2\theta}\, d\theta = \frac{1}{2}\pi\, {}_2F_1(-\tfrac{1}{2}, \tfrac{1}{2}; 1; k^2).$$

Jacobi polynomials

(33) $$P_n^{(\alpha,\beta)}(z) = (-1)^n P_n^{(\beta,\alpha)}(-z)$$
$$= \binom{\alpha + n}{n} {}_2F_1\left[\begin{matrix} -n, \alpha + \beta + n + 1; & \\ & \dfrac{1-z}{2} \\ \alpha + 1; & \end{matrix}\right].$$

Gegenbauer (or ultraspherical) polynomials

$$C_n^\nu(z) = \binom{n+\nu-\frac{1}{2}}{n}^{-1}\binom{n+2\nu-1}{n} P_n^{(\nu-(1/2),\,\nu-(1/2))}(z) \tag{34}$$

$$= \binom{n+2\nu-1}{n} {}_2F_1\left[\begin{matrix} -n, 2\nu+n; & \\ & \dfrac{1-z}{2} \\ \nu+\frac{1}{2}; & \end{matrix}\right].$$

Legendre polynomials

$$P_n(z) = P_n^{(0,0)}(z) = {}_2F_1[-n, n+1; 1; \tfrac{1}{2}(1-z)]. \tag{35}$$

Chebyshev polynomials of the first and second kinds

$$T_n(z) = \frac{n!}{(\frac{1}{2})_n} P_n^{(-1/2,\,-1/2)}(z) = {}_2F_1\left(-n, n; \frac{1}{2}; \frac{1-z}{2}\right); \tag{36}$$

$$U_n(z) = \frac{(n+1)!}{(\frac{3}{2})_n} P_n^{(1/2,\,1/2)}(z) = (n+1)\,{}_2F_1\left(-n, n+1; \frac{3}{2}; \frac{1-z}{2}\right). \tag{37}$$

II.3 The Confluent Hypergeometric Function

Kummer's Differential Equation

If in Gauss's hypergeometric equation II.2(1) we replace z by z/b, the resulting equation will have three regular singularities at

$$z = 0, b, \infty.$$

By letting $|b| \to \infty$, this transformed equation reduces to

$$z\frac{d^2w}{dz^2} + (c-z)\frac{dw}{dz} - aw = 0, \tag{1}$$

or, equivalently,

$$[\delta(\delta+c-1) - z(\delta+a)]w = 0, \qquad \delta = z(d/dz). \tag{2}$$

Equation (1) or (2) has a regular singularity at $z = 0$, and an irregular singularity at $z = \infty$ which is formed by the *confluence* of two regular singularities at b and ∞ of Gauss's equation II.2(1) with z replaced by z/b. Consequently, (1) or (2) is called the *confluent hypergeometric equation* or Kummer's differential equation after E. E. Kummer (1810–1893), who presented a detailed study of its solutions in 1836 [cf. Kummer (1836)].

The simplest solution of (1) or (2) is Kummer's function $F(a, c, z)$, which, in the notation of Section II.4, is

$$(3)\qquad {}_1F_1(a;c;z) = 1 + \frac{a}{1\cdot c}z + \frac{a(a+1)}{1\cdot 2\cdot c(c+1)}z^2 + \cdots$$

$$= \sum_{n=0}^{\infty} \frac{(a)_n}{(c)_n}\frac{z^n}{n!}, \qquad c \neq 0, -1, -2, \ldots, \quad |z| < \infty.$$

Alternatively, since

$$(4)\qquad \lim_{|\lambda|\to\infty}\left\{(\lambda)_n\left(\frac{z}{\lambda}\right)^n\right\} = \lim_{|\mu|\to\infty}\left\{\frac{(\mu z)^n}{(\mu)_n}\right\} = z^n,$$

for bounded z and $n = 0, 1, 2, \ldots,$ we have

$$(5)\qquad {}_1F_1(a;c;z) = \lim_{|b|\to\infty} {}_2F_1(a, b; c; z/b).$$

In view of the principle of confluence involved in (5), the solution (3) is also called the *confluent hypergeometric function.*

Other popular notations for the series solution (3) are (i) $M(a, c, z)$ used by Airey and Webb (1918), and (ii) $\Phi(a;c;z)$ introduced by Humbert (1921a, b).

Important Properties of Kummer's Function

The principle of confluence exhibited by (5) is useful to deduce properties of the confluent hypergeometric ${}_1F_1$ function from those of Gauss's hypergeometric ${}_2F_1$ function. Thus, from II.2(10) and II.2(19), we readily have

$$(6)\qquad {}_1F_1(a;c;z) = \frac{\Gamma(c)}{\Gamma(a)\Gamma(c-a)}\int_0^1 t^{a-1}(1-t)^{c-a-1}e^{zt}\,dt,$$

$$\mathrm{Re}(c) > \mathrm{Re}(a) > 0;$$

$$(7)\qquad {}_1F_1(a;c;z) = e^z\,{}_1F_1(c-a;c;-z).$$

The transformation formula (7) is known as *Kummer's first theorem,* while *Kummer's second theorem* may be recalled here in the form

$$(8)\quad e^{-z}\,{}_1F_1(a;2a;2z) = {}_0F_1(-;a+\tfrac{1}{2};\tfrac{1}{4}z^2), \qquad 2a \neq -1, -3, -5, \ldots,$$

where

$$(9)\qquad {}_0F_1(-;c;z) = \lim_{|a|\to\infty} {}_1F_1(a;c;z/a)$$

$$= \sum_{n=0}^{\infty}\frac{z^n}{n!(c)_n}, \qquad c \neq 0, -1, -2, \ldots; \quad |z| < \infty,$$

which, in view of (4), at once yields

$$ {}_1F_1(a;a;z) = {}_0F_0(-;-;z) = e^z. \tag{10} $$

Similarly, from II.2(3) we obtain the general solution of the confluent hypergeometric equation (1) valid in a neighborhood of the origin:

$$ w = A\,{}_1F_1(a;c;z) + Bz^{1-c}\,{}_1F_1(a-c+1;2-c;z), \tag{11} $$

where A and B are arbitrary constants and c is not an integer.

Finally, we record the following asymptotic estimates of ${}_1F_1(a;c;z)$:

$$ {}_1F_1(a;c;z) = \begin{cases} \dfrac{\Gamma(c)}{\Gamma(a)} e^z z^{a-c}\{1 + O(|z|^{-1})\}, & \operatorname{Re}(z) \to \infty \\ \dfrac{\Gamma(c)}{\Gamma(c-a)} (-z)^{-a}\{1 + O(|z|^{-1})\}, & \operatorname{Re}(z) \to -\infty, \end{cases} \tag{12} $$

where a and c are bounded complex numbers;

$$ {}_1F_1(a;c;z) = \begin{cases} 1 + O(|c|^{-1}), & a, z \text{ bounded}, c \to \infty \\ e^z\{1 + O(|c|^{-1})\}, & c - a, z \text{ bounded}, c \to \infty. \end{cases} \tag{13} $$

The Second Form of Solutions of Kummer's Equation

The function

$$ \Psi(a,c;z) = \frac{1}{\Gamma(a)} \int_0^\infty e^{-zt} t^{a-1} (1+t)^{c-a-1}\, dt, \tag{14} $$

$$ \operatorname{Re}(a) > 0, \qquad \operatorname{Re}(z) > 0, $$

defines an alternative form of solution of Kummer's equation (1) in the half-plane $\operatorname{Re}(z) > 0$. The Ψ function was introduced in 1927 by F. G. Tricomi who first denoted it by $G(a,c,z)$. It is related to the F_1 function used in 1933 by J. Meixner, to the hypergeometric ${}_2F_0(\alpha,\beta;-;z)$ function used in 1939 by A. Erdélyi (1908–1977), and to the $E(\alpha,\beta::z)$ function used in 1941 by T. M. MacRobert (1884–1962); in fact, we have

$$ F_1(a,c,z) = e^{i\pi a} \frac{\Gamma(c)}{\Gamma(c-a)} \Psi(a,c;z), \tag{15} $$

$$ {}_2F_0(\alpha,\beta;-;-1/z) = z^\alpha \Psi(\alpha, \alpha-\beta+1; z), \tag{16} $$

$$ E(\alpha,\beta::z) = \Gamma(\alpha)\Gamma(\beta) z^\alpha \Psi(\alpha,\alpha-\beta+1;z). \tag{17} $$

The following relation between the Φ (or ${}_1F_1$) and Ψ functions is worthy of note:

$$\Psi(a, c; z) = \frac{\Gamma(1-c)}{\Gamma(a-c+1)}\,{}_1F_1(a; c; z) + \frac{\Gamma(c-1)}{\Gamma(a)} z^{1-c}\,{}_1F_1(a-c+1; 2-c; z), \tag{18}$$

$$c \neq 0, \pm 1, \pm 2, \ldots .$$

Furthermore, we have

$$\Psi(a, c; z) = z^{1-c}\Psi(a-c+1, 2-c; z). \tag{19}$$

We remark in passing that $\Psi(a, c; z)$ is a many-valued function of z, for complex values of a, c, and z. We usually consider its principal branch which lies in the complex z plane cut along the negative real axis. [See, for details, Erdélyi *et al.* (1935, Vol. I. Chapter 6), and Slater (1960), who uses the notation $U(a; c; z)$ instead of $\Psi(a, c; z)$.]

The Laplace Transforms of Kummer's Functions

In the usual notation for Laplace transforms, we shall write

$$f(s) \equiv \mathscr{L}\{F(t): s\} = \int_0^\infty e^{-st}F(t)\,dt, \qquad \mathrm{Re}(s-\alpha) > 0, \tag{20}$$

where $F(t)$ is piecewise continuous on every finite closed interval $0 \leqq t \leqq R$, $R > 0$, and

$$F(t) = O(e^{\alpha t}), \qquad t \to \infty. \tag{21}$$

From the theory of the Laplace transformation, we find it worthwhile to recall here the following useful properties [cf., e.g., Erdélyi *et al.* (1954), Vol. I, Chapter 4) and Widder (1941)]:

$$\mathscr{L}\{F^{(n)}(t): s\} = s^n f(s) - s^{n-1}F(0) - s^{n-2}F'(0) - \cdots - sF^{(n-2)}(0) - F^{(n-1)}(0), \tag{22}$$

where, as usual,

$$F^{(n)}(t) = \frac{d^n}{dt^n}\{F(t)\}, \qquad n = 0, 1, 2, \ldots; \tag{23}$$

$$\mathscr{L}\{e^{-\lambda t}F(t): s\} = f(s+\lambda), \qquad \mathrm{Re}(s+\lambda-\alpha) > 0; \tag{24}$$

$$\mathscr{L}\{t^n F(t): s\} = (-1)^n f^{(n)}(s), \qquad n = 0, 1, 2, \ldots; \tag{25}$$

(26) $$\mathscr{L}\{t^{-n}F(t):s\} = \int_s^\infty \cdots \int_s^\infty f(s)(ds)^n, \qquad n = 0, 1, 2, \ldots;$$

(27) $$\mathscr{L}\left\{\int_0^t \cdots \int_0^t F(t)(dt)^n : s\right\} = s^{-n}f(s), \qquad n = 0, 1, 2, \ldots;$$

(28) $$\mathscr{L}\{F * G(t):s\} = f(s)g(s),$$

where

(29) $$F * G(t) = \int_0^t F(\tau)G(t-\tau)\,d\tau = G * F(t)$$

and, as before,

(30) $$g(s) = \mathscr{L}\{G(t):s\}.$$

Formula (28) is well known in the literature as the *convolution* (or *product*) *theorem* for the Laplace transform.

There are many pairs of Laplace transforms in which the confluent hypergeometric functions occur. If, for convenience, we write the second member of (20) simply as $\mathscr{L}\{F(t)\}$, we have from (14) that

(31) $$\mathscr{L}\{t^{a-1}(1+t)^{c-a-1}\} = \Gamma(a)\Psi(a, c; s),$$
$$\mathrm{Re}(a) > 0, \qquad \mathrm{Re}(s) > 0.$$

Also, by term-by-term integration it can easily be shown that

(32) $$\mathscr{L}\{t^{b-1}\,{}_1F_1(a;c;zt)\}$$
$$= \begin{cases} \dfrac{\Gamma(b)}{s^b}\,{}_2F_1\left(a, b; c; \dfrac{z}{s}\right), & |z| < |s|, \\[2ex] \dfrac{\Gamma(b)}{(s-z)^b}\,{}_2F_1\left[c-a, b; c; \dfrac{z}{z-s}\right], & |z| < |z-s|, \end{cases}$$

provided that

(33) $$\mathrm{Re}(b) > 0 \qquad \text{and} \qquad \mathrm{Re}(s) > \max\{0, \mathrm{Re}(z)\}.$$

Note that the last member of (32) follows from the second by using Euler's transformation II.2(19).

For $b = c$, (32) assumes the simpler form

(34) $$\mathscr{L}\{t^{c-1}\,{}_1F_1(a;c;zt)\} = \frac{\Gamma(c)}{s^c}\left(1 - \frac{z}{s}\right)^{-a}, \qquad |z| < |s|,$$

which holds true under the conditions in (33).

The Laplace transform formula (32) corresponds to the special case $p = q = 1$ of the general result [cf. Slater (1960, p. 43) and Erdélyi *et al.* (1954,

Vol. I, p. 219)]

$$\mathscr{L}\{t^{\nu-1}\,{}_pF_q[\alpha_1,\ldots,\alpha_p;\beta_1,\ldots,\beta_q;zt]\} = \frac{\Gamma(\nu)}{s^\nu}\,{}_{p+1}F_q[\nu,\alpha_1,\ldots,\alpha_p;\beta_1,\ldots,\beta_q;z/s], \tag{35}$$

where, for convergence, *either*

(i) $p < q$, $\mathrm{Re}(\nu) > 0$, and $\mathrm{Re}(s) > 0$, *or*
(ii) $p = q$, $\mathrm{Re}(\nu) > 0$, and $\mathrm{Re}(s) > \mathrm{Re}(z)$,

${}_pF_q$ being the generalized hypergeometric function of Section II.4.

In the special case $p = q - 1 = 0$ of (35), we at once have

$$\mathscr{L}\{t^{a-1}\,{}_0F_1(-;c;zt)\} = \frac{\Gamma(a)}{s^a}\,{}_1F_1\left(a;c;\frac{z}{s}\right), \tag{36}$$

$$\mathrm{Re}(a) > 0, \qquad \mathrm{Re}(s) > 0.$$

And since

$${}_0F_1(-;c;-z) = \Gamma(c)z^{(1/2)(1-c)}J_{c-1}(2\sqrt{z}), \tag{37}$$

where $J_\nu(z)$ is the Bessel function of order ν defined by

$$J_\nu(z) = \sum_{m=0}^{\infty} \frac{(-1)^m(\frac{1}{2}z)^{\nu+2m}}{m!\Gamma(\nu+m+1)}, \qquad |z| < \infty \tag{38}$$

$$= \frac{(\frac{1}{2}z)^\nu}{\Gamma(\nu+1)}\,{}_0F_1\left(-;\nu+1;-\frac{z^2}{4}\right),$$

the Laplace transform formula (36) may be rewritten in its equivalent form,

$$\mathscr{L}\{t^{a-(1+c)/2}J_{c-1}(2\sqrt{zt})\} = \frac{\Gamma(a)}{\Gamma(c)}\,s^{-a}z^{(c-1)/2}\,{}_1F_1\left(a;c;-\frac{z}{s}\right), \tag{39}$$

$$\mathrm{Re}(a) > 0, \qquad \mathrm{Re}(s) > 0,$$

which is *Weber's first exponential integral.*

Similarly, for the modified Bessel function $K_\nu(z)$ defined by

$$K_\nu(z) = \frac{\pi}{2\sin\nu\pi}[I_{-\nu}(z) - I_\nu(z)], \tag{40}$$

where

$$I_\nu(z) = \sum_{m=0}^{\infty} \frac{(\frac{1}{2}z)^{\nu+2m}}{m!\Gamma(\nu+m+1)} = \frac{(\frac{1}{2}z)^\nu}{\Gamma(\nu+1)}\,{}_0F_1\left(-;\nu+1;\frac{z^2}{4}\right) \tag{41}$$

$$= \exp(-\tfrac{1}{2}i\pi\nu)\,J_\nu(iz),$$

we have

$$\mathscr{L}\{t^{a-(1+c)/2}K_{c-1}(2\sqrt{zt})\} = \frac{\Gamma(a)\Gamma(a-c+1)}{2s^a z^{(1-c)/2}}\Psi\left(a, c; \frac{z}{s}\right), \tag{42}$$

$$\operatorname{Re}(a) > 0, \qquad \operatorname{Re}(a-c) > -1, \qquad \operatorname{Re}(s) > 0.$$

Further results involving the Ψ function include

$$\begin{aligned} (43)\quad & \mathscr{L}\{t^{c-1}\,{}_2F_1(a, a-b+1; c; -t)\} \\ & = s^{a-c}\Gamma(c)\Psi(a, c; s), \\ & \operatorname{Re}(c) > 0, \qquad \operatorname{Re}(s) > 0; \end{aligned}$$

$$\begin{aligned} (44)\quad & \mathscr{L}\{t^{b-1}\Psi(a, c; t)\} \\ & = \frac{\Gamma(b)\Gamma(b-c+1)}{\Gamma(a+b-c+1)}\,{}_2F_1(b, b-c+1; a+b-c+1; 1-s), \\ & \operatorname{Re}(b) > 0, \qquad \operatorname{Re}(c) < 1 + \operatorname{Re}(b), \qquad |1-s| < 1; \end{aligned}$$

$$\begin{aligned} (45)\quad & \mathscr{L}\{t^{b-1}\Psi(a, c; t)\} \\ & = \frac{\Gamma(b)\Gamma(b-c+1)}{\Gamma(a+b-c+1)}s^{-b}\,{}_2F_1\left(a, b; a+b-c+1; 1-\frac{1}{s}\right), \\ & \operatorname{Re}(s) > \tfrac{1}{2}. \end{aligned}$$

The Laplace transform formula (45) follows at once from (44) by appealing to Euler's transformation II.2(18).

Another transform pair of interest is

$$\begin{aligned} (46)\quad & \mathscr{L}\{t^{2c-2}\exp(-t^2)\,{}_1F_1(a; c; t^2)\} \\ & = \frac{\Gamma(2c-1)}{2^{2c-1}}\Psi\left(c-\tfrac{1}{2}, a+\tfrac{1}{2}; \frac{s^2}{4}\right), \qquad \operatorname{Re}(c) > \tfrac{1}{2}, \quad \operatorname{Re}(s) > 0. \end{aligned}$$

Finally, we apply (35) in conjunction with the convolution theorem (28), and we obtain

$$\begin{aligned} (47)\quad & \mathscr{L}\left\{\int_0^t \tau^{\lambda-1}\,{}_lF_m[\alpha_1, \ldots, \alpha_l; \beta_1, \ldots, \beta_m; w\tau]\right. \\ & \qquad \left.\cdot (t-\tau)^{\mu-1}\,{}_pF_q[\gamma_1, \ldots, \gamma_p; \delta_1, \ldots, \delta_q; z(t-\tau)]\,d\tau\right\} \\ & = \frac{\Gamma(\lambda)\Gamma(\mu)}{s^{\lambda+\mu}}\,{}_{l+1}F_m\left[\lambda, \alpha_1, \ldots, \alpha_l; \beta_1, \ldots, \beta_m; \frac{w}{s}\right] \\ & \qquad \cdot {}_{p+1}F_q\left[\mu, \gamma_1, \ldots, \gamma_p; \delta_1, \ldots, \delta_q; \frac{z}{s}\right], \end{aligned}$$

which holds true when *either*

(i) $l < m$, $p < q$, and $\min\{\mathrm{Re}(\mu), \mathrm{Re}(v), \mathrm{Re}(s)\} > 0$, *or*
(ii) $l = m$, $p = q$, $\mathrm{Re}(\mu) > 0$, $\mathrm{Re}(v) > 0$, $\mathrm{Re}(s) > \max\{\mathrm{Re}(w), \mathrm{Re}(z)\}$.

For $l = m = p = q = 1$, (47) yields the corresponding result for the product of two confluent hypergeometric ${}_1F_1$ functions. In this case, if we further let $\beta_1 = \lambda$, $\delta_1 = \mu$, and $w = z$ and appeal to the formula (34), we shall obtain

(48)
$$\mathscr{L}\left\{\int_0^t \tau^{\lambda-1}\,{}_1F_1[\alpha;\lambda;z\tau](t-\tau)^{\mu-1}\,{}_1F_1[\gamma;\mu;z(t-\tau)]\,d\tau\right\}$$
$$= B(\lambda,\mu)\mathscr{L}\{t^{\lambda+\mu-1}\,{}_1F_1[\alpha+\gamma;\lambda+\mu;zt]\},$$

whence we have the *integral addition theorem* for the ${}_1F_1$ (or Φ) function:

(49)
$$\int_0^t \tau^{\lambda-1}\,{}_1F_1[\alpha;\lambda;z\tau](t-\tau)^{\mu-1}\,{}_1F_1[\gamma;\mu;z(t-\tau)]\,d\tau$$
$$= B(\lambda,\mu)t^{\lambda+\mu-1}\,{}_1F_1[\alpha+\gamma;\lambda+\mu;zt], \quad \mathrm{Re}(\lambda) > 0, \quad \mathrm{Re}(\mu) > 0.$$

For $\gamma = 0$ and $\lambda + \mu = c$, (37) immediately yields the elegant form

(50)
$$\int_0^t \tau^{\lambda-1}(t-\tau)^{c-\lambda-1}\,{}_1F_1(\alpha;\lambda;z\tau)\,d\tau$$
$$= B(\lambda, c-\lambda)\,{}_1F_1(\alpha;c;zt), \qquad \mathrm{Re}(c) > \mathrm{Re}(\lambda) > 0.$$

Now, from definition in (41) and Kummer's second theorem (8) it is easily verified that [cf., e.g., Buchholz (1969, p. 7)]

(51)
$${}_1F_1(\alpha;2\alpha;z) = \Gamma(\alpha+\tfrac{1}{2})(\tfrac{1}{4}z)^{(1/2)-\alpha}e^{z/2}I_{\alpha-(1/2)}(\tfrac{1}{2}z),$$

which may be used in (38) with $t = 1$ to derive the formula

(52)
$${}_1F_1(a;c;z) = \frac{\Gamma(c)\sqrt{\pi}}{\Gamma(a)\Gamma(c-2a)}z^{(1/2)-a}$$
$$\cdot \int_0^1 \tau^{a-(1/2)}(1-\tau)^{c-2a-1}\exp(\tfrac{1}{2}z\tau)\,I_{a-(1/2)}(\tfrac{1}{2}z\tau)\,d\tau,$$
$$\mathrm{Re}(c) > 2\,\mathrm{Re}(a) > 0.$$

Functions Expressible as Φ and Ψ Functions

For convenience, we shall henceforth use Humbert's notation $\Phi(a;c;z)$ for the confluent hypergeometric function ${}_1F_1(a;c;z)$. We list below a large number of important functions which can be expressed in terms of the Φ and Ψ functions.

Bessel functions [cf. Eqs. (38) and (41)]

(53) $$J_\nu(z) = \frac{(\frac{1}{2}z)^\nu}{\Gamma(\nu+1)} \exp(\pm iz)\,\Phi(\nu+\tfrac{1}{2}; 2\nu+1; \mp 2iz);$$

(54) $$I_\nu(z) = \frac{(\frac{1}{2}z)^\nu}{\Gamma(\nu+1)} \exp(\pm z)\,\Phi(\nu+\tfrac{1}{2}; 2\nu+1; \mp 2z);$$

(55) $$K_\nu(z) = \sqrt{\pi}\,(2z)^\nu e^{-z}\Psi(\nu+\tfrac{1}{2}, 2\nu+1; 2z).$$

Whittaker functions

(56) $$M_{\kappa,\mu}(z) = z^{\mu+(1/2)}\exp(-\tfrac{1}{2}z)\,\Phi(\mu-\kappa+\tfrac{1}{2}; 2\mu+1; z)$$
$$= z^{\mu+(1/2)}\exp(\tfrac{1}{2}z)\,\Phi(\mu+\kappa+\tfrac{1}{2}; 2\mu+1; -z);$$

(57) $$W_{\kappa,\mu}(z) = z^{\mu+(1/2)}\exp(-\tfrac{1}{2}z)\,\Psi(\mu-\kappa+\tfrac{1}{2}, 2\mu+1; z)$$
$$= W_{\kappa,-\mu}(z).$$

Incomplete Gamma functions [see also II.1(52), (53), (55)]

(58) $$\gamma(\alpha, z) = \alpha^{-1}z^\alpha\Phi(\alpha; \alpha+1; -z);$$

(59) $$\Gamma(\alpha, z) = e^{-z}\Psi(1-\alpha, 1-\alpha; z).$$

Error functions [cf. II.1(70)]

(60) $$\mathrm{Erf}(z) = \frac{\sqrt{\pi}}{2}\mathrm{erf}(z) = \tfrac{1}{2}\gamma(\tfrac{1}{2}, z^2)$$
$$= z\Phi(\tfrac{1}{2}; \tfrac{3}{2}; -z^2);$$

(61) $$\mathrm{Erfc}(z) = \frac{\sqrt{\pi}}{2}\mathrm{erfc}(z) = \tfrac{1}{2}\Gamma(\tfrac{1}{2}, z^2)$$
$$= \tfrac{1}{2}\exp(-z^2)\,\Psi(\tfrac{1}{2}, \tfrac{1}{2}; z^2).$$

Parabolic cylinder (or Weber) functions

(62) $$D_\nu(z) = 2^{\nu/2}\exp(-\tfrac{1}{4}z^2)\,\Psi(-\tfrac{1}{2}\nu, \tfrac{1}{2}; \tfrac{1}{2}z^2)$$
$$= 2^{(\nu-1)/2}\exp(-\tfrac{1}{4}z^2)z\,\Psi(\tfrac{1}{2}-\tfrac{1}{2}\nu, \tfrac{3}{2}; \tfrac{1}{2}z^2).$$

Bateman's *k* function [see Bateman (1931) and Srivastava (1972)]

(63) $$k_\nu(z) = \frac{2}{\pi}\int_0^{\pi/2}\cos(z\tan\theta - \nu\theta)\,d\theta = \frac{1}{\Gamma(1+\frac{1}{2}\nu)}W_{\nu/2,\pm 1/2}(2z)$$
$$= \frac{e^{-z}}{\Gamma(1+\frac{1}{2}\nu)}\Psi(-\tfrac{1}{2}\nu, 0; 2z).$$

Hermite polynomials

(64) $$H_n(z) = \sum_{k=0}^{[n/2]} \frac{(-1)^k n!(2z)^{n-2k}}{k!(n-2k)!} = (2z)^n {}_2F_0(-\tfrac{1}{2}n, \tfrac{1}{2} - \tfrac{1}{2}n; -; -1/z^2);$$

(65) $$2^{n/2} He_n(z) \equiv H_n(z/\sqrt{2}) = 2^{n/2} \exp(\tfrac{1}{4}z^2)\, D_n(z)$$
$$= 2^n \Psi(-\tfrac{1}{2}n, \tfrac{1}{2}; \tfrac{1}{2}z^2),$$

where n is a nonnegative integer.

Laguerre functions and polynomials

(66) $$L_\nu^{(\alpha)}(z) = \frac{\Gamma(\nu + \alpha + 1)}{\Gamma(\nu + 1)\Gamma(\alpha + 1)} \Phi(-\nu; \alpha + 1; z);$$

(67) $$L_n^{(\alpha)}(z) = \sum_{k=0}^{n} \binom{n+\alpha}{n-k} \frac{(-z)^k}{k!}$$
$$= \binom{n+\alpha}{n} \Phi(-n; \alpha + 1; z)$$
$$= \frac{(-1)^n}{n!} \Psi(-n, \alpha + 1; z)$$
$$= \frac{(-1)^n}{n!} z^{-\alpha} \Psi(-n - \alpha, 1 - \alpha; z), \qquad n = 0, 1, 2, \ldots .$$

Poisson–Charlier polynomials

(68) $$c_n(x; \alpha) = \sum_{k=0}^{n} (-1)^k \binom{n}{k} \binom{x}{k} k!\alpha^{-k}, \qquad \alpha > 0, \quad x = 0, 1, 2, \ldots$$
$$= n!(-\alpha)^{-n} L_n^{(x-n)}(\alpha)$$
$$= (-\alpha)^{-n} (x - n + 1)_n \Phi(-n; x - n + 1; \alpha).$$

In an alternative notation, we have [see Szegö (1975, p. 35)]

(69) $$p_n(x) = \alpha^{-n/2} \frac{(x - n + 1)_n}{\sqrt{n!}} \Phi(-n; x - n + 1; \alpha).$$

II.4 The Generalized Hypergeometric Function

Generalized Gauss (and Kummer) Series

A natural generalization of the hypergeometric functions ${}_2F_1$, ${}_1F_1$, etc. (considered in the preceding sections), is accomplished by the introduction of an arbitrary number of numerator and denominator parameters. The resulting series,

$$(1)\qquad {}_pF_q\left[\begin{matrix}\alpha_1,\ldots,\alpha_p;\\ \beta_1,\ldots,\beta_q;\end{matrix}\; z\right] = \sum_{n=0}^{\infty}\frac{(\alpha_1)_n\cdots(\alpha_p)_n}{(\beta_1)_n\cdots(\beta_q)_n}\frac{z^n}{n!}$$
$$= {}_pF_q(\alpha_1,\ldots,\alpha_p;\beta_1,\ldots,\beta_q;z),$$

where $(\lambda)_n$ is the Pochhammer symbol defined by II.1(14), is known as the *generalized Gauss (and Kummer) series*, or simply the *generalized hypergeometric series*. Here p and q are positive integers or zero (interpreting an empty product as 1), and we assume that the variable z, the numerator parameters $\alpha_1,\ldots,\alpha_p$, and the denominator parameters $\beta_1,\ldots,\beta_q$ take on complex values, provided that

$$(2)\qquad \beta_j \neq 0, -1, -2, \ldots, \qquad j = 1,\ldots,q.$$

Thus, if a numerator parameter is a negative integer or zero, the ${}_pF_q$ series terminates in view of the identity II.1(23), and we are led to a (generalized hypergeometric) polynomial of the type

$$(3)\qquad {}_{p+1}F_q\left[\begin{matrix}-n,\alpha_1,\ldots,\alpha_p;\\ \beta_1,\ldots,\beta_q;\end{matrix}\; z\right]$$
$$= \sum_{k=0}^{n}\frac{(-n)_k(\alpha_1)_k\cdots(\alpha_p)_k}{(\beta_1)_k\cdots(\beta_q)_k}\frac{z^k}{k!}$$
$$= \frac{(\alpha_1)_n\cdots(\alpha_p)_n}{(\beta_1)_n\cdots(\beta_q)_n}(-z)^n$$
$$\cdot\, {}_{q+1}F_p\left[\begin{matrix}-n, 1-\beta_1-n,\ldots,1-\beta_q-n;\\ 1-\alpha_1-n,\ldots,1-\alpha_p-n;\end{matrix}\; \frac{(-1)^{p+q}}{z}\right],$$
$$n = 0, 1, 2, \ldots,$$

where we have reversed the order of the terms of the polynomial by using II.1(21), (22). [As an interesting application of (3), we can write II.3(67) in the form

$$(4)\qquad L_n^{(\alpha)}(z) = \binom{n+\alpha}{n}\, {}_1F_1(-n;\alpha+1;z)$$
$$= \frac{(-z)^n}{n!}\, {}_2F_0(-n;-\alpha-n;-;-1/z),$$

which follows also from II.3(16) and (67).]

If we suppose that none of the numerator parameters is zero or a negative integer (otherwise the question of convergence will not arise), and with the usual restriction (2), the ${}_pF_q$ series in (1)

$$\text{converges for} \quad |z| < \infty \qquad \text{if} \quad p \leqq q,$$

$$\text{converges for} \quad |z| < 1 \qquad \text{if} \quad p = q + 1,$$

and

$$\text{diverges for all} \quad z, \quad z \neq 0, \qquad \text{if} \quad p > q + 1.$$

Furthermore, if we set

$$\omega = \sum_{j=1}^{q} \beta_j - \sum_{j=1}^{p} \alpha_j, \tag{5}$$

it is known that the ${}_pF_q$ series, with $p = q + 1$, is

$$\text{absolutely convergent for} \quad |z| = 1 \qquad \text{if} \quad \operatorname{Re}(\omega) > 0,$$

$$\text{conditionally convergent for} \quad |z| = 1,\ z \neq 1, \qquad \text{if} \quad -1 < \operatorname{Re}(\omega) \leqq 0,$$

and

$$\text{divergent for} \quad |z| = 1 \qquad \text{if} \quad \operatorname{Re}(\omega) \leqq -1.$$

Analytic Continuation of the ${}_pF_q$ Function

The generalized hypergeometric ${}_pF_q$ function is defined by the (absolutely) convergent series (1) whenever

(i) $p \leqq q$ and $|z| < \infty$, or
(ii) $p = q + 1$ and $|z| < 1$.

As in the case of the ${}_2F_1$ function (see Section II.2), the generalized hypergeometric function ${}_pF_q$ (with $p = q + 1$) can be continued analytically to the domain $|\arg(1 - z)| < \pi$, that is, to the whole z plane cut along the real axis from $z = 1$ to $z = \infty$, by using (1) in conjunction with the Mellin–Barnes contour integral representation:

$$\frac{\Gamma(\alpha_1)\cdots\Gamma(\alpha_p)}{\Gamma(\beta_1)\cdots\Gamma(\beta_q)}\,{}_pF_q\left[\begin{matrix}\alpha_1,\ldots,\alpha_p; \\ \\ \beta_1,\ldots,\beta_q;\end{matrix}\ z\right] \tag{6}$$

$$= \frac{1}{2\pi i}\int_{-i\infty}^{i\infty} \frac{\Gamma(\alpha_1 + \zeta)\cdots\Gamma(\alpha_p + \zeta)\Gamma(-\zeta)}{\Gamma(\beta_1 + \zeta)\cdots\Gamma(\beta_q + \zeta)}(-z)^{\zeta}\,d\zeta,$$

$$\alpha_j \neq 0, -1, -2, \ldots; \qquad j = 1, \ldots, p; \quad |\arg(-z)| < \pi,$$

where the path of integration is the imaginary axis (in the complex ζ plane) which is indented, if necessary, in order to separate the poles of $\Gamma(\alpha_j + \zeta)$, $j = 1, \ldots, p$, from those of $\Gamma(-\zeta)$.

Euler, Laplace, and Mellin Transforms of the ${}_pF_q$ Function

By appealing to the Eulerian integral II.1(43) and the definition in (1), we can easily derive the Euler (or beta) transform of the ${}_pF_q$ function in the form

$$(7)\quad \int_0^1 t^{\nu-1}(1-t)^{\sigma-\nu-1}\,{}_pF_q\left[\begin{matrix}\alpha_1,\ldots,\alpha_p;\\ \beta_1,\ldots,\beta_q;\end{matrix}\; zt\right]dt$$

$$= \frac{\Gamma(\nu)\Gamma(\sigma-\nu)}{\Gamma(\sigma)}\,{}_{p+1}F_{q+1}\left[\begin{matrix}\nu,\alpha_1,\ldots,\alpha_p;\\ \sigma,\beta_1,\ldots,\beta_q;\end{matrix}\; z\right],$$

$$p \leqq q+1, \qquad \mathrm{Re}(\sigma) > \mathrm{Re}(\nu) > 0; \qquad |z| < 1 \quad \text{if} \quad p = q+1.$$

Next we apply the Γ-function formula II.1(1) and the definition in (1), and we obtain the Laplace transform of the ${}_pF_q$ function:

$$(8)\quad \int_0^\infty e^{-st}t^{\nu-1}\,{}_pF_q\left[\begin{matrix}\alpha_1,\ldots,\alpha_p;\\ \beta_1,\ldots,\beta_q;\end{matrix}\; zt\right]dt$$

$$= \frac{\Gamma(\nu)}{s^\nu}\,{}_{p+1}F_q\left[\begin{matrix}\nu,\alpha_1,\ldots,\alpha_p;\\ \beta_1,\ldots,\beta_q;\end{matrix}\; \frac{z}{s}\right],$$

$$p \leqq q, \qquad \mathrm{Re}(\nu) > 0; \qquad \mathrm{Re}(s) > 0 \quad \text{if} \quad p < q,$$

$$\mathrm{Re}(s) > \mathrm{Re}(z) \quad \text{if} \quad p = q.$$

The Mellin transform of the ${}_pF_q$ function can be derived from (8) by letting $s \to 0$, provided we make use of suitable limiting forms on the right-hand side of (8). Alternatively, we can apply the *Mellin inversion theorem* to the representation (6) above, and we finally obtain

$$(9)\quad \int_0^\infty t^{\zeta-1}\,{}_pF_q\left[\begin{matrix}\alpha_1,\ldots,\alpha_p;\\ \beta_1,\ldots,\beta_q;\end{matrix}\; -zt\right]dt$$

$$= \frac{\Gamma(\alpha_1-\zeta)\cdots\Gamma(\alpha_p-\zeta)\Gamma(\beta_1)\cdots\Gamma(\beta_q)}{\Gamma(\beta_1-\zeta)\cdots\Gamma(\beta_q-\zeta)\Gamma(\alpha_1)\cdots\Gamma(\alpha_p)}\Gamma(\zeta)z^{-\zeta},$$

$$z \neq 0, \qquad 0 < \mathrm{Re}(\zeta) < \min_{1\leqq j\leqq p}\{\mathrm{Re}(\alpha_j)\};$$

$$p = q-1, \qquad \mathrm{Re}(2\zeta-\omega) < \tfrac{1}{2}, \qquad z > 0;$$

$$p = q, \qquad |\arg(z)| < \tfrac{1}{2}\pi \quad \text{or} \quad |\arg(z)| = \tfrac{1}{2}\pi \quad \text{if} \quad \mathrm{Re}(\zeta-\omega) < 1;$$

$$p = q+1, \qquad |\arg(z)| < \pi.$$

Several special cases of (9) will be needed in the following chapters.

Functions Expressible in Terms of the ${}_pF_q$

The following list is in addition to the important functions (considered in the preceding sections) which can be expressed as a hypergeometric ${}_1F_1$ or ${}_2F_1$ function.

Elementary functions

(10) $$e^z = {}_0F_0[-;-;z];$$

(11) $$(1-z)^{-\alpha} = {}_1F_0[\alpha;-;z], \qquad |z|<1;$$

(12) $$\cos z = {}_0F_1[-;\tfrac{1}{2};-z^2/4];$$

(13) $$\sin z = z\,{}_0F_1[-;\tfrac{3}{2};-z^2/4].$$

Bessel functions [see also II.3(53), (54)]

(14) $$J_\nu(z) = \frac{(\frac{1}{2}z)^\nu}{\Gamma(\nu+1)}\,{}_0F_1\left[-;\nu+1;-\frac{z^2}{4}\right];$$

(15) $$I_\nu(z) = \frac{(\frac{1}{2}z)^\nu}{\Gamma(\nu+1)}\,{}_0F_1\left[-;\nu+1;\frac{z^2}{4}\right].$$

Lommel functions

(16) $$s_{\mu,\nu}(z) = \frac{z^{\mu+1}}{(\mu-\nu+1)(\mu+\nu+1)} \cdot {}_1F_2\left[\begin{matrix}1; \\ \frac{1}{2}(\mu-\nu+3), \frac{1}{2}(\mu+\nu+3);\end{matrix} -\frac{z^2}{4}\right];$$

(17) $$S_{\mu,\nu}(z) = s_{\mu,\nu}(z) + 2^{\mu-1}\Gamma\left(\frac{\mu-\nu+1}{2}\right)\Gamma\left(\frac{\mu+\nu+1}{2}\right) \cdot \{\sin[\tfrac{1}{2}(\mu-\nu)\pi]\,J_\nu(z) - \cos[\tfrac{1}{2}(\mu-\nu)\pi]\,Y_\nu(z)\},$$

where $J_\nu(z)$ is given by (14), and

(18) $$Y_\nu(z) = \csc(\nu\pi)\,[\cos(\nu\pi)\,J_\nu(z) - J_{-\nu}(z)].$$

Struve functions

(19) $$\mathbf{H}_\nu(z) = \frac{(\frac{1}{2}z)^{\nu+1}}{\Gamma(\frac{3}{2})\Gamma(\nu+\frac{3}{2})}\,{}_1F_2\left[\begin{matrix} 1; \\ \frac{3}{2}, \nu+\frac{3}{2}; \end{matrix}\; -\frac{z^2}{4}\right]$$

$$= \frac{2^{1-\nu}}{\sqrt{\pi}\,\Gamma(\nu+\frac{1}{2})} s_{\nu,\nu}(z)$$

$$= Y_\nu(z) + \frac{2^{1-\nu}}{\sqrt{\pi}\,\Gamma(\nu+\frac{1}{2})} S_{\nu,\nu}(z);$$

(20) $$\mathbf{L}_\nu(z) = \frac{(\frac{1}{2}z)^{\nu+1}}{\Gamma(\frac{3}{2})\Gamma(\nu+\frac{3}{2})}\,{}_1F_2\left[\begin{matrix} 1; \\ \frac{3}{2}, \nu+\frac{3}{2}; \end{matrix}\; \frac{z^2}{4}\right]$$

$$= \exp[-\tfrac{1}{2}(\nu+1)i\pi]\mathbf{H}_\nu(iz).$$

II.5 The *E*, *G*, and *H* Functions

MacRobert's E function and Meijer's G function arose from *independent* attempts to give a meaning to the symbol ${}_pF_q$ when $p > q + 1$. Meijer's interpretation leading to his G function is in complete agreement with the one provided by MacRobert's E function. Furthermore, all significant particular solutions of a generalized hypergeometric differential equation may be expressed in terms of the G function.

The G function is defined by [cf. II.4(6)]

(1) $$G_{p,q}^{m,n}\left(z\,\middle|\,\begin{matrix}\alpha_1,\ldots,\alpha_p\\ \beta_1,\ldots,\beta_q\end{matrix}\right) = \frac{1}{2\pi i}\int_L \frac{\prod_{j=1}^m \Gamma(\beta_j-\zeta)\prod_{j=1}^n \Gamma(1-\alpha_j+\zeta)}{\prod_{j=m+1}^q \Gamma(1-\beta_j+\zeta)\prod_{j=n+1}^p \Gamma(\alpha_j-\zeta)} z^\zeta\, d\zeta,$$

where $z \neq 0$; an empty product is interpreted as 1; m, n, p, and q are integers such that $0 \leqq m \leqq q$ and $0 \leqq n \leqq p$; and the α and β parameters are so constrained that no pole of $\Gamma(\beta_j - \zeta), j = 1, \ldots, m$, coincides with any pole of $\Gamma(1 - \alpha_j + \zeta), j = 1, \ldots, n$. The three different paths L of integration in (1) are discussed, among others, by Erdélyi *et al.* (1953, Vol. I, p. 207) and Luke (1969, Vol. I, p. 144). For example, L is a Mellin–Barnes contour of the type considered in II.4(6); it runs from $-i\infty$ to $i\infty$ with indentations, if necessary, so that all poles of $\Gamma(\beta_j - \zeta), j = 1, \ldots, m$, lie to the right of the contour, and all poles of $\Gamma(1 - \alpha_j + \zeta), j = 1, \ldots, n$, lie to the left of the contour. The integral (1) converges in this case if

$$(2) \qquad \Lambda \equiv m + n - \tfrac{1}{2}(p + q) > 0 \qquad \text{and} \qquad |\arg(z)| < \Lambda\pi.$$

If $|\arg(z)| = \Lambda\pi$, $\Lambda \geqq 0$, then the integral (1) converges absolutely when $p = q$ if $\mathrm{Re}(\omega) < -1$; and when $p \neq q$, if ξ is so chosen that, for $\eta \to \pm\infty$,

$$(3) \qquad (q - p)\xi > 1 - \tfrac{1}{2}(q - p) + \mathrm{Re}(\omega),$$

where ω is defined by II.4(5), ξ and η being, respectively, the real and imaginary parts of ζ occurring in the definition (1).

The G function is an analytic function of z with a branch point at the origin. It is symmetric in the parameters

$$\alpha_1, \ldots, \alpha_n; \qquad \alpha_{n+1}, \ldots, \alpha_p; \qquad \beta_1, \ldots, \beta_m; \qquad \text{and} \qquad \beta_{m+1}, \ldots, \beta_q.$$

The following elementary properties of the G function follow readily from the definition in (1):

$$(4) \qquad G_{p,q}^{m,n}\left(z \,\middle|\, \begin{matrix} \alpha_1, \ldots, \alpha_p \\ \beta_1, \ldots, \beta_{q-1}, \alpha_1 \end{matrix}\right) = G_{p-1,q-1}^{m,n-1}\left(z \,\middle|\, \begin{matrix} \alpha_2, \ldots, \alpha_p \\ \beta_1, \ldots, \beta_{q-1} \end{matrix}\right),$$

$$n, p, q \geqq 1;$$

$$(5) \qquad G_{p,q}^{m,n}\left(z \,\middle|\, \begin{matrix} \alpha_1, \ldots, \alpha_{p-1}, \beta_1 \\ \beta_1, \ldots, \beta_q \end{matrix}\right) = G_{p-1,q-1}^{m-1,n}\left(z \,\middle|\, \begin{matrix} \alpha_1, \ldots, \alpha_{p-1} \\ \beta_2, \ldots, \beta_q \end{matrix}\right),$$

$$m, p, q \geqq 1;$$

$$(6) \qquad G_{p,q}^{m,n}\left(z \,\middle|\, \begin{matrix} \alpha_1, \ldots, \alpha_p \\ \beta_1, \ldots, \beta_q \end{matrix}\right) = G_{q,p}^{n,m}\left(\frac{1}{z} \,\middle|\, \begin{matrix} 1 - \beta_1, \ldots, 1 - \beta_q \\ 1 - \alpha_1, \ldots, 1 - \alpha_p \end{matrix}\right),$$

which exhibits the fact that (in the discussion of the G function) we can, without loss of generality, suppose that $p \leqq q$;

$$(7) \qquad z^{\sigma} G_{p,q}^{m,n}\left(z \,\middle|\, \begin{matrix} \alpha_1, \ldots, \alpha_p \\ \beta_1, \ldots, \beta_q \end{matrix}\right) = G_{p,q}^{m,n}\left(z \,\middle|\, \begin{matrix} \alpha_1 + \sigma, \ldots, \alpha_p + \sigma \\ \beta_1 + \sigma, \ldots, \beta_q + \sigma \end{matrix}\right);$$

$$(8) \qquad G_{p,q}^{m,n}\left(z \,\middle|\, \begin{matrix} \alpha_1, \ldots, \alpha_p \\ \beta_1, \ldots, \beta_q \end{matrix}\right) = N^{\theta}(2\pi)^{(1-N)\Lambda}$$

$$\cdot\, G_{pN,qN}^{mN,nN}\left(N^{(p-q)N} z^N \,\middle|\, \begin{matrix} \Delta(N;\alpha_1), \ldots, \Delta(N;\alpha_p) \\ \Delta(N;\beta_1), \ldots, \Delta(N;\beta_q) \end{matrix}\right),$$

$$N = 1, 2, 3, \ldots; \qquad \theta = 1 + \tfrac{1}{2}(p - q) + \omega,$$

where, for convenience, $\Delta(N;\lambda)$ represents the array of N parameters $(\lambda+j-1)/N$, $j=1,\ldots,N$, Λ is defined by (2), and ω by II.4(5). [See, for details, Srivastava (1979, p. 195); see also Mathai and Saxena (1973, p. 6) and Luke (1969, Vol. I, p. 150).]

A comparison of (1) and II.4(6) yields the relation

$$\text{(9)}\qquad \frac{\Gamma(\alpha_1)\cdots\Gamma(\alpha_p)}{\Gamma(\beta_1)\cdots\Gamma(\beta_q)}\,{}_pF_q\left[\begin{matrix}\alpha_1,\ldots,\alpha_p;\\ \beta_1,\ldots,\beta_q;\end{matrix}\ z\right]$$

$$= G^{1,p}_{p,q+1}\left(-z\,\middle|\,\begin{matrix}1-\alpha_1,\ldots,1-\alpha_p\\ 0,1-\beta_1,\ldots,1-\beta_q\end{matrix}\right)$$

$$= G^{p,1}_{q+1,p}\left(-\frac{1}{z}\,\middle|\,\begin{matrix}1,\beta_1,\ldots,\beta_q\\ \alpha_1,\ldots,\alpha_p\end{matrix}\right)\qquad\text{[by (6)]},$$

$$p\leqq q\qquad(\text{or}\quad p=q+1\qquad\text{and}\qquad |z|<1).$$

Thus every special function expressible in terms of the ${}_pF_q$ function is automatically contained in the G function. The latter, however, can be specialized to a number of other interesting functions not contained in ${}_pF_q$. For example, we have MacRobert's E function

$$\text{(10)}\qquad E(p;\alpha_1,\ldots,\alpha_p:q;\beta_1,\ldots,\beta_q:z) = G^{p,1}_{q+1,p}\left(z\,\middle|\,\begin{matrix}1,\beta_1,\ldots,\beta_q\\ \alpha_1,\ldots,\alpha_p\end{matrix}\right);$$

the Bessel function [cf. II.4(18) and II.3(55)]

$$\text{(11)}\qquad z^\mu Y_\nu(z) = (-1)^m 2^\mu G^{2,0}_{1,3}\left(\frac{z^2}{4}\,\middle|\,\begin{matrix}\frac{1}{2}(\mu-\nu-1)-m\\ \frac{1}{2}(\mu+\nu),\frac{1}{2}(\mu-\nu),\frac{1}{2}(\mu-\nu-1)-m\end{matrix}\right),$$

$$m=0,\pm1,\pm2,\ldots,$$

$$\text{(12)}\qquad z^\mu K_\nu(z) = 2^{\mu-1}G^{2,0}_{0,2}\left(\frac{z^2}{4}\,\middle|\,\begin{matrix}\overline{\qquad}\\ \frac{1}{2}(\mu+\nu),\frac{1}{2}(\mu-\nu)\end{matrix}\right)$$

$$= \sqrt{\pi}\,2^{-\mu}e^z G^{2,0}_{1,2}\left(2z\,\middle|\,\begin{matrix}\mu+\frac{1}{2}\\ \mu+\nu,\mu-\nu\end{matrix}\right)$$

$$= \frac{\cos(\nu\pi)}{\sqrt{\pi}}2^{-\mu}e^z G^{2,1}_{1,2}\left(2z\,\middle|\,\begin{matrix}\mu+\frac{1}{2}\\ \mu+\nu,\mu-\nu\end{matrix}\right);$$

the Whittaker function [cf. II.3(57)]

$$(13)\qquad z^l W_{k,m}(z) = \exp(\tfrac{1}{2}z)\, G^{2,0}_{1,2}\left(z \,\middle|\, \begin{matrix} l-k+1 \\ l+m+\tfrac{1}{2}, l-m+\tfrac{1}{2} \end{matrix} \right)$$

$$= \frac{\exp(-\tfrac{1}{2}z)}{\Gamma(\tfrac{1}{2}+m-k)\Gamma(\tfrac{1}{2}-m-k)} \cdot G^{2,1}_{1,2}\left(z \,\middle|\, \begin{matrix} k+l+1 \\ l+m+\tfrac{1}{2}, l-m+\tfrac{1}{2} \end{matrix} \right);$$

and the Lommel function [cf. II.4(17)]

$$(14)\qquad (\tfrac{1}{2}z)^\lambda S_{\mu,\nu}(z) = \frac{2^{\mu-1}}{\Gamma[\tfrac{1}{2}(1-\mu+\nu)]\Gamma[\tfrac{1}{2}(1-\mu-\nu)]} \cdot G^{3,1}_{1,3}\left(\frac{z^2}{4} \,\middle|\, \begin{matrix} \tfrac{1}{2}(\nu+\lambda+1) \\ \tfrac{1}{2}(\lambda+\nu), \tfrac{1}{2}(\lambda-\nu), \tfrac{1}{2}(\nu+\lambda+1) \end{matrix} \right).$$

[For an extensive list of important special functions, their suitable products and combinations, which can be expressed as G functions, see Erdélyi *et al.* (1953, Vol. I, pp. 215–222) and Luke (1969, Vol. I, pp. 225–234).]

Many of the properties of the G function extend appropriately to the H function, which indeed contains the G function as a special case. A study of one form or the other of the H function, which was initiated as long ago as 1888 by S. Pincherle, appeared in the works of E. W. Barnes in 1908, H. Mellin in 1910, A. L. Dixon and W. L. Ferrar in 1936, S. Bochner in 1958, and several others [see, for details, Erdélyi *et al.* (1953, Vol. I, p. 49, §1.19)]. A first systematic discussion of the properties of the H function as a symmetric Fourier kernel was incorporated in a recent paper by C. Fox (1897–1977), whose name seems to have been associated with this function ever since [cf. Fox (1961)].

We recall the definition of the H function in the form

$$(15)\quad H^{m,n}_{p,q}\left[z \,\middle|\, \begin{matrix} (\alpha_1, A_1), \ldots, (\alpha_p, A_p) \\ (\beta_1, B_1), \ldots, (\beta_q, B_q) \end{matrix} \right] = \frac{1}{2\pi i} \int_L \frac{\prod_{j=1}^m \Gamma(\beta_j - B_j\zeta) \prod_{j=1}^n \Gamma(1-\alpha_j + A_j\zeta)}{\prod_{j=m+1}^q \Gamma(1-\beta_j + B_j\zeta) \prod_{j=n+1}^p \Gamma(\alpha_j - A_j\zeta)} z^\zeta \, d\zeta,$$

where L is a suitable contour of the Mellin–Barnes type [as, for example, in II.4(6) and II.5(1)] separating the poles of

$$\Gamma(\beta_j - B_j\zeta), \qquad j = 1, \ldots, m,$$

from those of

$$\Gamma(1-\alpha_j+A_j\zeta), \qquad j=1,\ldots,n,$$

an empty product is interpreted as 1, the integers m, n, p, and q satisfy the inequalities

$$0 \leqq m \leqq q \qquad \text{and} \qquad 0 \leqq n \leqq p,$$

the coefficients $A_1,\ldots,A_p$ and $B_1,\ldots,B_q$ are positive real numbers, and the complex parameters $\alpha_1,\ldots,\alpha_p$ and $\beta_1,\ldots,\beta_q$ are so constrained that no poles of the integrand coincide. If we set

$$\Omega = \sum_{j=1}^{n} A_j - \sum_{j=n+1}^{p} A_j + \sum_{j=1}^{m} B_j - \sum_{j=m+1}^{q} B_j, \tag{16}$$

then, for $\Omega > 0$, the integral in (15) is absolutely convergent and defines the H function, which is analytic in the sector

$$|\arg(z)| < \tfrac{1}{2}\Omega\pi, \tag{17}$$

the point $z = 0$ being tacitly excluded.

The H function makes sense also when *either* [see Braaksma (1964)]

$$\delta \equiv \sum_{j=1}^{q} B_j - \sum_{j=1}^{p} A_j > 0 \qquad \text{and} \qquad 0 < |z| < \infty \tag{18}$$

or

$$\delta = 0 \qquad \text{and} \qquad 0 < |z| < R \equiv \prod_{j=1}^{p} A_j^{-A_j} \prod_{j=1}^{q} B_j^{B_j}. \tag{19}$$

It is easy to observe that the H function would reduce to the G function in the *trivial* cases when $A_j = B_k = C$, $C > 0$, $j = 1,\ldots,p$ and $k = 1,\ldots,q$, since the definitions (1) and (15) readily yield the relation

$$H_{p,q}^{m,n}\left[z\left|\begin{matrix}(\alpha_1,C),\ldots,(\alpha_p,C)\\ \\ (\beta_1,C),\ldots,(\beta_q,C)\end{matrix}\right.\right] = \frac{1}{C}G_{p,q}^{m,n}\left(z^{1/C}\left|\begin{matrix}\alpha_1,\ldots,\alpha_p\\ \\ \beta_1,\ldots,\beta_q\end{matrix}\right.\right), \quad C>0, \tag{20}$$

or when the A_j and B_k are *positive* rational numbers [see Boersma (1962)]. Thus the H function contains, as its special cases, all of the functions that are expressible in terms of the G function. More important, it contains the generalized hypergeometric function ${}_p\Psi_q$ and the generalized Bessel function $\phi(\alpha,\beta;z)$ or J_ν^μ, considered by Wright (1933, 1934, 1935, 1940), the Mittag–Leffler function E_α and its such generalizations as $E_{\alpha,\beta}$ studied in the literature [Erdélyi *et al.* (1955, pp. 206–212); see also Srivastava (1968)], and so on. These (additional) special cases of the H function, evidently not contained in the G function, are given below:

(21)
$$
{}_p\Psi_q\left[\begin{matrix}(\alpha_1, A_1), \ldots, (\alpha_p, A_p); \\ (\beta_1, B_1), \ldots, (\beta_q, B_q);\end{matrix}\ z\right]
= \sum_{n=0}^{\infty} \frac{\prod_{j=1}^{p} \Gamma(\alpha_j + A_j n)}{\prod_{j=1}^{q} \Gamma(\beta_j + B_j n)} \frac{z^n}{n!}
= H^{1,p}_{p,q+1}\left[-z \left| \begin{matrix}(1-\alpha_1, A_1), \ldots, (1-\alpha_p, A_p) \\ (0,1), (1-\beta_1, B_1), \ldots, (1-\beta_q, B_q)\end{matrix}\right.\right];
$$

(22)
$$
J_\nu^\mu(z) = \sum_{n=0}^{\infty} \frac{(-z)^n}{n!\,\Gamma(\mu n + \nu + 1)} = \phi(\mu, \nu + 1; -z)
= H^{1,0}_{0,2}\left[z \left| \begin{matrix}\overline{} \\ (0,1), (-\nu, \mu)\end{matrix}\right.\right];
$$

(23)
$$
E_\alpha(z) = \sum_{n=0}^{\infty} \frac{z^n}{\Gamma(n\alpha + 1)} = H^{1,1}_{1,2}\left[-z \left| \begin{matrix}(0,1) \\ (0,1), (0,\alpha)\end{matrix}\right.\right];
$$

(24)
$$
E_{\alpha,\beta}(z) = z^{(\beta-1)/\alpha} \sum_{n=0}^{\infty} \frac{z^n}{\Gamma(n\alpha + \beta)}
= z^{(\beta-1)/\alpha} H^{1,1}_{1,2}\left[-z \left| \begin{matrix}(0,1) \\ (0,1), (1-\beta, \alpha)\end{matrix}\right.\right].
$$

Indeed, the last three functions are expressible also in terms of the Wright function ${}_p\Psi_q$. Furthermore, by using an obvious generalization of (6) in the form

(25)
$$
H^{m,n}_{p,q}\left[z \left| \begin{matrix}(\alpha_1, A_1), \ldots, (\alpha_p, A_p) \\ (\beta_1, B_1), \ldots, (\beta_q, B_q)\end{matrix}\right.\right]
= H^{n,m}_{q,p}\left[\frac{1}{z} \left| \begin{matrix}(1-\beta_1, B_1), \ldots, (1-\beta_q, B_q) \\ (1-\alpha_1, A_1), \ldots, (1-\alpha_p, A_p)\end{matrix}\right.\right],
$$

which follows from the definition in (15), we can obtain alternative expressions for each of the above functions in terms of the H function.

For several interesting properties and applications of the H function, including its Hankel, Laplace, Mellin, and other transforms, see, for example, Fox (1961, pp. 408–429), Braaksma (1964), Gupta and Jain (1966), Srivastava and Daoust (1969a, pp. 451–453), Srivastava (1972a,b), Srivastava and Singhal (1972), Srivastava and Panda (1973, 1976a), Srivastava and Buschman (1974, 1976), Srivastava *et al.* (1975), Buschman and Srivastava (1975), Mathai and Saxena (1978), H. M. Srivastava (1980), and Srivastava and Raina (1981).

II.6 Hypergeometric Functions of Two Variables

Appell Functions

The great success of the theory of hypergeometric functions of a single variable has stimulated the development of a corresponding theory in two and more variables. In 1880, P. Appell considered the product of two Gaussian functions, viz.,

(1)
$$
\begin{aligned}
&{}_2F_1(a,b;c;x)\,{}_2F_1(a',b';c';y)\\
&\quad = \sum_{m,n=0}^{\infty} \frac{(a)_m(a')_n(b)_m(b')_n}{(c)_m(c')_n}\frac{x^m}{m!}\frac{y^n}{n!}.
\end{aligned}
$$

This double series, in itself, yields nothing new, but if one or more of the three pairs of products

$$(a)_m(a')_n, \qquad (b)_m(b')_n, \qquad (c)_m(c')_n$$

be replaced by the corresponding expressions

$$(a)_{m+n}, \qquad (b)_{m+n}, \qquad (c)_{m+n},$$

we are led to five distinct possibilities of getting new functions. One such possibility, however, gives us the double series

$$\sum_{m,n=0}^{\infty} \frac{(a)_{m+n}(b)_{m+n}}{(c)_{m+n}}\frac{x^m}{m!}\frac{y^n}{n!},$$

which is simply the Gaussian series for

$${}_2F_1(a,b;c;x+y),$$

since it is easily verified that [cf., e.g., Srivastava (1971, p. 4)]

(2)
$$\sum_{m,n=0}^{\infty} f(m+n)\frac{x^m}{m!}\frac{y^n}{n!} = \sum_{N=0}^{\infty} f(N)\frac{(x+y)^N}{N!},$$

or, more generally, that [Srivastava (1971, p. 4, Eq. (12))]

(3)
$$
\begin{aligned}
&\sum_{m_1,\ldots,m_n=0}^{\infty} f(m_1+\cdots+m_n)\frac{x_1^{m_1}}{m_1!}\cdots\frac{x_n^{m_n}}{m_n!}\\
&\quad = \sum_{m=0}^{\infty} f(m)\frac{(x_1+\cdots+x_n)^m}{m!}.
\end{aligned}
$$

The remaining four possibilities lead to the four Appell functions of two variables, which are defined below:

(4) $F_1[a, b, b'; c; x, y]$

$$= \sum_{m,n=0}^{\infty} \frac{(a)_{m+n}(b)_m(b')_n}{(c)_{m+n}} \frac{x^m}{m!} \frac{y^n}{n!}$$

$$= \sum_{m=0}^{\infty} \frac{(a)_m(b)_m}{(c)_m} {}_2F_1\left[\begin{matrix} a+m, b'; \\ c+m; \end{matrix} y\right] \frac{x^m}{m!}, \qquad \max\{|x|, |y|\} < 1;$$

(5) $F_2[a, b, b'; c, c'; x, y]$

$$= \sum_{m,n=0}^{\infty} \frac{(a)_{m+n}(b)_m(b')_n}{(c)_m(c')_n} \frac{x^m}{m!} \frac{y^n}{n!}$$

$$= \sum_{m=0}^{\infty} \frac{(a)_m(b)_m}{(c)_m} {}_2F_1\left[\begin{matrix} a+m, b'; \\ c'; \end{matrix} y\right] \frac{x^m}{m!}, \qquad |x| + |y| < 1;$$

(6) $F_3[a, a', b, b'; c; x, y]$

$$= \sum_{m,n=0}^{\infty} \frac{(a)_m(a')_n(b)_m(b')_n}{(c)_{m+n}} \frac{x^m}{m!} \frac{y^n}{n!}$$

$$= \sum_{m=0}^{\infty} \frac{(a)_m(b)_m}{(c)_m} {}_2F_1\left[\begin{matrix} a', b'; \\ c+m; \end{matrix} y\right] \frac{x^m}{m!}, \qquad \max\{|x|, |y|\} < 1;$$

(7) $F_4[a, b; c, c'; x, y]$

$$= \sum_{m,n=0}^{\infty} \frac{(a)_{m+n}(b)_{m+n}}{(c)_m(c')_n} \frac{x^m}{m!} \frac{y^n}{n!}$$

$$= \sum_{m=0}^{\infty} \frac{(a)_m(b)_m}{(c)_m} {}_2F_1\left[\begin{matrix} a+m, b+m; \\ c'; \end{matrix} y\right] \frac{x^m}{m!}, \qquad \sqrt{|x|} + \sqrt{|y|} < 1;$$

here, as usual, the denominator parameters c and c' are neither zero nor negative integers.

The standard work on the theory of Appell functions is the monograph by Appell and Kampé de Fériet (1926), which contains an extensive bibliography of all relevant papers up to 1926 (by, for example, L. Pochhammer, J. Horn, E. Picard, and E. Goursat). [See Erdélyi *et al.* (1953, Vol. I, pp. 222–245) for a review of the subsequent work on the subject, and Srivastava (1969, 1973, 1974a).]

Particular Cases of the Appell Functions

All four Appell functions reduce to the Gauss hypergeometric function

$$ {}_2F_1(a, b; c; x) $$

when $y = 0$. The first two functions also reduce to the Gauss function when $b' = 0$, while the third does so when a' or b' is zero. Apart from these obvious particular cases of Appell's functions, we note the following:

(8) $$F_1[a, b, b'; a; x, y] = (1 - x)^{-b}(1 - y)^{-b'};$$

(9) $$F_2[a, b, b'; b, b'; x, y] = (1 - x - y)^{-a};$$

(10) $$F_2[a, b, b'; a, b'; x, y] = (1 - y)^{b-a}(1 - x - y)^{-b};$$

(11) $$F_2[a, b, b'; b, a; x, y] = (1 - x)^{b'-a}(1 - x - y)^{-b'};$$

(12) $$xF_2[1, 1, b'; a, b'; x, y] + yF_2[1, b, 1; b, a; x, y] = -\ln(1 - x - y);$$

and

(13) $$F_3[a, b, 1, 1; a + b; x, y] = (x + y - xy)^{-1}\left\{x\,{}_2F_1\left[\begin{matrix} a, 1; \\ a + b; \end{matrix}\; x\right] + y\,{}_2F_1\left[\begin{matrix} b, 1; \\ a + b; \end{matrix}\; y\right]\right\},$$

which is due to Srivastava (1972b, p. 1259, Eq. (2.2)).

Another result of interest is the reduction formula [Appell and Kampé de Fériet (1926), p. 23, Eq. (25)]

(14) $$F_1[a, b, b'; c; x, x] = {}_2F_1(a, b + b'; c; x),$$

which, in view of Gauss's summation theorem [cf. II.2(7)], yields the well-known result

(15) $$F_1[a, b, b'; c; 1, 1] = \frac{\Gamma(c)\Gamma(c - a - b - b')}{\Gamma(c - a)\Gamma(c - b - b')},$$

$$c \neq 0, -1, -2, \ldots; \qquad \mathrm{Re}(c - a - b - b') > 0.$$

For the Appell function F_4, we have

(16) $$F_4[a, b; c, c'; x, x] = {}_4F_3\left[\begin{matrix} a, b, \tfrac{1}{2}(c + c'), \tfrac{1}{2}(c + c' - 1); \\ c, c', c + c' - 1; \end{matrix}\; 4x\right],$$

which was obtained by Burchnall (1942, p. 101) while investigating solutions of equivalent systems of hypergeometric differential equations; and

(17) $$F_4[a, b; c, c; x, -x] = {}_4F_3\left[\begin{matrix} \tfrac{1}{2}a, \tfrac{1}{2}(a + 1), \tfrac{1}{2}b, \tfrac{1}{2}(b + 1); \\ c, \tfrac{1}{2}c, \tfrac{1}{2}(c + 1); \end{matrix}\; -4x^2\right],$$

which was given by Srivastava (1973, p. 296, Eq. (9)). In fact, Srivastava (1973, p. 295, Eq. (4) and (5); p. 297, Eq. (16) and (17)) proved much more general reduction formulas than (14), (16), and (17) contained in the following theorem.

Theorem [Srivastava (1973)] *Let $\{c_n\}$ be a sequence of arbitrary complex numbers. Then*

(18) $$\sum_{m,n=0}^{\infty} \frac{c_{m+n}}{(\nu)_m(\sigma)_n} \frac{x^{m+n}}{m!n!} = \sum_{n=0}^{\infty} \frac{c_n(\nu+\sigma+n-1)_n}{(\nu)_n(\sigma)_n} \frac{x^n}{n!},$$

(19) $$\sum_{m,n=0}^{\infty} (-1)^n \frac{c_{m+n}}{(\nu)_m(\nu)_n} \frac{x^{m+n}}{m!n!} = \sum_{n=0}^{\infty} (-1)^n \frac{c_{2n}}{(\nu)_n(\nu)_{2n}} \frac{x^{2n}}{n!},$$

(20) $$\sum_{m,n=0}^{\infty} c_{m+n}(\nu)_m(\sigma)_n \frac{x^{m+n}}{m!n!} = \sum_{n=0}^{\infty} c_n(\nu+\sigma)_n \frac{x^n}{n!},$$

and

(21) $$\sum_{m,n=0}^{\infty} (-1)^n c_{m+n}(\nu)_m(\nu)_n \frac{x^{m+n}}{m!n!} = \sum_{n=0}^{\infty} c_{2n}(\nu)_n \frac{x^{2n}}{n!},$$

provided that each of the series involved is absolutely convergent.

This last identity (21) applies to Appell's function F_1 to yield a reduction formula corresponding to (17). [See also H. M. Srivastava (1981) for various generalizations of the above identities.]

Horn Functions

In 1931 J. Horn defined ten hypergeometric functions of two variables and denoted them by $G_1, G_2, G_3, H_1, \ldots, H_7$; he thus completed the set of all possible second-order (complete) hypergeometric functions of two variables in the terminology of Appell and Kampé de Fériet (1926, p. 143 et seq.). [See also Erdélyi *et al.* (1953, Vol. I, pp. 224–225).]

The Horn function H_3, for example, is defined by

(22) $$H_3[\alpha, \beta; \gamma; x, y] = \sum_{m,n=0}^{\infty} \frac{(\alpha)_{2m+n}(\beta)_n}{(\gamma)_{m+n}} \frac{x^m}{m!} \frac{y^n}{n!},$$

where, for convergence, $|x|$ and $|y|$ are sufficiently small [see, for details, Erdélyi *et al.* (1953, Vol. I, pp. 227–228)]. An interesting result involving Appell's F_2 and Horn's H_3 functions was given by Srivastava (1969, p. 681, Eq. (2.2)):

(23) $$\sum_{n=0}^{\infty} \frac{(\nu)_n}{n!} F_2[\alpha, -n, -n; \mu, \nu; x, y]z^n$$
$$= (1-z)^{\alpha-\nu}\zeta^{-\alpha} H_3[\alpha, \mu-\nu; \mu; xyz/\zeta^2, xz/\zeta],$$

where $\zeta = 1 - z + xz + yz$. This result was subsequently applied by Srivastava and Panda (1973a, pp. 860–861) to obtain the following relations between Appell's F_4 and Horn's H_3:

(24) $$F_4[\alpha, \beta; \gamma, \beta; x, y] = (1 - x - y)^{-\alpha} H_3[\alpha, \gamma - \beta; \gamma; xy/(x + y - 1)^2, x/(x + y - 1)],$$

(25) $$F_4[\alpha, \beta; \alpha, \gamma; x, y] = (1 - x - y)^{-\beta} H_3[\beta, \gamma - \alpha; \gamma; xy/(x + y - 1)^2, y/(x + y - 1)],$$

or, inversely,

(26) $$H_3[\alpha, \beta; \gamma; xy, y] = (1-x-y)^{-\alpha} F_4[\alpha, \gamma-\beta; \gamma-\beta, \gamma; x/(x+y-1), y/(x+y-1)],$$

which exhibits the fact that the Horn function H_3 can always be expressed in terms of Appell's function F_4.

Confluent Hypergeometric Functions of Two Variables

Seven confluent forms of the four Appell functions were defined in 1920 by P. Humbert, and he denoted these confluent hypergeometric functions of two variables by

$$\Phi_1, \Phi_2, \Phi_3, \Psi_1, \Psi_2, \Xi_1, \Xi_2.$$

In addition, there exist 13 confluent forms of the Horn functions, which are denoted by

$$\Gamma_1, \Gamma_2, H_1, \ldots, H_{11}.$$

The work of Humbert is described fairly fully by Appell and Kampé de Fériet (1926, pp. 124–135), and the series definitions and convergence conditions of all of these twenty confluent hypergeometric functions of two variables are given also in Erdélyi *et al.* (1953, Vol. I, pp. 225–228). The definition of Φ_1, given in Erdélyi *et al.* (1953, p. 225, Eq. (20)), is in error; we recall the corrected definition in the form

(27) $$\Phi_1[\alpha, \beta; \gamma; x, y] = \sum_{m,n=0}^{\infty} \frac{(\alpha)_{m+n}(\beta)_m}{(\gamma)_{m+n}} \frac{x^m}{m!} \frac{y^n}{n!},$$

$$|x| < 1, \qquad |y| < \infty,$$

so that

(28) $$\Phi_1[\alpha, \beta; \gamma; x, y] = \lim_{|\beta'| \to \infty} F_1[\alpha, \beta, \beta'; \gamma; x, y/\beta'] = \lim_{\varepsilon \to 0} F_1[\alpha, \beta, 1/\varepsilon; \gamma; x, \varepsilon y],$$

which follows from (4) and (27) in view of II.3(4).

Another confluent hypergeometric function of two variables, which is of special interest here, is the Ψ_2 function defined by

(29) $$\Psi_2[\alpha;\gamma,\gamma';x,y] = \sum_{m,n=0}^{\infty} \frac{(\alpha)_{m+n}}{(\gamma)_m(\gamma')_n} \frac{x^m}{m!} \frac{y^n}{n!},$$

$$|x| < \infty, \qquad |y| < \infty,$$

which is a confluent form of Appell's F_2; in fact, we havc

(30) $$\Psi_2[\alpha;\gamma,\gamma';x,y] = \lim_{|\beta|\to\infty} F_2[\alpha,\beta,\beta;\gamma,\gamma';x/\beta,y/\beta]$$
$$= \lim_{\varepsilon\to 0} F_2[\alpha,1/\varepsilon,1/\varepsilon;\gamma,\gamma';\varepsilon x,\varepsilon y].$$

We remark in passing that the reduction formulas (18) and (19) apply to Ψ_2 when $y = \pm x$. [See Srivastava (1973), p. 296, Eq. (12); p. 297, Eq. (13).]

II.7 Hypergeometric Functions of Several Variables

Lauricella Functions

Lauricella (1893) further generalized the four Appell functions $F_1, \ldots, F_4$ to functions of n variables and defined his functions as follows:

(1) $$F_A^{(n)}[a,b_1,\ldots,b_n;c_1,\ldots,c_n;x_1,\ldots,x_n]$$
$$= \sum_{m_1,\ldots,m_n=0}^{\infty} \frac{(a)_{m_1+\cdots+m_n}(b_1)_{m_1}\cdots(b_n)_{m_n}}{(c_1)_{m_1}\cdots(c_n)_{m_n}} \frac{x_1^{m_1}}{m_1!}\cdots\frac{x_n^{m_n}}{m_n!},$$
$$|x_1| + \cdots + |x_n| < 1;$$

(2) $$F_B^{(n)}[a_1,\ldots,a_n,b_1,\ldots,b_n;c;x_1,\ldots,x_n]$$
$$= \sum_{m_1,\ldots,m_n=0}^{\infty} \frac{(a_1)_{m_1}\cdots(a_n)_{m_n}(b_1)_{m_1}\cdots(b_n)_{m_n}}{(c)_{m_1+\cdots+m_n}} \frac{x_1^{m_1}}{m_1!}\cdots\frac{x_n^{m_n}}{m_n!},$$
$$\max\{|x_1|,\ldots,|x_n|\} < 1;$$

(3) $$F_C^{(n)}[a,b;c_1,\ldots,c_n;x_1,\ldots,x_n]$$
$$= \sum_{m_1,\ldots,m_n=0}^{\infty} \frac{(a)_{m_1+\cdots+m_n}(b)_{m_1+\cdots+m_n}}{(c_1)_{m_1}\cdots(c_n)_{m_n}} \frac{x_1^{m_1}}{m_1!}\cdots\frac{x_n^{m_n}}{m_n!},$$
$$\sqrt{|x_1|} + \cdots + \sqrt{|x_n|} < 1;$$

(4) $$F_D^{(n)}[a, b_1, \ldots, b_n; c; x_1, \ldots, x_n]$$

$$= \sum_{m_1, \ldots, m_n = 0}^{\infty} \frac{(a)_{m_1 + \cdots + m_n} (b_1)_{m_1} \cdots (b_n)_{m_n}}{(c)_{m_1 + \cdots + m_n}} \frac{x_1^{m_1}}{m_1!} \cdots \frac{x_n^{m_n}}{m_n!},$$

$$\max\{|x_1|, \ldots, |x_n|\} < 1.$$

Clearly, we have

(5) $$F_A^{(2)} = F_2, \qquad F_B^{(2)} = F_3, \qquad F_C^{(2)} = F_4, \qquad F_D^{(2)} = F_1,$$

where $F_1, \ldots, F_4$ are the Appell functions defined by II.6(4)–(7).

Lauricella (1893) also gave several elementary properties of these functions including, for instance, integral representations of the Eulerian type, transformations and reducible cases, and the systems of partial differential equations associated with them. A summary of Lauricella's work is given by Appell and Kampé de Fériet (1926, Chapter VII, pp. 114–120). [See also the recent papers by Carlson (1963), Carlitz and Srivastava (1976), and Srivastava and Exton (1973, 1974, 1979).]

We recall here the following integral formula:

(6) $$\int_0^{\infty} t^{a-1} e^{-st} \, {}_1F_1\begin{bmatrix} b_1; & \\ & x_1 t \\ c_1; & \end{bmatrix} \cdots {}_1F_1\begin{bmatrix} b_n; & \\ & x_n t \\ c_n; & \end{bmatrix} dt$$

$$= s^{-a} \Gamma(a) F_A^{(n)}[a, b_1, \ldots, b_n; c_1, \ldots, c_n; x_1/s, \ldots, x_n/s],$$

$$\mathrm{Re}(a) > 0, \qquad \mathrm{Re}(s) > \mathrm{Re}(x_1) + \cdots + \mathrm{Re}(x_n),$$

which was given, in a slightly different form, by Erdélyi (1936). Another result of interest is the following generalization of the reduction formula II.6(14):

(7) $$F_D^{(n)}[a, b_1, \ldots, b_n; c; x, \ldots, x] = {}_2F_1[a, b_1 + \cdots + b_n; c; x],$$

which is due to Lauricella himself [Lauricella (1893, p. 150); Appell and Kampé de Fériet (1926, p. 116, Eq. (11))].

For $x = 1$, the second member of (7) is summable by Gauss's theorem II.2(7), and we have the following generalization of II.6(15):

(8) $$F_D^{(n)}[a, b_1, \ldots, b_n; c; 1, \ldots, 1] = \frac{\Gamma(c)\Gamma(c - a - b_1 - \cdots - b_n)}{\Gamma(c - a)\Gamma(c - b_1 - \cdots - b_n)},$$

$$c \neq 0, -1, -2, \ldots; \qquad \mathrm{Re}(c - a - b_1 - \cdots - b_n) > 0,$$

which was also given by Lauricella (1893, p. 150); see also Appell and Kampé de Fériet (1926, p. 117).

The reduction formula (7) follows at once from a much more general result, given by Singhal (1974, p. 163, Eq. (2)) and Panda (1974, p. 166, Eq. (6)), which indeed is a multiple-series extension of Srivastava's formula II.6(20); it states that

$$(9)\qquad \sum_{m_1,\ldots,m_n=0}^{\infty} f(m_1+\cdots+m_n)(\nu_1)_{m_1}\cdots(\nu_n)_{m_n}\frac{x^{m_1+\cdots+m_n}}{m_1!\cdots m_n!} = \sum_{N=0}^{\infty} f(N)(\nu_1+\cdots+\nu_n)_N\frac{x^N}{N!},$$

provided that the series involved converge absolutely.

Confluent Forms of Lauricella Functions

Two important confluent hypergeometric functions of n variables are the functions $\Phi_2^{(n)}$ and $\Psi_2^{(n)}$ defined by [see Srivastava (1976)]

$$(10)\qquad \Phi_2^{(n)}[b_1,\ldots,b_n;c;x_1,\ldots,x_n] = \sum_{m_1,\ldots,m_n=0}^{\infty}\frac{(b_1)_{m_1}\cdots(b_n)_{m_n}}{(c)_{m_1+\cdots+m_n}}\frac{x_1^{m_1}}{m_1!}\cdots\frac{x_n^{m_n}}{m_n!}$$

and

$$(11)\qquad \Psi_2^{(n)}[a;c_1,\ldots,c_n;x_1,\ldots,x_n] = \sum_{m_1,\ldots,m_n=0}^{\infty}\frac{(a)_{m_1+\cdots+m_n}}{(c_1)_{m_1}\cdots(c_n)_{m_n}}\frac{x_1^{m_1}}{m_1!}\cdots\frac{x_n^{m_n}}{m_n!},$$

so that

$$(12)\qquad \Phi_2^{(n)}[b_1,\ldots,b_n;c;x_1,\ldots,x_n] = \lim_{|a|\to\infty} F_D^{(n)}[a,b_1,\ldots,b_n;c;x_1/a,\ldots,x_n/a] = \lim_{\min\{|a_1|,\ldots,|a_n|\}\to\infty} F_B^{(n)}[a_1,\ldots,a_n,b_1,\ldots,b_n;c;x_1/a_1,\ldots,x_n/a_n]$$

and

$$(13)\qquad \Psi_2^{(n)}[a;c_1,\ldots,c_n;x_1,\ldots,x_n] = \lim_{|b|\to\infty} F_C^{(n)}[a,b;c_1,\ldots,c_n;x_1/b,\ldots,x_n/b] = \lim_{\min\{|b_1|,\ldots,|b_n|\}\to\infty} F_A^{(n)}[a,b_1,\ldots,b_n;c_1,\ldots,c_n;x_1/b_1,\ldots,x_n/b_n].$$

Clearly, we have

$$(14)\qquad \Phi_2^{(2)}=\Phi_2,\qquad \Psi_2^{(2)}=\Psi_2,$$

where Φ_2 and Ψ_2 are Humbert's confluent hypergeometric functions of two variables.

In terms of $\Psi_2^{(n)}$, the multivariable extension of Whittaker's $M_{\kappa,\mu}$ function was defined by Humbert (1920) in the form

$$(15)\quad M_{\kappa,\mu_1,\ldots,\mu_n}(x_1,\ldots,x_n) = x_1^{\mu_1+(1/2)}\cdots x_n^{\mu_n+(1/2)}\exp[-\tfrac{1}{2}(x_1+\cdots+x_n)] \cdot \Psi_2^{(n)}[\mu_1+\cdots+\mu_n-\kappa+\tfrac{1}{2}n; 2\mu_1+1,\ldots,2\mu_n+1; x_1,\ldots,x_n].$$

As a matter of fact, Humbert (1920) also studied the corresponding $W_{\kappa,\mu_1,\ldots,\mu_n}(x_1,\ldots,x_n)$ function of n variables [cf., e.g., Appell and Kampé de Fériet (1926, p. 135, Eq. (38))].

Further Generalizations of the Appell and Lauricella Functions

Just as the Gaussian ${}_2F_1$ function was generalized to ${}_pF_q$ by increasing the number of the numerator and denominator parameters, the four Appell functions were unified and generalized by Kampé de Fériet (1921), who defined a general hypergeometric function of two variables [see Appell and Kampé de Fériet (1926, p. 150, Eq. (29)); see also Srivastava and Panda (1976, p. 423, Eq. (26))]. A further generalization of the Kampé de Fériet function is due to Srivastava and Daoust (1969), who indeed defined an extension of Wright's ${}_p\Psi_q$ function in two variables.

A similar extension of the Wright function ${}_p\Psi_q$ in several variables is due to Srivastava and Daoust (1969a, p. 454) who defined the generalized Lauricella function of several variables. See also Srivastava and Daoust (1972); Srivastava and Panda (1975, p. 1127, Eq. (4.1)); Exton (1976, Section 3.7); Exton (1978, Section 1.4); Srivastava and Panda (1980, Section 3); R. Srivastava (1980).]

Finally, it should be mentioned that the literature also contains studies of various generalizations of the G and H functions in two and more variables. A reasonable background of these generalizations may be found in Srivastava and Panda (1976a), who defined a general H function of n variables. [See also Buschman (1977, 1978, 1979); Srivastava and Panda (1975a; 1976b,c; 1978; 1979); Panda (1977a,b,c); Srivastava and Srivastava (1978); Prasad and Singh (1977); Siddiqui (1979); Tandon (1980a,b); R. Srivastava (1981); and Srivastava and Raina (1981).]

II.8 The Classical Orthogonal Polynomials

In Sections II.2 and II.3 we have given explicit hypergeometric representations of the classical orthogonal polynomials such as the Jacobi polynomials [II.2(33)], Gegenbauer (or ultraspherical) polynomials [II.2(34)], Legendre (or spherical) polynomials [II.2(35)], Chebyshev polynomials of the first and second kinds [II.2(36) and (37)], Hermite polynomials [II.3(64) and (65)], and Laguerre polynomials [II.3(67)]. We shall list here their orthogonality properties, generating functions, Rodrigues formulas, recurrence relations, and the differential equations satisfied by them. For further details and applications, the interested reader may refer to Szegö (1975), Erdélyi *et al.* (1953, Vol. II, Chapter 10), Lebedev (1965, Chapter 4), Luke (1969, Vol. II, Chapter 8), McBride (1971), Beckmann (1973), Askey (1975), Danese (1976–1977), Carlson (1977, Chapter 7), and Chihara (1978).

Jacobi polynomials

(1)
$$\int_{-1}^{1} (1-x)^{\alpha}(1+x)^{\beta} P_n^{(\alpha,\beta)}(x) P_m^{(\alpha,\beta)}(x)\,dx = \frac{2^{\alpha+\beta+1}\Gamma(\alpha+n+1)\Gamma(\beta+n+1)}{n!\,(\alpha+\beta+2n+1)\Gamma(\alpha+\beta+n+1)}\delta_{mn},$$
$$\min\{\mathrm{Re}(\alpha), \mathrm{Re}(\beta)\} > -1,$$

where, and in what follows, δ_{mn} is the Kronecker delta;

(2)
$$\sum_{n=0}^{\infty} P_n^{(\alpha,\beta)}(x) z^n = 2^{\alpha+\beta} R^{-1}(1-z+R)^{-\alpha}(1+z+R)^{-\beta},$$
$$|z| < 1, \qquad R = (1-2xz+z^2)^{1/2}$$

[for a generalization of (2) see Srivastava and Singhal (1973); see also Askey (1978) and Srivastava (1979b)];

(3)
$$P_n^{(\alpha,\beta)}(x) = \frac{(-1)^n(1-x)^{-\alpha}(1+x)^{-\beta}}{2^n n!} D_x^n\{(1-x)^{\alpha+n}(1+x)^{\beta+n}\};$$

(4)
$$2(n+1)(\alpha+\beta+n+1)(\alpha+\beta+2n)P_{n+1}^{(\alpha,\beta)}(x)$$
$$= (\alpha+\beta+2n+1)[(\alpha+\beta+2n)(\alpha+\beta+2n+2)x+\alpha^2-\beta^2]P_n^{(\alpha,\beta)}(x)$$
$$-2(\alpha+n)(\beta+n)(\alpha+\beta+2n+2)\,P_{n-1}^{(\alpha,\beta)}(x);$$

(5)
$$(1-x^2)\frac{d^2y}{dx^2} + [\beta-\alpha-(\alpha+\beta+2)x]\frac{dy}{dx} + (\alpha+\beta+n+1)ny = 0,$$
$$y \equiv P_n^{(\alpha,\beta)}(x).$$

Gegenbauer (or ultraspherical) polynomials

(6) $$\int_{-1}^{1} (1-x^2)^{\nu-(1/2)} C_n^{\nu}(x) C_m^{\nu}(x)\,dx = \frac{(2\nu)_n \Gamma(\frac{1}{2})\Gamma(\nu+\frac{1}{2})}{n!(\nu+n)\Gamma(\nu)}\delta_{mn}, \qquad \mathrm{Re}(\nu) > -\tfrac{1}{2};$$

(7) $$\sum_{n=0}^{\infty} C_n^{\nu}(x) z^n = (1-2xz+z^2)^{-\nu}, \qquad |z| < 1;$$

(8) $$C_n^{\nu}(x) = \frac{(-1)^n (2\nu)_n (1-x^2)^{(1/2)-\nu}}{2^n n! (\nu+\frac{1}{2})_n} D_x^n \{(1-x^2)^{\nu+n-(1/2)}\};$$

(9) $$(n+1)C_{n+1}^{\nu}(x) = 2(\nu+n)xC_n^{\nu}(x) - (2\nu+n-1)C_{n-1}^{\nu}(x);$$

(10) $$(1-x^2)\frac{d^2y}{dx^2} - (2\nu+1)x\frac{dy}{dx} + (2\nu+n)ny = 0, \qquad y \equiv C_n^{\nu}(x).$$

The Gegenbauer polynomials $C_n^{\nu}(x)$ are also known as *ultraspherical* polynomials and are often denoted by $P_n^{(\nu)}(x)$. Except for the generating function (7), the properties of Gegenbauer (or ultraspherical) polynomials can alternatively be deduced as special cases of the corresponding properties of Jacobi polynomials by setting

$$\alpha = \beta = \nu - \tfrac{1}{2}$$

and appealing to the relation II.2(34).

Legendre (or spherical) polynomials Since

(11) $$P_n(x) = P_n^{(0,0)}(x) = C_n^{1/2}(x),$$

the properties of Legendre (or spherical) polynomials follow at once from (1)–(5) when

$$\alpha = \beta = 0$$

or from (6)–(10) when $\nu = \frac{1}{2}$. The details may be omitted.

Chebyshev polynomials of the first and second kinds The Chebyshev polynomials $T_n(x)$ and $U_n(x)$ of the first and second kinds, respectively, are special ultraspherical polynomials. These are often introduced by the relations

(12) $$T_n(\cos\theta) = \cos n\theta, \qquad U_n(\cos\theta) = \frac{\sin(n+1)\theta}{\sin\theta}.$$

The various properties of these polynomials can easily be derived from those of the Jacobi polynomials by setting

$$\alpha = \beta = -\tfrac{1}{2} \qquad \text{and} \qquad \alpha = \beta = \tfrac{1}{2}$$

and using II.2(36) and II.2(37). Additionally, we have the generating functions

(13) $$\sum_{n=0}^{\infty} T_n(x)z^n = (1 - xz)(1 - 2xz + z^2)^{-1}, \qquad |z| < 1;$$

(14) $$\sum_{n=0}^{\infty} U_n(x)z^n = (1 - 2xz + z^2)^{-1}, \qquad |z| < 1,$$

which readily yield the following connections between the two kinds of Chebyshev polynomials:

(15) $$T_n(x) = U_n(x) - xU_{n-1}(x), \qquad n \geqq 1;$$

(16) $$(1 - x^2)U_n(x) = xT_{n+1}(x) - T_{n+2}(x).$$

Hermite polynomials

(17) $$\int_{-\infty}^{\infty} \exp(-x^2)\, H_n(x)H_m(x)\, dx = 2^n n! \sqrt{\pi}\, \delta_{mn};$$

(18) $$\int_{-\infty}^{\infty} \exp(-\tfrac{1}{2}x^2)\, He_n(x)He_m(x)\, dx = n! \sqrt{2\pi}\, \delta_{mn}.$$

The orthogonal polynomials $H_n(x)$ and $He_n(x)$ are both known in the literature as the Hermite polynomials. They are related to each other by

(19) $$He_n(x) = 2^{-n/2} H_n(x/\sqrt{2}), \qquad H_n(x) = 2^{n/2} He_n(x\sqrt{2}).$$

We shall state the other properties of Hermite polynomials in their relatively more popular notation $H_n(x)$; the reader will have no difficulty in restating each property in terms of the alternative notation $He_n(x)$:

(20) $$\sum_{n=0}^{\infty} H_n(x)\frac{z^n}{n!} = \exp(2xz - z^2);$$

(21) $$H_n(x) = (-1)^n \exp(x^2)\, D_x^n\{\exp(-x^2)\};$$

(22) $$H_{n+1}(x) - 2xH_n(x) + 2nH_{n-1}(x) = 0, \qquad n \geqq 1;$$

(23) $$\frac{d^2y}{dx^2} - 2x\frac{dy}{dx} + 2ny = 0, \qquad y \equiv H_n(x).$$

Laguerre polynomials

(24) $$\int_0^\infty x^\alpha e^{-x} L_n^{(\alpha)}(x) L_m^{(\alpha)}(x)\,dx = \frac{\Gamma(\alpha+n+1)}{n!}\delta_{mn}, \qquad \mathrm{Re}(\alpha) > -1;$$

(25) $$\sum_{n=0}^\infty L_n^{(\alpha)}(x) z^n = (1-z)^{-\alpha-1}\exp\left(\frac{xz}{z-1}\right), \quad |z| < 1;$$

(26) $$L_n^{(\alpha)}(x) = \frac{x^{-\alpha}e^x}{n!} D_x^n\{x^{\alpha+n}e^{-x}\};$$

(27) $$(n+1)L_{n+1}^{(\alpha)}(x) = (\alpha+2n-x+1)L_n^{(\alpha)}(x) - (\alpha+n)L_{n-1}^{(\alpha)}(x);$$

(28) $$x\frac{d^2y}{dx^2} + (\alpha - x + 1)\frac{dy}{dx} + ny = 0, \qquad y \equiv L_n^{(\alpha)}(x).$$

The Laguerre polynomials are, in fact, limiting cases of the Jacobi polynomials, and we have the relation

(29) $$L_n^{(\alpha)}(x) = \lim_{|\beta|\to\infty}\left\{P_n^{(\alpha,\beta)}\left(1 - \frac{2x}{\beta}\right)\right\},$$

which provides an alternative method of derivation of (24)–(28) from (1)–(5).

Other limiting relations of type (29) include

(30) $$H_n(x) = n!\lim_{|\nu|\to\infty}\{\nu^{-n/2}C_n^\nu(x/\sqrt{\nu})\}$$

between Hermite and Gegenbauer polynomials, and

(31) $$H_n(x/\sqrt{2}) = (-1)^n 2^{n/2} n! \lim_{|\alpha|\to\infty}\{\alpha^{-n/2}L_n^{(\alpha)}(\alpha + x\sqrt{\alpha})\}$$

between Hermite and Laguerre polynomials. A limit formula of a special type,

(32) $$\lim_{n\to\infty}\{n^{-\alpha}L_n^{(\alpha)}(x/n)\} = x^{-\alpha/2}J_\alpha(2\sqrt{x}),$$

relates Laguerre polynomials with the Bessel function [cf. II.3(38)].

The even and odd Hermite polynomials are essentially Laguerre polynomials, since

(33) $$H_{2n+\varepsilon}(x) = (-1)^n 2^{2n+\varepsilon} n!\, x^\varepsilon L_n^{(\varepsilon-(1/2))}(x^2), \qquad \varepsilon = 0 \quad \text{or} \quad \varepsilon = 1.$$

II.9 Other Polynomial Systems and Generalizations

The Bessel polynomials [Krall and Frink (1949); see also Grosswald (1978)]

(1) $$y_n(x,\alpha,\beta) = {}_2F_0[-n, \alpha+n-1; -; -x/\beta]$$

provide another example of a hypergeometric polynomial which has received considerable attention in the literature. Indeed, there are several useful systems of polynomials of a *discrete* variable in addition to the Poisson–Charlier polynomials defined by II.3(68). For the sake of completeness, we give their definitions below.

Krawtchouk polynomials

(2) $$K_n(x;p,N) = {}_2F_1[-n, -x; -N; p^{-1}], \quad 0 < p < 1, \qquad x = 0, 1, \ldots, N.$$

Meixner polynomials

(3) $$M_n(x;\beta,c) = {}_2F_1[-n, -x; \beta; 1 - c^{-1}], \quad \beta > 0, \qquad 0 < c < 1, \qquad x = 0, 1, 2, \ldots .$$

Hahn polynomials

(4) $$Q_n(x;\alpha,\beta,N) = \sum_{k=0}^{n} \frac{(-n)_k(-x)_k(\alpha+\beta+n+1)_k}{(-N)_k(\alpha+1)_k\,k!}, \quad \alpha,\beta > -1, \qquad n = 0, 1, \ldots, N.$$

When $x = 0, 1, \ldots, N$, one can look upon the right-hand side of (4) as a hypergeometric function and write

(5) $$Q_n(x;\alpha,\beta,N) = {}_3F_2[-n, -x, \alpha+\beta+n+1; -N, \alpha+1; 1], \quad \alpha,\beta > -1, \qquad n, x = 0, 1, \ldots, N.$$

Of the various generalizations (and unifications) of the classical orthogonal polynomials, we recall here the Gould–Hopper generalization of the Hermite polynomials in the form [Gould and Hopper (1962), p. 58, Eq. (6.2)]

(6) $$g_n^m(x,h) = \sum_{k=0}^{[n/m]} \frac{n!}{k!(n-mk)!} h^k x^{n-mk} = x^n\, {}_mF_0[\Delta(m; -n); -; h(-m/x)^m],$$

where m is a positive integer and, as in II.5(8), $\Delta(m; -n)$ abbreviates the array of m parameters

$$-\frac{n}{m}, -\frac{n-1}{m}, \ldots, -\frac{n-m+1}{m}, \qquad m \geqq 1.$$

These polynomials are generated by [cf. 6.8(87)]

(7) $$\sum_{n=0}^{\infty} g_n^m(x,h)\frac{z^n}{n!} = \exp(xz + hz^m),$$

so that, by comparison with II.8(20), we have

(8) $$H_n(x) = g_n^2(2x, -1).$$

The Gould–Hopper polynomials are, in fact, contained in certain general hypergeometric polynomials defined earlier by Brafman (1957, p. 186):

(9) $$\mathscr{B}_n^m[\alpha_1, \ldots, \alpha_r; \beta_1, \ldots, \beta_s : x] = {}_{m+r}F_s[\Delta(m; -n), \alpha_1, \ldots, \alpha_r; \beta_1, \ldots, \beta_s; x]$$

where the α and β are arbitrary (complex) parameters independent of n.

The generalized Hermite polynomials $H_{n,m}(\alpha, \beta)$ used by Gupta and Jain (1974) are precisely the Gould–Hopper polynomials $g_n^m(\alpha, \beta)$. This fact does not seem to have been known to Gupta and Jain (1974), who consider $H_{n,m}(\alpha, \beta)$ as a "new" generalization of the polynomials $H_{n,m,v}(x)$ defined by Lahiri (1971, p. 118, Eq. (3.2)) by

(10) $$H_{n,m,v}(x) = (vx)^n \, {}_mF_0[\Delta(m; -n); -; -(-m/vx)^m],$$

so that [see Srivastava (1976a, p. 458, Eq. (1.7))]

(11) $$H_{n,m,v}(x) = v^n g_n^m(x, -1) = g_n^m(vx, -1),$$

or, in the notation of Gupta and Jain (1974),

(12) $$H_{n,m,v}(x) = v^n H_{n,m}(x, -1) = H_{n,m}(vx, -1).$$

Other known generalizations (and unifications) of the various polynomials mentioned above include the general systems introduced and studied, for example, by Srivastava and Singhal (1971, 1972a), Srivastava and Joshi (1974), Srivastava and Lavoie (1975), Srivastava and Panda (1975b, 1979a), Srivastava *et al.* (1979), Panda (1977d), and R. Srivastava (1980a), where references to earlier works may be found.

CHAPTER 1

Poisson Queues I

In this chapter we give a brief description of the uses of special functions in obtaining various results for queuing systems with Poisson arrivals and negative-exponential service times. We include the work on the transient solutions of $M/M/1$ and $M/M/c$ queues by Heathcote and Moyal (1959), Saaty (1960), and Chaudhry and Templeton (1973), and on double queues by Kashyap (1965a). We also describe the study of sojourn time problems by Gibson and Conolly (1971), certain limit theorems for the $M/M/c$ queue by Tomko (1972), and a discussion of the optimal initial number in the simulation of $M/M/1$ by Madansky (1976). We close the chapter with the study of the Poisson queue with bounded waiting time by Gavish and Schweitzer (1977).

1.1 A Random Walk Approach

Consider a random walk in continuous time [Heathcote and Moyal (1959)], where a particle moves on the real axis, taking one step to the right or left, respectively, with probabilities λ and μ per unit time. Let i be the initial position of the particle at time $t = 0$. The probabilities $P_{in}(t)$ of a transition from position i to n in time t are easily seen to satisfy the following "backward" equation [Bartlett (1955); Feller (1968)]:

$$\frac{d}{dt} P_{in}(t) = -(\lambda + \mu)P_{in}(t) + \lambda P_{i+1,n}(t) + \mu P_{i-1,n}(t). \tag{1}$$

Using the generating function (g.f.)

$$G_i(z, t) = \sum_n P_{in}(t)z^n$$

and the Laplace transform (L.T.)

$$g_i(z, s) = \int_0^\infty e^{-st} G_i(z, t)\, dt, \qquad \mathrm{Re}(s) > 0,$$

we have the difference equation

(2) $$(s + \lambda + \mu) g_i(z, s) = \lambda g_{i+1}(z, s) + \mu g_{i-1}(z, s) + z^i.$$

It can be easily seen that the solution of (2) is

(3) $$g_i(z, s) = z^i \phi(z, s) + A(z, s)\alpha_1^i(s) + B(z, s)\alpha_2^i(s),$$

where

(4) $$\phi(z, s) = (s + \lambda + \mu - \lambda z - \mu z^{-1})^{-1}$$

and α_1, α_2 are the zeros of $\phi(z, s)$, given by

(5) $$\alpha_1(s) = (2\lambda)^{-1}[s + \lambda + \mu - \{(s + \lambda + \mu)^2 - 4\lambda\mu\}^{1/2}],$$

$$\alpha_2(s) = (2\lambda)^{-1}[s + \lambda + \mu + \{(s + \lambda + \mu)^2 - 4\lambda\mu\}^{1/2}],$$

and A, B are arbitrary functions determined from boundary conditions.

If we consider the random walk with reflecting barriers at $n = 0$ and $n = N$, this corresponds to the queuing system $M/M/1 : N$. In this case we have the following boundary conditions:

(6) $$(s + \lambda) g_0 = \lambda g_1 + 1,$$

$$(s + \mu) g_N = \mu g_{N-1} + z^N.$$

Using these conditions to evaluate A and B, we finally obtain

(7) $$g_i(z, s) = \phi(z, s)\,[z^i + (1 - z^{-1})\alpha_1^{i+1}(1 - \alpha_1)^{-1}$$

$$+ X_i\{z^N - z^{N+1} + (1 - z^{-1})\alpha_1^{N+1}\}]$$

(8) $$\to \phi(z, s)[z^i + (1 - z^{-1})\alpha_1^{i+1}(1 - \alpha_1)^{-1}] \qquad \text{as} \quad N \to \infty,$$

where

(9) $$X_i = \lambda s^{-1}[(\alpha_2^{i+1} - \alpha_1^{i+1}) - \alpha_1\alpha_2(\alpha_2^i - \alpha_1^i)]\{\alpha_2^{N+1} - \alpha_1^{N+1}\}^{-1}.$$

Expanding the right-hand member of (7) in powers of z and inverting the coefficient of z^n with the help of the known results [Erdélyi *et al.* (1954, Vol. I, p. 240, Eq. (28); p. 130, Eq. (9))], we have, for the probability that there are n

units in the system at time t,

$$(10)\quad P_{i,n}(t) = \beta^{n-i}e^{-\alpha t}I_{i-n} + \beta^{n-i}e^{-\alpha t}\sum_{j=0}^{\infty}\{I_{2(j+1)(N+1)+i-n} + I_{2(j+1)(N+1)-i-n}\}$$
$$+ \lambda\beta^{n-i}\sum_{j=0}^{\infty}\int_0^t e^{-\alpha\tau}\{I_{2(j+1)(N+1)-i-n-2} - 2\beta I_{2(j+1)(N+1)-i-n-1}$$
$$+ \beta^{-2}I_{2(j+1)(N+1)-i-n} + I_{2j(N+1)+i+n+2}$$
$$- 2\beta^{-1}I_{2j(N+1)+i+n+1} + \beta^{-2}I_{2j(N+1)+i+n}\}\,d\tau,$$
$$(n = 0, 1, 2, \ldots, N),$$

where

$$(10')\qquad \alpha = \lambda + \mu, \qquad \beta = (\lambda/\mu)^{1/2},$$

and I_ν stands for $I_\nu(2\sqrt{\lambda\mu}\,t)$ or $I_\nu(2\sqrt{\lambda\mu}\,\tau)$ (as is clear from the context), the modified Bessel function of the first kind and order ν.

For the stationary case we use the Tauberian theorem [Widder (1941, Chapter V)],

$$(11)\qquad \lim_{s\downarrow 0}\{sg_i(z, s)\} = \lim_{t\to\infty} G_i(z, t),$$

and obtain the steady-state probability of finding n units in the system:

$$(12)\qquad P_n = \lim_{t\to\infty} P_{in}(t) = \begin{cases} \dfrac{(\lambda - \mu)(\mu/\lambda)^{N-n}}{\lambda\{1 - (\mu/\lambda)^{N+1}\}} & \text{if } \lambda \neq \mu \\[2ex] \dfrac{1}{N+1} & \text{if } \lambda = \mu. \end{cases}$$

The expected value of the position n at time t, conditional on initial position i, is given by

$$(13)\qquad E(n|i, t) = \frac{\partial}{\partial z}G_i(z, t)\bigg|_{z=1},$$

and its L.T. is given by

$$(14)\qquad \mathscr{L}[E(n|i, t)] = \frac{\partial}{\partial z}g_i(z, s)\bigg|_{z=1}.$$

It can easily be seen that the expected length of the waiting line L at time t, conditional on initial length $i - 1$, is given by

$$(15)\qquad E(L|i - 1, t) = E(n|i, t) + P_{i0}(t) - 1,$$

and that the steady-state expected queue length[1] is, on letting $t \to \infty$ in (15),

$$[N - \{1 - (\mu/\lambda)^N\}(1 - \mu/\lambda)^{-1}]\{1 - (\mu/\lambda)^{N+1}\}^{-1}. \tag{16}$$

For the distribution of the busy period, we consider the random walk with absorbing barrier at the origin and reflecting barrier at N, for which case the boundary conditions are

$$g_0 = 1/s, \qquad (s + \mu)g_N = \mu g_{N-1} + z^N. \tag{17}$$

Using these equations to evaluate A and B and substituting these values of A and B in (3), we have

$$g_i = s^{-1}V_i + \phi \cdot \{(z^N - z^{N+1})W_i + z^i - V_i\}, \tag{18}$$

where

$$V_i = \{\alpha_1^i \alpha_2^N(\alpha_2 - 1) + \alpha_1^N \alpha_2^i(1 - \alpha_1)\}\{\alpha_2^N(\alpha_2 - 1) + \alpha_1^N(1 - \alpha_1)\}^{-1} \tag{19}$$

and

$$W_i = (\alpha_2^i - \alpha_1^i)\{\alpha_2^N(\alpha_2 - 1) + \alpha_1^N(1 - \alpha_1)\}^{-1}. \tag{20}$$

The term independent of z in the right-hand member of (18) clearly gives the L.T. of the probability that the particle will be absorbed at the origin at time t or earlier, and this probability is, by definition, the distribution function (D.F.) of the duration of a busy period initiated by i customers. Denoting this D.F. by $B_i(t)$, and the corresponding density function (d.f.) by $b_i(t)$, we have

$$\mathscr{L}[B_i(t)] = V_i/s, \tag{21}$$

and, therefore,

$$\mathscr{L}[b_i(t)] = V_i(s), \tag{22}$$

where $\mathscr{L}[\cdot]$ stands for the L.T.

Letting $N \to \infty$ in the above, we get results for the queue $M/M/1/\infty$. The *probability generating function* (p.g.f.) for this case is given by (8), and the transition probabilities can be seen to be

$$P_{in}(t) = \beta^{n-i}\left[e^{-\alpha t}I_{i-n} + \lambda \int_0^t e^{-\alpha\tau}\{\beta^{-2}I_{i+n} - 2\beta^{-1}I_{i+n+1} + I_{i+n+2}\}\,d\tau\right] \quad (n = 0, 1, 2, \ldots), \tag{23}$$

where the notation is as in (10).

[1] Heathcote and Moyal (1959) denote the expression in (16) by $E(L|i - 1, \infty)$, giving the impression that the steady-state queue length is a function of the initial value $i - 1$, which is not the case.

Equation (23) is a well-known result obtained by several authors using different methods [Ledermann and Reuter (1954, pp. 365–366); Bailey (1954); Clarke (1956); Champernowne (1956); Conolly (1958a)].

The conditional expected value of n at time t is found as before by differentiating $g_i(z, s)$ at $z = 1$ and inverting the result, which gives

$$\text{(24)} \quad E(n|i, t) = i + (\lambda - \mu)t + \sum_{j=0}^{\infty} \beta^{-(i+1+j)}(i + 1 + j) \int_0^t \tau^{-1} e^{-\alpha\tau} I_{i+1+j} \, d\tau.$$

The expected length of the waiting line at time t is given by (15), (23), and (24).

The density function of the duration of a busy period initiated by i customers is found from (22) and (19), letting $N \to \infty$, to be

$$\text{(25)} \qquad b_i(t) = \mathscr{L}^{-1}\alpha_1^i = i\beta^{-i}t^{-1}e^{-\alpha t}I_i,$$

with the help of Erdélyi *et al.* (1954, Vol. I, p. 240, Eq. (28)), where in (24) and (25) the notation is as in (10).

For the queue $M/M/c : \infty$, corresponding to (2) we have

$$\text{(26)} \qquad (s + \lambda + i\mu)g_i = \lambda g_{i+1} + \mu i g_{i-1} + z^i \qquad (i = 0, 1, \ldots, c),$$

$$\text{(27)} \qquad (s + \lambda + c\mu)g_i = \lambda g_{i+1} + \mu c g_{i-1} + z^i \qquad (i = c, c + 1, \ldots).$$

The solutions of (26) and (27) are, respectively,

$$\text{(28)} \qquad g_i(z, s) = \psi_i(z, s) + A(z, s)\beta_i(s) \qquad (i = 0, 1, \ldots, c),$$

$$\text{(29)} \qquad g_i(z, s) = z^i\phi(z, s) + B(z, s)\alpha_1^i(s) \qquad (i = c, c + 1, \ldots),$$

where $\phi(z, s)$ and $\alpha_1(s)$ are given by (4) and (5) with μc substituted for μ; $A(z, s)$ and $B(z, s)$ are arbitrary functions;

$$\text{(30)} \qquad \psi_i(z, s) = \mathscr{L}\left[\{1 - (1 - z)e^{-\mu t}\}^i \exp\left\{-\left(\frac{\lambda}{\mu}\right)(1 - z)(1 - e^{-\mu t})\right\}\right]$$

$$= \exp\left[-\left(\frac{\lambda}{\mu}\right)(1 - z)\right] \sum_{j=0}^{i} \binom{i}{j} \frac{(z - 1)^j}{\mu}$$

$$\cdot {}_1F_1\left(\frac{s}{\mu} + j; \frac{s}{\mu} + j + 1; \left(\frac{\lambda}{\mu}\right)(1 - z)\right)$$

and

$$\text{(31)} \qquad \beta_i(s) = \int_0^1 \exp\left(-\frac{\lambda x}{\mu}\right) x^i (1 - x)^{(s/\mu) - 1} \, dx$$

$$= B(i + 1, s/\mu) \, {}_1F_1\left(i + 1; i + 1 + \frac{s}{\mu}; -\frac{\lambda}{\mu}\right),$$

$B(\alpha, \beta)$ and ${}_1F_1(a; b; x)$ being the beta and the confluent hypergeometric functions, respectively.

The Laplace transform of the expected number of customers in the system at time t, given i initially, is found as before to be

$$(32)\quad \mathscr{L}[E(n|i,t)] = \begin{cases} \dfrac{\lambda}{\mu s} + \dfrac{\mu i - \lambda}{\mu(s+\mu)} + R_i\left\{\dfrac{\mu c - \lambda}{\lambda\mu(\alpha_2 - 1)} + \dfrac{1}{\lambda(\alpha_2 - 1)^2} - \dfrac{(1-\alpha_1)(c - \lambda/\mu) + \alpha_1}{s+\mu}\right\}, \\ \qquad (i = 0, 1, \ldots, c); \\ \dfrac{i}{s} + \dfrac{\lambda - \mu}{s^2} + S_i\left\{\dfrac{(\lambda-\mu)(\beta_{c-1} - \beta_c)}{s^2} - \dfrac{\mu\{(c-1-\lambda/\mu)\beta_c - (c - c\mu)\beta_{c-1}\}}{s(s+\mu)}\right\}, \\ \qquad (i = c, c+1, \ldots), \end{cases}$$

where

$$(33)\quad R_i = \beta_i[\beta_c - \alpha_1\beta_{c-1}]^{-1}, \qquad S_i = \alpha_1^{i+1-c}[\beta_c - \alpha_1\beta_{c-1}]^{-1}.$$

For the steady state, use of (11), (28), and (29) yields

$$(34)\quad \lim_{t\to\infty} G_i(z,t) = P_0\left\{\sum_{j=0}^{c-1} \frac{(\lambda/\mu)^j}{j!} z^j + \sum_{j=c}^{\infty} \frac{(\lambda/\mu)^j}{c!\,c^{j-c}} z^j\right\}, \qquad \mu c > \lambda,$$

where

$$(35)\quad P_0 = \left\{\sum_{j=0}^{c-1} \frac{(\lambda/\mu)^j}{j!} + \sum_{j=c}^{\infty} \frac{(\lambda/\mu)^j}{c!\,c^{j-c}}\right\}^{-1}.$$

Equation (34) is a well-known result for $M/M/c$. [Cf., e.g., Prabhu (1965a, Eqs. (1.83), (1.84)).]

The density function $b_i(t)$ of the duration of a busy period initiated by i customers is found as before by introducing an absorbing barrier at $i = 0$, and is given by

$$(36)\quad b_i(t) = \frac{d}{dt} P_{i0}(t) = \mathscr{L}^{-1}\{s\bar{g}_i(0,s)\},$$

where $\bar{g}_i(0,s)$ stands for the solution, evaluated at $z = 0$, of the system (26), (27) with the absorbing barrier and, for $i \geqq c$, is given by

$$(37)\quad \bar{g}_i(0,s) = \frac{\alpha_1^{i+1-c}\{(\gamma_0\beta_c - \gamma_c\beta_0)(\gamma_{c-1} + s\gamma_0\psi_{c-1} - s\gamma_{c-1}\psi_0)}{s\gamma_0\{(\gamma_0\beta_c - \gamma_c\beta_0) - \alpha_1(\gamma_0\beta_{c-1} - \gamma_{c-1}\beta_0)\}} - \frac{(\gamma_0\beta_{c-1} - \gamma_{c-1}\beta_0)(\gamma_c + s\gamma_0\psi_c - s\gamma_c\psi_0)\}}{s\gamma_0\{(\gamma_0\beta_c - \gamma_c\beta_0) - \alpha_1(\gamma_0\beta_{c-1} - \gamma_{c-1}\beta_0)\}},$$

where

$$\gamma_i(s) = \sum_{j=0}^{i} \binom{i}{j} \left(\frac{\mu}{\lambda}\right)^j \Gamma\left(j + \frac{s}{\mu}\right) \tag{38}$$

and the function $\psi(z, s)$ appearing in (37) is evaluated at $z = 0$.

1.2 The Many-Server Queue

Consider the queuing system $M/M/c : \infty$ with parameters λ, μ of the arrival and service time distributions, respectively (Saaty, 1960). Let there be i units in the system at $t = 0$; then writing $P_n(t)$ in place of

$$P_{in}(t) = \text{Prob}\{n \text{ units in system at } t \mid i \text{ units at } 0\},$$

we have the following forward equations describing the system:

$$\frac{d}{dt} P_0(t) = -\lambda P_0(t) + \mu P_1(t), \tag{1}$$

$$\frac{d}{dt} P_n(t) = -(\lambda + n\mu) P_n(t) + \lambda P_{n-1}(t) + (n+1)\mu P_{n+1}(t) \quad (1 \leqq n < c), \tag{2}$$

$$\frac{d}{dt} P_n(t) = -(\lambda + c\mu) P_n(t) + \lambda P_{n-1}(t) + c\mu P_{n+1}(t) \qquad (n \geqq c), \tag{3}$$

with the initial condition

$$P_n(0) = \delta_{in}, \quad \text{the Kronecker delta.} \tag{4}$$

Using the generating function

$$P(z, t) = \sum_{n=0}^{\infty} P_n(t) z^n, \tag{5}$$

and Laplace transforms, denoted, typically, by

$$f^*(s) = \int_0^{\infty} e^{-st} f(t)\, dt, \qquad \text{Re}(s) \geqq 0, \tag{6}$$

we have

$$P^*(z, s) = \frac{z^{i+1} - \mu(1 - z) \sum_{n=0}^{c-1} (c - n) z^n P_n^*(s)}{sz - (1 - z)(c\mu - \lambda z)}. \tag{7}$$

The zeros of the denominator of the right-hand member of (7) are given by

$$\begin{aligned} \alpha_1 &= (2\lambda)^{-1}[\lambda + c\mu + s + \sqrt{(\lambda + c\mu + s)^2 - 4c\lambda\mu}], \\ \alpha_2 &= (2\lambda)^{-1}[\lambda + c\mu + s - \sqrt{(\lambda + c\mu + s)^2 - 4c\lambda\mu}], \end{aligned} \tag{8}$$

and by usual analyticity arguments the numerator must vanish at α_2, which is inside the unit circle $|z| = 1$. Thus, we have

$$\text{(9)} \qquad \sum_{n=0}^{c-1} (c-n)\alpha_2^n P_n^*(s) = \frac{\alpha_2^{i+1}}{\mu(1-\alpha_2)}.$$

In order to determine the c unknowns $P_n^*(s)$ $(0 \leqq n \leqq c-1)$, we require c equations involving them, of which (9) is one and we take the first $(c-1)$ equations given by (1)–(2), viz., the equations corresponding to $n = 0, 1, 2, \ldots, c-2$.

Let us first consider the case $c > 2$. Using the generating function

$$\text{(10)} \qquad Q(z,t) = \sum_{n=0}^{c-2} P_n(t) z^n,$$

we have, from these last-mentioned $(c-1)$ equations,

$$\text{(11)} \qquad \frac{\partial Q}{\partial t} = \mu(1-z)\frac{\partial Q}{\partial z} - \lambda(1-z)Q - \lambda z^{c-1}P_{c-2} + \mu(c-1)z^{c-2}P_{c-1},$$

which is a first-order linear partial differential equation. The associated simultaneous equations are

$$\text{(12)} \qquad \frac{dt}{-1} = \frac{dz}{\mu(1-z)} = \frac{dQ}{\lambda(1-z)Q + \lambda z^{c-1}P_{c-2} - \mu(c-1)z^{c-2}P_{c-1}}.$$

The first and second members give

$$\text{(13)} \qquad 1 - z = c_1 e^{\mu t}.$$

Taking first and third members and using (13), we have

$$\text{(14)} \qquad \frac{\partial Q}{\partial t} + \lambda c_1 e^{\mu t} Q = \mu(c-1)(1 - c_1 e^{\mu t})^{c-2} P_{c-1} - \lambda(1 - c_1 e^{\mu t})^{c-1} P_{c-2},$$

whose solution is

$$\text{(15)} \qquad Q \exp\left[\frac{\lambda}{\mu} c_1 e^{\mu t}\right] = \int_0^t [\mu(c-1)(1 - c_1 e^{\mu x})^{c-2} P_{c-1}(x) - \lambda(1 - c_1 e^{\mu x})^{c-1} P_{c-2}(x)] \exp\left[\frac{\lambda}{\mu} c_1 e^{\mu x}\right] dx + c_2.$$

From (13) and (15) the solution of (11) is

$$\text{(16)} \qquad Q(z,t) \exp\left[\frac{\lambda}{\mu} c_1 e^{\mu t}\right] - \int_0^t [\mu(c-1)(1 - c_1 e^{\mu x})^{c-2} P_{c-1}(x) - \lambda(1 - c_1 e^{\mu x})^{c-1} P_{c-2}(x)] \exp\left[\frac{\lambda}{\mu} c_1 e^{\mu x}\right] dx = f[(1-z)e^{-\mu t}],$$

where f is an arbitrary function to be determined from initial conditions and c_1 is given by (13).

Now, if $i \leqq c - 2$, then

$$Q(z, 0) = z^i. \tag{17}$$

Using (13) and (17), (16) gives

$$f(1 - z) = z^i \exp\left[\frac{\lambda}{\mu}(1 - z)\right], \tag{18}$$

whence, setting $1 - z = y$,

$$f(y) = (1 - y)^i \exp\left[\frac{\lambda}{\mu} y\right]$$

and, therefore,

$$f[(1 - z)e^{-\mu t}] = [1 - (1 - z)e^{-\mu t}]^i \exp\left[\frac{\lambda}{\mu}(1 - z)e^{-\mu t}\right]. \tag{19}$$

From (16), on substituting for c_1 from (13) and for f from (19), we have

$$\begin{aligned} Q(z, t) = \exp\left[-\frac{\lambda}{\mu}(1 - z)\right] \Bigg[\int_0^t & [\mu(c - 1)\{1 - (1 - z) \\ & \cdot \exp(-\mu(t - x))\}^{c-2} P_{c-1}(x) \\ & - \lambda\{1 - (1 - z)\exp(-\mu(t - x))\}^{c-1} P_{c-2}(x)] \\ & \cdot \exp\left[\frac{\lambda}{\mu}(1 - z)\exp(-\mu(t - x))\right] dx \\ & + \exp\left[-\frac{\lambda}{\mu}(1 - z)(1 - e^{-\mu t})\right]\{1 - (1 - z)e^{-\mu t}\}^i. \end{aligned} \tag{20}$$

If $i > c - 2$, then $Q(z, 0) = 0$, and therefore $f(y) = 0$; so the last term in (20) will vanish. Taking Laplace transforms in (20), for the case $i > c - 2$, and noting that [Erdélyi *et al.* (1954, Vol. I, p. 147, Eq. (40))]

$$\begin{aligned} & \int_0^\infty e^{-sw}[1 - (1 - z)e^{-\mu w}]^{c-2} \exp\left[\frac{\lambda}{\mu}(1 - z)e^{-\mu w}\right] dw \\ & = \int_0^\infty e^{-(s/\mu)y}[1 - (1 - z)e^{-y}]^{c-2} \exp\left[\frac{\lambda}{\mu}(1 - z)e^{-y}\right] \cdot \frac{1}{\mu}\, dy \\ & = \frac{1}{s}\Phi_1\left[\frac{s}{\mu}, -(c - 2); 1; 1 - z, \frac{\lambda}{\mu}(1 - z)\right], \end{aligned} \tag{21}$$

where [see also II.6(27)]

$$\text{(22)}\quad \Phi_1(\alpha, \beta; \gamma; x, y) = \sum_{m,n=0}^{\infty} \frac{(\alpha)_{m+n}(\beta)_m}{(\gamma)_{m+n}} \frac{x^m}{m!} \frac{y^n}{n!}, \qquad |x| < 1, \quad |y| < \infty,$$

we have

$$\text{(23)}\quad Q^*(z, s) = \exp\left[-\frac{\lambda}{\mu}(1 - z)\right]\left\{(c - 1)P^*_{c-1}(s) \cdot \frac{\mu}{s}\Phi_1\left[\frac{s}{\mu}, -(c - 2); 1; 1 - z, \frac{\lambda}{\mu}(1 - z)\right] - P^*_{c-2}(s) \cdot \frac{\lambda}{s}\Phi_1\left[\frac{s}{\mu}, -(c - 1); 1; 1 - z, \frac{\lambda}{\mu}(1 - z)\right]\right\}.$$

Now

$$\text{(24)}\qquad P^*_{c-2}(s) = \frac{1}{(c - 2)!} \left.\frac{\partial^{c-2}Q^*}{\partial z^{c-2}}\right|_{z=0}.$$

Thus, from (23) by differentiation we obtain an equation in terms of P^*_{c-2} and P^*_{c-1}, which we solve for P^*_{c-2}. Then, using (24) and using Leibniz's theorem for successive differentiation, we obtain, from (23),

$$\text{(25)}\quad P^*_{c-2}(s) = \frac{1}{(c - 2)!}\left\{(c - 1)P^*_{c-1}(s) \cdot \frac{\mu}{s}\sum_{k=0}^{c-2}\binom{c - 2}{k}\left(\frac{\lambda}{\mu}\right)^{c-k-2} e^{-\lambda/\mu} \cdot \frac{d^k}{dz^k}\Phi_1\left[\frac{s}{\mu}, -(c - 2); 1; 1 - z, \frac{\lambda}{\mu}(1 - z)\right]\right|_{z=0} - \frac{\lambda}{s}P^*_{c-2}(s)\sum_{k=0}^{c-2}\binom{c - 2}{k}\left(\frac{\lambda}{\mu}\right)^{c-k-2} e^{-\lambda/\mu} \cdot \frac{d^k}{dz^k}\Phi_1\left[\frac{s}{\mu}, -(c - 1); 1; 1 - z, \frac{\lambda}{\mu}(1 - z)\right]\left.\right|_{z=0}\Bigg\},$$

which gives

$$\text{(26)}\quad P^*_{c-2}(s) = \frac{\dfrac{c-1}{(c-2)!}\dfrac{\mu}{s}P^*_{c-1}(s)\displaystyle\sum_{k=0}^{c-2}\binom{c-2}{k}\left(\frac{\lambda}{\mu}\right)^{c-k-2} e^{-\lambda/\mu}\frac{d^k}{dz^k}\Phi_1\left[\frac{s}{\mu}, -(c-2); 1; 1-z, \frac{\lambda}{\mu}(1-z)\right]\Bigg|_{z=0}}{1+\dfrac{\lambda}{s}P^*_{c-2}(s)\displaystyle\sum_{k=0}^{c-2}\binom{c-2}{k}\left(\frac{\lambda}{\mu}\right)^{c-k-2} e^{-\lambda/\mu}\frac{d^k}{dz^k}\Phi_1\left[\frac{s}{\mu}, -(c-1); 1; 1-z, \frac{\lambda}{\mu}(1-z)\right]\Bigg|_{z=0}}.$$

We also have, for $n \leqq c-2$,

$$(27)\qquad P_n^*(s) = \frac{1}{n!}\frac{\mu}{s}\Bigg\{(c-1)P_{c-1}^*(s)\sum_{k=0}^{n}\binom{n}{k}\left(\frac{\lambda}{\mu}\right)^{n-k} e^{-\lambda/\mu} \cdot \frac{d^k}{dz^k}\Phi_1\left[\frac{s}{\mu}, -(c-2); 1; 1-z, \frac{\lambda}{\mu}(1-z)\right]\Bigg|_{z=0} - \frac{\lambda}{\mu}P_{c-2}^*(s)\sum_{k=0}^{n}\binom{n}{k}\left(\frac{\lambda}{\mu}\right)^{n-k} e^{-\lambda/\mu} \cdot \frac{d^k}{dz^k}\Phi_1\left[\frac{s}{\mu}, -(c-1); 1; 1-z, \frac{\lambda}{\mu}(1-z)\right]\Bigg|_{z=0}\Bigg\}.$$

Substituting from (27) in (9) and solving for $P_{c-1}^*(s)$, we have[2]

$$(28)\quad P_{c-1}^*(s) = \Bigg\{\frac{\alpha_2^{i+1}}{\mu(1-\alpha_2)} - 2\alpha_2^{c-2}P_{c-2}^*(s) + \sum_{n=0}^{c-3}(c-n)\alpha_2^n \cdot \frac{1}{n!}\frac{\lambda}{s}P_{c-2}^*(s)\sum_{k=0}^{n}\binom{n}{k}\left(\frac{\lambda}{\mu}\right)^{n-k} e^{-\lambda/\mu} \cdot \frac{d^k}{dz^k}\Phi_1\left[\frac{s}{\mu}, -(c-1); 1; 1-z, \frac{\lambda}{\mu}(1-z)\right]\Bigg|_{z=0}\Bigg\} \div \Bigg\{\alpha_2^{c-1} + \sum_{n=0}^{c-3}(c-n)\alpha_2^n \cdot \frac{1}{n!}\frac{\mu}{s}(c-1)\sum_{k=0}^{n}\binom{n}{k}\left(\frac{\lambda}{\mu}\right)^{n-k} e^{-\lambda/\mu} \cdot \frac{d^k}{dz^k}\Phi_1\left[\frac{s}{\mu}, -(c-2); 1; 1-z, \frac{\lambda}{\mu}(1-z)\right]\Bigg|_{z=0}\Bigg\}.$$

Equations (28) and (25) completely determine $P_{c-1}^*(s)$. Then, (27) gives $P_n^*(s)$ $(n = 0, 1, \ldots, c-3)$, and $P_n^*(s)$ $(n \geqq c)$ can be found by taking coefficient of z^n in the right-hand member of (7).

The case $c = 2$ is considered separately, and in this case Saaty (1960) has been able to obtain the inverse Laplace transform in terms of the modified Bessel functions of the first kind $I_\nu(z)$.

Further, Saaty (1960) considers the busy period distribution for the $M/M/2$ system. He, in fact, makes a mistake in writing his equation for $P^*(z, s)$ in this case [Saaty (1960, p. 769, Eq. (40))], which should read

$$(29)\qquad P^*(z,s) = \frac{z^{i+1} - \mu(1-z)[2P_0^* + zP_1^*] + \lambda z(1-z)P_0^*}{-\lambda(z-\alpha_1)(z-\alpha_2)}.$$

Then, P_0^* and P_1^* are given by the following equation, obtained from the

[2] The corresponding expression in Saaty (1960, p. 762, Eq. (22)) seems to be in error.

analyticity consideration of $P^*(z, s)$,

(30) $$(2\mu - \lambda\alpha_2)P_0^* + \mu\alpha_2 P_1^* = \frac{\alpha_2^{i+1}}{1 - \alpha_2},$$

and the Laplace transform of the modified form of (1), viz.,

(31) $$sP_0^* = \mu P_1^*.$$

From (30) and (31) we obtain

(32) $$P_1^*(s) = \frac{s\alpha_2^{i+1}}{\mu(1 - \alpha_2)(2\mu - \lambda\alpha_2 + s\alpha_2)} = \frac{\lambda(1 - \alpha_1)\alpha_2^{i+1}}{\mu(2 - \alpha_2)(\lambda\alpha_2 - 2\mu)}.$$

For finding the distribution of an *ordinary* busy period as different from the *initial* one, we set $i = 1$ in (32) and, after further simplification, obtain

(33) $$\mu P_1^*(s) = \alpha_2/(2 - \alpha_2),$$

which agrees with the corresponding result obtained by Chaudhry and Templeton (1973), and which can be inverted to give the following form for the busy period density function [cf. 1.6(16), (17)]:

(34) $$\frac{d}{dt}P_0(t) = \mu P_1(t) = e^{-(\lambda + 2\mu)t}t^{-1}\sum_{n=1}^{\infty}\left\{\frac{\mu}{2\lambda}\right\}^{n/2} I_n(2\sqrt{2\lambda\mu}\, t).$$

1.3 The Double-Ended Queue

Kendall (1951) discussed queues of taxis and passengers at a taxistand, where at times there are passengers waiting for taxis, and at other times there may be taxis waiting for passengers. Such systems, sometimes called double queues, where either the units wait in queue for service or idle servers queue up for customers, have varied applications in inventory and production processes and have been studied by various authors [e.g., Dobbie (1961); Sasieni (1961); Kashyap (1965a,b,c, 1966)]. Similar processes have also been discussed by Brigham (1955), Foster (1959) and Latouche (1979) among others. In this system there are two types of units, and interchangeably either may be called passengers (customers) and the other taxis (servers).

As in Section 1.1, consider a random walk on the entire real line, where the particle moves one step to the right or left with probabilities λ, μ per unit of time, respectively, and there are barriers at positions $-M$ and N. The first barrier reflects the particle with probability r_1 and absorbs it with probability $1 - r_1$. The second barrier reflects and absorbs with probabilities r_2 and $1 - r_2$ (Kashyap, 1965a).

Define

(1) $P_n(t) = \text{Prob}\{\text{particle is at position } n \text{ at time } t\} \qquad (-M \leqq n \leqq N).$

Then, we have

$$\frac{d}{dt}P_{-M}(t) = -r_1\lambda P_{-M}(t) + \mu P_{-M+1}(t), \tag{2}$$

$$\frac{d}{dt}P_{-M+1}(t) = -(\lambda+\mu)P_{-M+1}(t) + r_1\lambda P_{-M}(t) + \mu P_{-M+2}(t), \tag{3}$$

$$\frac{d}{dt}P_n(t) = -(\lambda+\mu)P_n(t) + \lambda P_{n-1}(t) + \mu P_{n+1}(t) \tag{4}$$

$$(-M+2 \leqq n \leqq N-2),$$

$$\frac{d}{dt}P_{N-1}(t) = -(\lambda+\mu)P_{N-1}(t) + r_2\mu P_N(t) + \lambda P_{N-2}(t), \tag{5}$$

$$\frac{d}{dt}P_N(t) = -\mu r_2 P_N(t) + \lambda P_{N-1}(t). \tag{6}$$

Defining the generating functions

$$F(z,t) = \sum_{n=-M}^{-1} P_n(t)z^n, \tag{7}$$

$$G(z,t) = \sum_{n=0}^{N} P_n(t)z^n,$$

and using Laplace transforms (L.T.), denoted by the corresponding lowercase letters, typically,

$$\mathscr{L}[F(t)] = f(s) = \int_0^\infty e^{-st}F(t)\,dt, \tag{8}$$

we obtain

$$f(z,s) = \frac{z^{-M}[\lambda z(1-r_1)-\mu](1-z)p_{-M}(s) - \lambda z p_{-1}(s) + \mu p_0(s)}{-[\lambda z^2 - (s+\lambda+\mu)z + \mu]}, \tag{9}$$

$$g(z,s) = \frac{z^{i+1} + z^N[\lambda z - \mu(1-r_2)](1-z)p_N(s) + \lambda z p_{-1}(s) - \mu p_0(s)}{-[\lambda z^2 - (s+\lambda+\mu)z + \mu]}, \tag{10}$$

where it has been assumed that initially at time $t = 0$ the particle is at the position $n = i \geqq 0$. The four unknowns in (9) and (10), viz., $p_n(s)$, $n = -M, -1, 0, N$, are determined by requiring that the numerators of the right-hand members of both (9) and (10) vanish at the zeros of the denominator, which are the same as given by 1.1(5). Thus $f(z,s)$ and $g(z,s)$ are completely determined. We also define the generating function

$$(11) \qquad H(z,t) = \sum_{n=-M}^{N} P_n(t) z^n,$$

and its L.T. is given by

$$(12) \quad h(z,s) = f(z,s) + g(z,s)$$

$$= \frac{z^{i+1} + z^N\{\lambda z - \mu(1-r_2)\}(1-z)p_N(s) + z^{-M}\{\lambda z(1-r_1) - \mu\}(1-z)p_{-M}(s)}{-[\lambda z^2 - (s+\lambda+\mu)z + \mu]}.$$

By setting r_1 and r_2 at each of the values 0 and 1, we obtain the particular cases where both the barriers are absorbing (reflecting), and one absorbing and the other reflecting, and the corresponding results of Heathcote and Moyal (1959) can be obtained on setting $M = 0$.

The case $r_1 = r_2 = 1$, with reflecting barriers at $-M$ and N, corresponds to the double-ended queuing model with limited waiting space for N passengers or M taxis on defining

$$(13) \qquad P_n(t) = \begin{cases} \text{Prob}\{n \text{ passengers waiting at time } t\} & (n \geqq 0) \\ \text{Prob}\{-n \text{ taxis waiting at time } t\} & (n < 0). \end{cases}$$

Initially at $t = 0$, there are i passengers waiting. In this case, evaluating $p_{-M}(s)$ and $p_N(s)$ as indicated above, we have

$$(14) \quad h(z,s) = \phi(z,s) \cdot \Bigg[z^i + \frac{\lambda}{s} z^N (1-z)$$

$$\cdot \frac{\alpha_2^{M+i+1} - \alpha_1^{M+i+1} - \alpha_1\alpha_2(\alpha_2^{M+i} - \alpha_1^{M+i})}{\alpha_2^{M+N+1} - \alpha_1^{M+N+1}}$$

$$+ \frac{\lambda}{s} z^{-M} \left(1 - \frac{1}{z}\right) (\alpha_1\alpha_2)^{M+i+1} \cdot \frac{\alpha_2^{N-i}(\alpha_2 - 1) + \alpha_1^{N-i}(1-\alpha_1)}{\alpha_2^{M+N+1} - \alpha_1^{M+N+1}} \Bigg],$$

where $\phi(z,s)$ is given by 1.1(4).

Inverting the coefficient of z^n by applying Erdélyi *et al.* (1954, Vol. I, p. 240, Eqs. (28), (30); and p. 130, Eq. (9)), we have

$$(15) \quad P_n(t) = \beta^{n-i} \Bigg[e^{-\alpha t} \Bigg\{ I_{i-n} + \sum_{j=0}^{\infty} (I_{2(M+N+1)(j+1)+i-n}$$

$$+ I_{2(M+N+1)(j+1)-i+n}) \Bigg\} + \lambda \sum_{j=0}^{\infty} \int_0^t e^{-\alpha\tau} \{ I_{2(M+N+1)j+2N-i-n}$$

$$- 2\beta^{-1} I_{2(M+N+1)j+2N-i-n+1}$$

$$+ \beta^{-2} I_{2(M+N+1)j+2N-i-n+2} + I_{2(M+N+1)j+2M+i+n+2}$$

$$- 2\beta^{-1} I_{2(M+N+1)j+2M+i+n+1} + \beta^{-2} I_{2(M+N+1)j+2M+i+n} \} \, d\tau \Bigg]$$

$$(n = -M, -M+1, \ldots, N),$$

where I_ν has the same meaning as in 1.1(10). Equation (15) agrees with 1.1(10) on setting $M = 0$.

Setting $r_1 = 1$ in (9), we have

$$f(z, s) = \frac{\mu z^{-M}(z-1)p_{-M}(s) - \lambda z p_{-1}(s) + \mu p_0(s)}{-[\lambda z^2 - (s + \lambda + \mu)z + \mu]}, \tag{16}$$

whence

$$f(1, s) = \frac{\mu p_0(s) - \lambda p_{-1}(s)}{s}, \tag{17}$$

which, on inversion, gives

$$F(1, t) = \int_0^t [\mu P_0(\tau) - \lambda P_{-1}(\tau)]\, d\tau. \tag{18}$$

Clearly, (18) gives the probability that at time t there is a queue of taxis waiting, $P_0(\tau)$ and $P_{-1}(\tau)$ being given by (15). Also, $l_T(s)$, the L.T. of $L_T(t)$, the mean queue length of taxis at time t, is given by

$$\begin{aligned} l_T(s) &= -\frac{\partial}{\partial z} f(z, s)\bigg|_{z=1} \\ &= \frac{\lambda - \mu}{s^2}\{-\mu p_0(s) + \lambda p_{-1}(s)\} + \frac{\mu}{s}\{-p_{-M}(s) + p_0(s)\}. \end{aligned} \tag{19}$$

Inverting (19), we have

$$\begin{aligned} L_T(t) = \mu \int_0^t \{P_0(\tau) - P_{-M}(\tau)\}\, d\tau \\ + (\lambda - \mu)\int_0^t (t - \tau)\{\lambda P_{-1}(\tau) - \mu P_0(\tau)\}\, d\tau. \end{aligned} \tag{20}$$

Setting $r_2 = 1$, $z = 1$ in (10), we have

$$g(1, s) = \frac{1 + \lambda z p_{-1}(s) - \mu p_0(s)}{s}, \tag{21}$$

whence, on inversion,

$$G(1, t) = 1 + \int_0^t [\lambda P_{-1}(\tau) - \mu P_0(\tau)]\, d\tau. \tag{22}$$

$G(1, t) - P_0(t)$ is the probability that there is a queue of customers at time t.

The L.T. of $L_C(t)$, the mean number of customers waiting at time t, is given by

$$(23)\qquad l_C(s) = \left.\frac{\partial g(z,s)}{\partial z}\right|_{z=1} = \frac{\lambda-\mu}{s^2}\{1+\lambda p_{-1}(s)-\mu p_0(s)\} + \frac{1}{s}\{i-\lambda p_N(s)+\mu p_0(s)\}.$$

On inversion, we have

$$(24)\qquad L_C(t) = i + (\lambda-\mu)t + \int_0^t \{\mu P_0(\tau) - \lambda P_N(\tau)\}\,d\tau + (\lambda-\mu)\int_0^t (t-\tau)\{\lambda P_{-1}(\tau) - \mu P_0(\tau)\}\,d\tau.$$

Busy Periods

We divide busy periods for the double-ended queue into (1) customer periods and (2) taxi periods, which have obvious interpretations. For customer periods, we consider an initial state $n = 1$, a reflecting barrier at $n = N$, and an absorbing barrier at $n = 0$. Then, if $\Gamma_C(t)$ is the probability density of customer periods with L.T. $\gamma_C(s)$, we have

$$(25)\qquad \Gamma_C(t) = \frac{d}{dt}P_0(t)$$

and, therefore,

$$(26)\qquad \gamma_C(s) = sp_0(s),$$

where $p_0(s)$ is the value of $p_{-M}(s)$ in (12), with $M = 0, i = 1, r_1 = 0$, and $r_2 = 1$. Hence, we have, on simplification,

$$(27)\qquad \gamma_C(s) = \frac{(\alpha_1\alpha_2)[\alpha_2^{N-1}(\alpha_2-1)+\alpha_1^{N-1}(1-\alpha_1)]}{\alpha_2^N(\alpha_2-1)+\alpha_1^N(1-\alpha_1)}.$$

Relaxing the limited waiting space condition by letting $N \to \infty$, we have

$$(28)\qquad \gamma_C(s) = \alpha_1,$$

whence by using Erdélyi *et al.* (1954, Vol. I, p. 240, Eq. (28)), we have

$$(29)\qquad \Gamma_C(t) = \sqrt{\mu/\lambda}\, t^{-1} e^{-(\lambda+\mu)t} I_1(2\sqrt{\lambda\mu}\, t),$$

which is a well-known result for the $M/M/1$ queue. [See, e.g., Saaty (1961a, p. 128, Problem 15).] Similarly, the L.T. of $\Gamma_T(t)$, the probability density function of taxi periods, is given by

$$\gamma_{\mathrm{T}}(s) = 1/\alpha_2, \tag{30}$$

whence

$$\Gamma_{\mathrm{T}}(t) = \sqrt{\lambda/\mu}\, t^{-1} e^{-(\lambda+\mu)t} I_1(2\sqrt{\lambda\mu}\, t). \tag{31}$$

Steady-State Case of the Queue

Let $P_n(t)$ be the steady-state probability given by

$$P_n = \lim_{t\to\infty} P_n(t). \tag{32}$$

That this limit and hence the steady state exist can be shown by using (15) and the asymptotic behavior of the Bessel function as follows. We know that, for large z,

$$I_v(z) \sim \exp(z)/\sqrt{2\pi z}, \tag{33}$$

which is independent of v. Hence, for large t,

$$e^{-(\lambda+\mu)t} I_v(2\sqrt{\lambda\mu}\, t) \sim \exp[-(\sqrt{\lambda} - \sqrt{\mu})^2 t]/\sqrt{2\pi(2\sqrt{\lambda\mu}\, t)}, \tag{34}$$

which approaches 0 as $t \to \infty$. As the above result holds for all v, all the terms outside the integral sign in (15) vanish as $t \to \infty$.

Also, as $t \to \infty$, the integrals occurring in (15) all take the form

$$\int_0^\infty e^{-(\lambda+\mu)\tau} I_v(2\sqrt{\lambda\mu}\,\tau)\, d\tau$$

which on using Erdélyi *et al.* (1954, Vol. I, p. 240, Eq. (30)) equals

$$\frac{(2\sqrt{\lambda\mu})^v}{\sqrt{(\lambda+\mu)^2 - 4\lambda\mu}\,[\lambda + \mu + \sqrt{(\lambda+\mu)^2 - 4\lambda\mu}]^v} \tag{35}$$

$$= \begin{cases} \dfrac{1}{\lambda - \mu}\left(\sqrt{\dfrac{\mu}{\lambda}}\right)^v & \text{if } \lambda > \mu \\[2ex] \dfrac{1}{\mu - \lambda}\left(\sqrt{\dfrac{\lambda}{\mu}}\right)^v & \text{if } \lambda < \mu. \end{cases}$$

Thus in both cases, when $\lambda > \mu$ and $\lambda < \mu$, the infinite sums of the integrals in (15) give convergent series, since $M + N + 1$ is positive.

Using the Tauberian theorem [Eq. 1.1(11)], we have

$$H(z) = \lim_{t\to\infty} H(z, t) = \lim_{s\downarrow 0} sh(z, s), \tag{36}$$

where $h(z, s)$ is given by (14). Now, taking the coefficient of z^n in (36), we have

$$P_n = \frac{(1-\rho)\rho^{M+n}}{1-\rho^{M+N+1}} \quad \text{if} \quad \rho \neq 1 \tag{37}$$

and, by proceeding to limits

$$P_n = \frac{1}{M+N+1} \quad \text{if} \quad \rho = 1. \tag{38}$$

1.4 The Unrestricted Linear Random Walk

Conolly (1971a) considers the random walk discussed in Sections 1.1 and 1.3 without any barriers at either end, so that the random walk continues unrestricted between $n = -\infty$ and $n = +\infty$. This has been described as a "randomized" random walk by Feller (1966, p. 58) and has been denoted by $\infty^2/M/M$ by Gibson (1978), and Conolly (1971a), using an obvious extension of the well-known queuing notation.

Let $S(t)$ be the position of the particle at time t. Thus $S(t)$ is given by two independent Poisson streams with parameters λt and μt for motion to the right and left, respectively. Let

$$P_n(t) = \text{Prob}\{S(t) = n \mid S(0) = 0\}. \tag{1}$$

Then, for $n \geqq 0$,

$$P_n(t) = \sum_{r \geqq 0} e^{-\lambda t} \frac{(\lambda t)^{n+r}}{(n+r)!} e^{-\mu t} \frac{(\mu t)^r}{r!}, \tag{2}$$

$$P_{-n}(t) = \sum_{r \geqq 0} e^{-\mu t} \frac{(\mu t)^{n+r}}{(n+r)!} e^{-\lambda t} \frac{(\lambda t)^r}{r!}. \tag{3}$$

The evenness in n of the modified Bessel function $I_n(x)$ of the first kind enables us to write

$$P_n(t) = \rho^{n/2} e^{-(\lambda+\mu)t} I_n(wt) \tag{4}$$

for all integers n, where

$$\rho = \lambda/\mu, \tag{5}$$

$$w = 2\sqrt{\lambda\mu}. \tag{6}$$

Also, clearly we have

$$\text{Prob}\{S(t) = n \mid S(0) = m\} = \rho^{(n-m)/2} e^{-(\lambda+\mu)t} I_{|n-m|}(wt). \tag{7}$$

First Passages

A first passage from m to n ($n > m$) is said to occur at time t if

$$S(0) = m, \qquad S(\tau) < n \quad (0 < \tau < t), \qquad S(t) = n \tag{8}$$

and will be assumed to have probability density function (p.d.f.) $f_{mn}(t)$ with L.T. $\phi_{mn}(s)$. Without loss of generality, we may consider $f_{0n}(t)$ ($n \geqq 1$), and because of the "lack of memory" property of the negative exponential distribution we have

$$f_{0n}(t) = f_{01}^{(n)}(t), \tag{9}$$

where $a^{(n)}(t)$ denotes the n-fold convolution of $a(t)$ with itself. By ordinary probability arguments, we have

$$f_{01}(t) = \lambda e^{-(\lambda+\mu)t} + \mu e^{-(\lambda+\mu)t} * f_{01}^{(2)}(t), \tag{10}$$

where the asterisk denotes convolution of the two functions it separates. Taking Laplace transforms in (10), we have

$$\mu\phi_{01}^2(s) - (s + \lambda + \mu)\phi_{01}(s) + \lambda = 0. \tag{11}$$

The roots of (11) are

$$\xi, \eta = \frac{(s + \lambda + \mu) \pm \sqrt{(s + \lambda + \mu)^2 - w^2}}{2\mu}, \tag{12}$$ [3]

ξ corresponding to the + sign.

Since $f_{01}(t)$ is a p.d.f., we have

$$\phi_{01}(s) = \eta, \tag{13}$$

and, by symmetry,

$$\phi_{10}(s) = \xi^{-1}. \tag{14}$$

Equations (13) and (14) yield, on inversion,

$$f_{01}(t) = \rho^{1/2} e^{-(\lambda+\mu)t} I_1(wt)/t, \tag{15}$$

$$f_{10}(t) = \rho^{-1/2} e^{-(\lambda+\mu)t} I_1(wt)/t. \tag{16}$$

Also, using (9) we have

$$f_{0n}(t) = \rho^{n/2} n e^{-(\lambda+\mu)t} I_n(wt)/t, \tag{17}$$

$$f_{n0}(t) = \rho^{-n/2} n e^{-(\lambda+\mu)t} I_n(wt)/t. \tag{18}$$

It may be noted that (15) and (16) are in agreement with 1.3(31) and 1.3(29), respectively. In fact, $\xi = \alpha_2^{-1}$ and $\eta = \alpha_1^{-1}$.

[3] The μ in the denominator is missing in Conolly (1971a).

From (4) and (17), we have

(19) $$f_{0n}(t)/n = P_n(t)/t,$$

which is the continuous-time counterpart of Feller's (1966) general discrete time theorem about ladder epochs.

Now consider

(20) $$F_{01}(t) = 1 - \int_0^t f_{01}(\tau)\,d\tau,$$

the probability that the first passage from 0 to 1 occurs after time t. Let $\Phi_{01}(s)$ be the L.T. of $F_{01}(t)$. Then

(21) $$\Phi_{01}(s) = \frac{1-\eta}{s} = \frac{1}{\mu\xi(1 - 1/\xi)} = \frac{\phi_{10}(s)}{\mu[1 - \phi_{10}(s)]}.$$

Using the obvious relation between ξ and η, we get

(22) $$(1 - \xi)(1 - \eta) = -s/\mu.$$

Similarly, we have

(23) $$\Phi_{10}(s) = \frac{\phi_{01}(s)}{\lambda[1 - \phi_{01}(s)]}.$$

From (23), by expansion of the denominator and term-by-term inversion, and then using the relation

(24) $$\frac{2n}{z} I_n(z) = I_{n-1}(z) - I_{n+1}(z),$$

we obtain

(25) $$F_{10}(t) = \frac{1}{\rho} e^{-(\lambda+\mu)t}[I_0(wt) + \rho^{1/2} I_1(wt)] - \left(\frac{1}{\rho} - 1\right) J(\mu t, \lambda t),$$

where

(26) $$J(x, y) = e^{-(x+y)} \sum_{m \geqq 0} \left(\frac{y}{x}\right)^{m/2} I_m(2\sqrt{xy}).$$

$F_{10}(t)$ can also be obtained by using the "reflection principle" for random walk paths (Feller, 1968, p. 72). Thus

(27) $$F_{10}(t) = \text{Prob}\{S(\tau) \geqq 1 \mid S(0) = 1\}, \qquad 0 < \tau \leqq t,$$

so that we are concerned with all possible paths from $S(0) = 1$ to $S(t) = r\,(r \geqq 1)$ that do not ever reach 0 during the passage. By the reflection principle and (7) it follows that

$$(28)\qquad F_{10}(t) = \sum_{r \geqq 1} \left[P_{r-1}(t) - \frac{1}{\rho} P_{r+1}(t) \right]$$
$$= (1 - 1/\rho)\,\text{Prob}\{S(t) \geqq 0\} + (1/\rho)\{P_0(t) + P_1(t)\}.$$

Comparison of (25) and (28) provides a probabilistic interpretation of $J(\mu t, \lambda t)$. $F_{01}(t)$ can be obtained by interchanging λ and μ.

The function $J(x, y)$ given by (26) whose probabilistic interpretation is shown above was introduced by Goldstein (1953) in the analysis of a one-dimensional diffusion process. Goldstein points out that this function also occurs in two-dimensional diffusion, certain solutions of the telegrapher's equation, and a probabilistic application in gunnery where one is interested in the probability of scoring a hit in an offset circle with a bivariate normal probability distribution of aiming errors. Conolly (1971b) and Hadidi (1969) also find the same function in connection with the waiting time problem for the $M/M/1$ queue with a certain kind of state-dependent service.

Returns to the Origin

We say that a first return to the origin occurs at time t if

$$(29)\qquad S(0) = 0, \qquad S(\tau) \neq 0 \quad (0 < \tau < t), \qquad S(t) = 0.$$

Let the distribution for the first return have a density function $g_1(t)$. Then the density function relating to the rth return to zero is given by

$$(30)\qquad g_r(t) = g_1^{(r)}(t).$$

Gibson (1968) shows that the L.T. $\gamma(s)$ of $g_1(t)$ is

$$(31)\qquad \gamma(s) = \frac{s + \lambda + \mu - \sqrt{(s + \lambda + \mu)^2 - w^2}}{s + \lambda + \mu},$$

from which, using (30) and the convolution theorem for Laplace transforms,

$$(32)\qquad g_r(t) = \frac{r w^r}{(r-1)!} e^{-(\lambda+\mu)t} \int_0^t (t - \tau)^{r-1} I_r(w\tau) \frac{d\tau}{\tau}.$$

Setting $\tau = ut$ and using the definition II.3(41) of $I_r(z)$, (32) simplifies to

$$(33)\quad g_r(t) = \frac{(2\lambda\mu)^r t^{2r-1}}{(2r-1)!} e^{-(\lambda+\mu)t}\, {}_2F_3\left[\frac{r}{2}, \frac{r+1}{2}; r, r + \frac{1}{2}, r + 1; \lambda\mu t^2\right],$$

from which asymptotic behavior can be determined. Gibson (1968) also obtains a general formula for $k_n(t)$, the probability of exactly n returns to zero in $(0, t)$. In the case $\lambda = \mu$ this reduces to

$$(34) \qquad k_n(t) = 2\mu e^{-2\mu t} \int_0^t I_n(2\mu\tau) \left[\frac{\{2\mu(t-\tau)\}^{n-1}}{(n-1)!} + \frac{\{2\mu(t-\tau)\}^n}{n!} \right] d\tau,$$

which, as before, simplifies to

$$(35) \quad k_n(t) = e^{-2\mu t} \left[\frac{2^n(\mu t)^{2n}}{(2n)!} {}_2F_3\left(\frac{n}{2} + \frac{1}{2}, \frac{n}{2} + 1; n + \frac{1}{2}, n+1, n+1; \mu^2 t^2 \right) \right.$$
$$\left. + \frac{2^{n+1}(\mu t)^{2n+1}}{(2n+1)!} {}_2F_3\left(\frac{n}{2} + \frac{1}{2}, \frac{n}{2} + 1; n+1, n+1, n + \frac{3}{2}; \mu^2 t^2 \right) \right].$$

Gibson (1968) provides numerical tables and notes that as $t \to \infty$, the expected number of returns to zero behaves like $t^{1/2}$, when $\lambda = \mu$, a result which corresponds to its discrete-time analogue.

For the queuing system $M/M/1$ the result corresponding to (32) is

$$(36) \qquad g_r(t) = \frac{r(\lambda\mu)^{r/2}}{(r-1)!} e^{-\lambda t} \int_0^t e^{-\mu\tau}(t-\tau)^{r-1} I_r(w\tau) \frac{d\tau}{\tau}.$$

Two-State Sojourn Time Problems

Let $S(0) = 0$. We say that the particle is in state A when $S(t) > 0$; otherwise it is in state B (Conolly, 1971a). Consider a random variable $\chi(t)$ which assumes the value 1 when the particle is in state A, and 0 when it is in state B. Assuming the existence of a random variable $\sigma_A(t)$ which measures the sojourn time in state A during the interval $(0, t)$ and is defined by

$$(37) \qquad \sigma_A(t) = \int_0^t \chi(\tau)\, d\tau,$$

we propose to find an expression for

$$(38) \qquad p_A(\tau, t)\, d\tau = \text{Prob}\{\tau < \sigma_A(t) < \tau + d\tau\}, \qquad 0 < \tau < t.$$

This problem has been addressed in some generality by Takács (1957a,b), who cites earlier work by himself, Kolmogorov, Dobrusin, Karpelevich, and Uspensky. A later and independent paper is by Good (1961).

If the particle has just passed into state B, its immediate sojourn there is governed by the first passage time from 0 to 1 when it reaches A. It then remains in A until it makes a first passage back to 0. Clearly, we have

$$(39) \qquad P_A(0) = \text{Prob}\{\sigma_A(t) = 0\} = F_{01}(t).$$

Also, from a heuristic argument depending on enumerating all possible state sequences during $(0, t)$, we have

(40) $$p_A(\tau, t) = f_{01}(t-\tau)F_{10}(\tau) + \sum_{n \geqq 1} f_{01}^{(n+1)}(t-\tau)f_{10}^{(n)}(\tau) * F_{10}(\tau)$$
$$+ \sum_{n \geqq 1} f_{10}^{(n)}(\tau)f_{01}^{(n)}(t-\tau) * F_{01}(t-\tau),$$

from which, after taking double L.T. and inverting, we have for $\infty^2/M/M$,

(41) $$p_A(\tau, t) = \lambda F_{10}(\tau)F_{01}(t-\tau).$$

For $M/M/1$, $f_{10}(t)$ is exactly as for $\infty^2/M/M$, but $f_{01}(t) = \lambda e^{-\lambda t}$ since state B consists only of the origin. So we have, from (40),

(42) $$p_A(\tau, t) = \mu e^{-\lambda t - \mu\tau}[\rho I_0(x) + (\rho t/\tau)^{1/2} I_1(x)]$$
$$- \mu(1-\rho)e^{(\mu-\lambda)(t-\tau)}J(\mu t, \lambda\tau),$$

where

(43) $$x = 2(\lambda\mu\tau t)^{1/2}.$$

Conolly (1971a) proceeds to investigate asymptotic results for sojourn time problems by evaluating

(44) $$\int_{\alpha t}^{t} p_A(\tau, t)\, d\tau = \text{Prob}\{\alpha t < \sigma_A(t) < t\}$$

for fixed α, $0 < \alpha < 1$, as $t \to \infty$ in the following four cases:

(i) $\infty^2/M/M: \lambda < \mu$;
(ii) $\infty^2/M/M: \lambda = \mu$;
(iii) $M/M/1: \lambda < \mu$;
(iv) $M/M/1: \lambda = \mu$.

Case (i) – Unsymmetrical $\infty^2/M/M$ It is shown that

(45) $$\text{Prob}\{\alpha t < \sigma_A(t) < t\} \underset{\text{D}}{=} \frac{\rho^{1/4}(1+\rho^{1/2})}{\pi^{1/2}}[\psi(\alpha t) - \psi(t)],$$

where

(46) $$\psi(t) = \frac{e^{-\Lambda t}}{(\Lambda t)^{1/2}} - 2\pi^{1/2}\,\text{Erfc}(2\Lambda t)^{1/2},$$

(47) $$\Lambda = \mu(1-\rho^{1/2})^2,$$

(48) $$\text{Erfc}(t) = \frac{1}{(2\pi)^{1/2}}\int_t^{\infty} e^{-x^2/2}\, dx,$$

and the D beneath the equal sign indicates that only dominant terms have been retained. One expects to find that for any $0 < \alpha < 1$, $\text{Prob}\{\alpha t < \sigma_A(t) < t\} \to 0$ as $t \to \infty$ in view of the long drift to the left since $\lambda < \mu$; and this is indeed the case since $\psi(t)$ given by (46) behaves more and more like $t^{3/2}$ as $t \to \infty$. Dependence on Λ shows that (45) will be a poor result even for large t, when Λt is small, as can happen for $\rho \simeq 1$. In that case one should consider a modification of Case (ii).

Case (ii)–Symmetrical $\infty^2/M/M$ Since $\lambda = \mu$, F_{10} and F_{01} have the common value

(49) $$F(t) = e^{-2\mu t}[I_0(2\mu t) + I_1(2\mu t)] \underset{\text{D}}{=} (\pi\mu t)^{-1/2}$$

as $t \to \infty$. Therefore, using (41), we see that as $t \to \infty$, $\tau \to \infty$, $t - \tau \to \infty$,

(50) $$p_A(\tau, t) \underset{\text{D}}{=} \frac{1}{\pi\tau^{1/2}(t-\tau)^{1/2}},$$

which leads as $t \to \infty$, for fixed $0 < \alpha < 1$, to

(51) $$\text{Prob}\{\alpha t < \sigma_A(t) < t\} \underset{\text{D}}{=} 1 - \frac{2}{\pi}\arcsin \alpha^{1/2};$$

and this confirms the arc sine law predicted in Takács's work (1957b). Equation (51) gives an insight into the probability for one of two evenly matched teams to be in the lead throughout the whole game in, for example, a long enough soccer game, the characteristic of which is a change in score of ± 1 at negative-exponentially distributed time intervals.

Case (iii)–Unsymmetrical M/M/1 From (42), we have as $t \to \infty$

(52) $$\text{Prob}\{\alpha t < \sigma_A(t) < t\} = \frac{1}{\sqrt{2\pi}} \int_{-\sqrt{2\mu t}\,(1-\rho^{1/2})}^{\sqrt{2\mu t}\,(\rho^{1/2}-\alpha^{1/2})} e^{-y^2/2} \cdot \left[1 + \frac{(1+\rho)y}{2(1-\rho)\sqrt{2\lambda t}}\right] dy + O\left(\frac{1}{t}\right).$$

Takács's work (1957b, Theorem 2, and Example 2, p. 185) is applicable in this case and gives

(53) $$\text{Prob}\{\sigma_A(t) > \alpha t\} \underset{\text{D}}{=} \frac{1}{\sqrt{2\pi}} \int_{-\infty}^{\sqrt{\mu t/2}\,(\rho^{1/2}-\alpha/\rho^{1/2})} e^{-y^2/2}\, dy.$$

Conolly discusses the merits of (52) and (53) with the help of numerical results calculated from the two formulas by numerical integration.

Case (iv)–Symmetrical M/M/1 The result is of operational queuing interest since it enables one to make asymptotic statements about busy time under heavy traffic conditions. From (42), with $\rho = 1$, we have as $t \to \infty$

$$\text{(54)} \qquad \text{Prob}\{\alpha t < \sigma_A(t) < t\} = \sqrt{\frac{2}{\pi}} \int_0^{\sqrt{2\mu t}\,(1 - \alpha^{1/2})} e^{-y^2/2}\, dy + O\left(\frac{1}{t}\right).$$

The corresponding result from Takács (1957b) can be put in the form

$$\text{(55)} \qquad \text{Prob}\{\alpha t < \sigma_A(t) < t\} = \sqrt{\frac{2}{\pi}} \int_0^{\sqrt{\mu t/2}\,(1 - \alpha)} e^{-y^2/2}\, dy,$$

so that there remains a slight difference between (54) and (55). Conolly compares the two results with the help of numerical results.

Three-State Sojourn Times

Gibson and Conolly (1971) consider a three-state sojourn time problem by defining a real-valued stochastic process

$$\text{(56)} \qquad \{S(t),\ 0 \leqq t < \infty\},$$

whose range is an arbitrary space X, and consider in particular the case where

$$\text{(57)} \qquad X = A \cup B_1 \cup B_2.$$

Under certain assumptions, including that from states B_1 and B_2 only transitions to state A are permitted, the joint probability density function of $\sigma_A(t)$, $\sigma_{B_1}(t)$, and $\sigma_{B_2}(t)$ is derived, where $\sigma_A(t)$, etc., are defined as in (37). The results are then specialized to the case of a "randomized" random walk on the entire real line, and by defining $A = \{0\}$ and B_1 and B_2 as positive and negative integers, respectively. Defining the probability distribution functions

$$\text{(58)} \qquad P(r, s, t) = \text{Prob}\{\sigma_A(t) \leqq r,\ \sigma_{B_1}(t) \leqq s\},$$

$$\text{(59)} \qquad P_i(r, t) = \text{Prob}\{\sigma_A(t) \leqq r,\ \sigma_{B_i}(t) = 0\},$$

and denoting the corresponding density functions by $p(r, s, t)$ and $p_i(r, t)$, it is shown that

$$\text{(60)} \qquad p_1(r, t) = e^{-(\lambda + \mu)t}\left\{\lambda I_0[l(r)] + \left[\frac{\lambda\mu(t - r)}{t}\right]^{1/2} I_1[l(r)]\right\} + (\mu - \lambda)e^{-\mu r} J[\lambda t, \mu(t - r)]$$

and

$$\text{(61)} \qquad p(r, s, t) = e^{-(\lambda + \mu)t} M(r, s) N(r, s, t) + e^{-(\lambda + \mu)s} p_2(r, t - s) M(r, s) + e^{-(\lambda + \mu)(t - r - s)} p_2(r, r + s) N(r, s, t),$$

$$0 < r < t - s, \quad 0 < s < t,$$

where

(62) $$l(r) = 2[\lambda\mu t(t-r)]^{1/2},$$

(63) $$M(r,s) = \left[\frac{\lambda\mu}{s(r+s)}\right]^{1/2} rI_1[l_1(r,s)],$$

(64) $$N(r,s,t) = \left[\frac{\lambda\mu}{(t-s)(t-s-r)}\right]^{1/2} rI_1[l_2(r,s,t)],$$

(65) $$l_1(r,s) = 2[\lambda\mu s(r+s)]^{1/2},$$

(66) $$l_2(r,s,t) = 2[\lambda\mu(t-s)(t-s-r)]^{1/2},$$

and $J(x,y)$ is given by (26). $p_2(r,t)$ can be obtained from $p_1(r,t)$ by interchanging λ and μ.

1.5 Ergodic Waiting Time Distributions

Consider the queuing system $M/M/m$ (Tomko, 1972) with mean arrival and service rates of λ and 1, respectively, and with n places for the waiting customers, so that the traffic intensity is $\rho = \lambda/m$. Denote by $\eta^{(n)}$ the waiting time of a customer enabled to enter the system, by $\nu^{(n)}$ the number of customers in the system, and suppose that the stationary case does hold from the initial moment. Then the ergodic waiting time distribution,

(1) $$1 - W_n(x) = \text{Prob}\{\eta^{(n)} \leqq x \mid \nu^{(n)} < m+n\},$$

is given by the formula[4]

(2) $$W_n(x) = \frac{P_m}{1-P_{m+n}} \sum_{r=0}^{n-1} \rho^r \sum_{j=0}^{r} \frac{(mx)^j}{j!} e^{-mx}$$
$$= \frac{P_m}{1-P_{m+n}} \sum_{j=0}^{n-1} \frac{(mx)^j}{j!} e^{-mx}(\rho^j + \rho^{j+1} + \cdots + \rho^{n-1}),$$

where P_k $(0 \leqq k \leqq m+n)$ is the stationary probability that at an arbitrary moment there are k customers in the system and is given by

(3) $$P_k = \begin{cases} \dfrac{\lambda^k}{k!} P_0 & (1 \leqq k \leqq m) \\[2ex] \dfrac{\lambda^k}{m!\,m^{k-m}} P_0 & (m < k \leqq m+n) \end{cases}$$

[4] The lower limit of the r summation is $r = 1$ in Tomko's paper, which seems to be in error.

and

$$P_0 = \left[\sum_{k=0}^{m} \frac{\lambda^k}{k!} + \frac{\lambda^m}{m!} \sum_{k=1}^{n} \rho^k \right]^{-1}. \tag{4}$$

If $\rho < 1$, the ergodic waiting time distribution exists also for infinite queue capacity and equals $\lim_{n\to\infty}[1 - W_n(x)]$. Therefore, for $\rho < 1$,

$$W(x) \equiv \lim_{n\to\infty} W_n(x) = \frac{\lambda^m P_0^* e^{-(m-\lambda)x}}{m!(1-\rho)}, \tag{5}$$

where

$$P_0^* = \left[\sum_{k=0}^{m} \frac{\lambda^k}{k!} + \frac{\lambda^m}{m!} \frac{\rho}{1-\rho} \right]^{-1}. \tag{6}$$

Now, if one approximates $W_n(x)$ by the right-hand side of (5), it is of interest to estimate the accuracy of the approximation. This is done by Tomko (1972), and we state his result as the following theorem.

Theorem 1 *If $\rho < 1$, then for every $x \geqq 0$ and integer $n \geqq 1$,*

$$|W_n(x) - W(x)| \leqq c(\lambda, m)\rho^n, \tag{7}$$

where $c(\lambda, m)$ is a constant.

In the case $\rho = 1$, the ergodic waiting time distribution becomes spread over the positive real axis as $n \to \infty$. Since the conditional distribution

$$\text{Prob}\{\nu^{(n)} = m + k \mid m \leqq \nu^{(n)} < m + n\} \qquad (k = 0, 1, \ldots, n-1) \tag{8}$$

is, for this case, a uniform distribution, one may reasonably infer that with a suitable normalization, the asymptotic distribution of the waiting time will also be uniform. This is shown in the following theorem.

Theorem 2 *If $\rho = 1$ $(\lambda = m)$, then for every real y and integer $n \geqq 1$,*

$$\left| \text{Prob}\left\{ \frac{m\eta^{(n)}}{n-1} \leqq y \,\middle|\, \nu^{(n)} < m + n \right\} - U(y) \right| \leqq \frac{C(m)}{\sqrt{n}}, \tag{9}$$

where $U(\cdot)$ is the distribution function of uniform random variable on [0, 1] *and $C(m)$ is a constant independent of y and n.*

The proof of Theorem 2 is outlined below.

Using (2) for $\rho = 1$ and the identity

$$1 - \sum_{k=0}^{n} \frac{z^k}{k!} e^{-z} = \frac{1}{n!} \int_0^z e^{-t} t^n \, dt, \tag{10}$$

we have

$$(11)\quad 1 - W_n\left(\frac{(n-1)y}{m}\right) = \frac{C_m}{C_m + n} + \frac{P_m}{1 - P_{m+n}} \cdot \int_0^{(n-1)y} \left(1 - \frac{1}{(n-1)!}\int_0^t e^{-x}x^{n-1}\,dx\right)dt,$$

where

$$(12)\quad C_m = \frac{m!}{m^m}\sum_{k=0}^{m-1}\frac{m^k}{k!}$$

and it can be shown that

$$(13)\quad \frac{\sum_{k=0}^{m-1} P_k}{1 - P_{m+n}} = \frac{C_m}{C_m + n},$$

$$(14)\quad 1 - \frac{(n-1)P_m}{1 - P_{m+n}} = \frac{C_m + 1}{C_m + n}.$$

Using the substitution $t = (n-1)z$ in the integral, we see that, for $y \leqq 1$, the left-hand member of (9) does not exceed the sum

$$(15)\quad \frac{C_m}{C_m + n} + \frac{C_m + 1}{C_m + n} + \frac{n-1}{C_m + n} \cdot \frac{1}{(n-1)!}\int_0^1\int_0^{(n-1)z} e^{-x}x^{n-1}\,dx\,dz.$$

The theorem is then proved by showing that the integral

$$(16)\quad I_n = \frac{(n-1)^n}{(n-1)!}\int_0^1\int_0^1 e^{-(n-1)z\tau}(z\tau)^{n-1}z\,dz\,d\tau = O\left(\frac{1}{\sqrt{n}}\right).$$

Considering the substitution $z\tau = t$, $z = w$ and a partial integration, we obtain

$$(17)\quad I_n = \frac{e^{-(n-1)}(n-1)^{n-1}}{(n-1)!} - \frac{1}{n-1} \cdot \frac{1}{(n-1)!}\int_0^{n-1} e^{-\tau}\tau^{n-1}\,d\tau.$$

The desired estimation then follows from the asymptotic relation, deducible from the central limit theorem,

$$(18)\quad \sum_{k=0}^{n}\frac{n^k}{k!}e^{-n} \sim \frac{1}{2}, \qquad n \to \infty,$$

by using (10) and Stirling's formula. The result is similarly proved for $y > 1$.

The case $\rho > 1$ is considered in the following theorem.

Theorem 3 *If $\rho > 1$, there exists a constant $C(\lambda, m)$ such that for every $n \geqq 1$ and $-\infty < x < \infty$,*

$$\left| \operatorname{Prob}\left\{ \frac{\eta^{(n)} - n/m}{\sqrt{n/m}} \leqq x \,\middle|\, \nu^{(n)} < m + n \right\} - \phi(x) \right| \leqq C(\lambda, m)/\sqrt{n}, \tag{19}$$

where $\phi(x)$ is the standard normal distribution function.

This is proved by using (10), Stirling's formula and the classical Berry–Esseen result

$$I(x\sqrt{n} + n, n) - \phi(x) = O(1/\sqrt{n}), \qquad n \to \infty, \tag{20}$$

where

$$I(z, n) = \gamma(n, z)/(n-1)!, \tag{21}$$

$\gamma(n, z)$ being the incomplete gamma function [see also II.1(52) and II.1(55)]

$$\gamma(n, z) = \int_0^z e^{-t} t^{n-1}\, dt. \tag{22}$$

1.6 Busy Period Distribution for $M/M/c$

The busy period during which k channels are busy in an $M/M/c$ queuing system (Chaudhry and Templeton, 1973) is defined as the time from the instant of arrival of a unit that makes k channels busy to the first subsequent instant when one of the k channels becomes empty. Let

$$P_n(t) = \operatorname{Prob}\{n \text{ units in system at time } t\}. \tag{1}$$

We postulate that

$$P_n(0) = \delta_{nk}, \tag{2}$$

where δ_{nk} is the Kronecker delta, and that the state $n = k - 1$ is an absorbing state; then

$$\frac{d}{dt} P_{k-1}(t) = \text{the density function of the busy period defined above.} \tag{3}$$

We have the following transition equations for the system,

$$\frac{d}{dt} P_{k-1}(t) = k\mu P_k(t), \tag{4}$$

$$\frac{d}{dt} P_n(t) = -(\lambda + n\mu) P_n(t) + \lambda P_{n-1}(t) + (n+1)\mu P_{n+1}(t), \qquad k \leqq n < c, \tag{5}$$

$$\frac{d}{dt} P_n(t) = -(\lambda + c\mu) P_n(t) + \lambda P_{n-1}(t) + c\mu P_{n+1}(t), \qquad n \geqq c, \tag{6}$$

where $\lambda P_{n-1}(t)$ is to be excluded from (5) when $n = k$ or from (6) when $n = k = c$, in which case (5) becomes redundant, and λ and μ are the mean arrival and service rates, respectively. Let $p_n(s)$ be the L.T. of $P_n(t)$, given by

(7) $$p_n(s) = \int_0^\infty e^{-st} P_n(t)\, dt.$$

Then, from Eqs. (4)–(6), we have

(8) $$sp_{k-1}(s) = k\mu p_k(s),$$

(9) $$(s + \lambda + n\mu)p_n(s) = \delta_{nk} + \lambda p_{n-1}(s) + (n+1)\mu p_{n+1}(s), \qquad k \leqq n < c,$$

(10) $$(s + \lambda + c\mu)p_n(s) = \delta_{nk} + \lambda p_{n-1}(s) + c\mu p_{n+1}(s), \qquad n \geqq c.$$

Using the generating function

(11) $$p(z, s) = \sum_{n=k}^{\infty} p_n(s) z^n,$$

we have, from (9) and (10),

(12) $$p(z, s) = \frac{z^{k+1} + \mu(z-1)\sum_{n=k}^{c-1}(c-n)p_n(s)z^n - k\mu p_k(s)z^k}{-\lambda z^2 + (s + \lambda + c\mu)z - c\mu}.$$

The numerator of (12) should vanish at $z = \alpha_2$, where α_1, α_2 are the zeros of the denominator, given by 1.2(8). Thus, we have

(13) $$(1 - \alpha_2)\sum_{n=k}^{c-1}(c-n)p_n(s)\alpha_2^n + kp_n(s)\alpha_2^k = \frac{\alpha_2^{k+1}}{\mu}.$$

Equation (13) involves $c - k$ unknowns $p_n(s)$, $k \leqq n \leqq c - 1$, which are determined by solving (10) and the first $c - k - 1$ equations of (9). Then, (8) gives the L.T. of the busy period density function. We have the following *particular cases*:

Case I If $c = 1, k = 1$, we have for the $M/M/1$ system, from (6),

(14) $$\mu p_1(s) = \alpha_2 = \frac{\mu}{\lambda \alpha_1},$$

whence, on inversion, we have the well-known result [cf. 1.3(31)]

(15) $$\frac{d}{dt} P_0(t) = \mu P_1(t) = \sqrt{\frac{\mu}{\lambda}}\, e^{-(\lambda+\mu)t} t^{-1} I_1(2\sqrt{\lambda\mu}\, t).$$

Case II If $c = 2, k = 1$, i.e., in the $M/M/2$ system for busy period starting with at least one server busy, (6) gives [cf. 1.2(33), (34)]

(16) $$\mu p_1(s) = \frac{\alpha_2}{2 - \alpha_2},$$

which, on inversion, yields

$$(17)\qquad \frac{d}{dt}P_0(t) = e^{-(\lambda+2\mu)t}t^{-1}\sum_{n=0}^{\infty}(n+1)\left(\frac{2\lambda}{\mu}\right)^{-(n+1)/2} I_{n+1}(2\sqrt{2\lambda\mu}\,t).$$

Case III If $c = k = 2$, we have from (6), for $M/M/2$ and busy periods lasting while both servers are busy,

$$(18)\qquad 2\mu p_2(s) = \alpha_2 = \frac{2\mu}{\lambda}\cdot\frac{1}{\alpha_1}$$

and, on inversion,

$$(19)\qquad \frac{d}{dt}P_1(t) = 2\mu P_2(t) = e^{-(\lambda+2\mu)t}t^{-1}\sqrt{\frac{2\mu}{\lambda}}\,I_1(2\sqrt{2\lambda\mu}\,t).$$

Case IV If $k = c$, i.e., for $M/M/c$ and busy period lasting when all servers are busy, we have

$$(20)\qquad c\mu p_c(s) = \alpha_2 = \frac{c\mu}{\lambda}\cdot\frac{1}{\alpha_1},$$

and, on inversion,

$$(21)\qquad \frac{d}{dt}P_{c-1}(t) = c\mu P_c(t) = e^{-(\lambda+c\mu)t}t^{-1}\sqrt{\frac{c\mu}{\lambda}}\,I_1(2\sqrt{c\lambda\mu}\,t).$$

It may be noted that Cases I and III are particular cases of Case IV, and it is evident from above that the busy period distribution for the case of all servers busy for $M/M/c$ can be derived from the corresponding $M/M/1$ result by replacing μ by $c\mu$, whose interpretation is obvious.

1.7 Optimal Initial Conditions for a Simulation Problem

Madansky (1976) considers the problem of finding the optimal value of n, the number in the system initially at time $t = 0$ when simulation is used for estimating the steady-state mean number in the system. Consider an $M/M/1$ system and suppose that we run N simulations of the queue, with n in the system initially, each to time t. Let

(1) $k_i(t;n) =$ the number in system at time t on the ith replication,

and let us use as an estimator of the steady-state mean

$$(2)\qquad \bar{k}_N(t;n) = \frac{1}{N}\sum_{i=1}^{N} k_i(t;n).$$

The parameter of interest is the steady-state mean, given by

(3) $$L = \rho/(1-\rho),$$

where, in the usual notation, $\rho = \lambda/\mu$, and for considerations of optimality we focus on the mean-squared error

(4) $$\omega_{n,N}(t) = [\bar{k}_N(t;n) - \rho/(1-\rho)]^2.$$

Consider first the case $N = 1$. When $t = 0$, clearly the optimal n is $\rho/(1-\rho)$, the steady-state mean. In what follows it is shown that for sufficiently large t, the initial condition that minimizes the mean-squared error is $n = 0$, thus verifying the conjecture of Conway (1963) that the optimal initial number in the system is the mode of the distribution and not the steady-state mean. Let $M_n(t)$ and $V_n(t)$ denote, respectively, the mean and variance of the transient distribution of the number in the system at time t, given n in the system at time 0. Let

(5) $$M = \lim_{t\to\infty} M_n(t), \qquad V = \lim_{t\to\infty} V_n(t).$$

Then

(6) $$\omega_{n,1}(t) = V_n(t) + [M_n(t) - M(t)]^2.$$

Using the results for the transient distribution and its mean (Takács, 1962, pp. 23, 27), we have

(7) $$\omega_{n,1}(t) = \frac{\rho}{(1-\rho)^2} + \rho^{-n/2}\int_{\mu t}^{\infty} \exp[-(1+\rho)z]\,z^{-1}\left[(1-\rho)(z-\mu t)^2 - \left(\frac{1+\rho}{1-\rho}\right)(z-\mu t)\right][nI_n(2\alpha z) - \alpha(n+1)I_{n+1}(2\alpha z)]\,dz,$$

where

(8) $$\alpha = \rho^{1/2}.$$

Using the asymptotic result 1.3(33) for $I_v(z)$, we obtain

(9) $$\omega_{n,1}(t) \sim \rho/(1-\rho)^2 + \rho^{-n/2}[n - \alpha(n+1)]2^{-3/2}\pi^{-1/2}\rho^{-1/4} \cdot \int_{\mu t}^{\infty} \exp[-(1-2\alpha+\rho)z]\,z^{-3/2}\left[(1-\rho)(z-\mu t)^2 - \left(\frac{1+\rho}{1-\rho}\right)(z-\mu t)\right]dz.$$

This integral is the sum of incomplete gamma functions, and so, noting that

$$\int_{\mu t}^{\infty} \exp[-(1-2\alpha+\rho)z]\,z^{-1/2}\,dz = \pi^{1/2}(1-\rho)^{-1}[1 - \Phi(1-\sqrt{\rho})\sqrt{\mu t}],$$

where Φ is the error function, and using standard expansions of this function (Abramowitz and Stegun, 1964) and retaining terms to $O(t^{-3/2})$, we see that

$$\omega_{n,1}(t) \sim \rho/(1-\rho)^2 + \exp\{-\mu t(1-\alpha)^2\}(\alpha^2+4\alpha+1) \cdot \alpha^{-n}[n-\alpha(n+1)][2(\pi\alpha)^{1/2}(1-\alpha)^2(1-\alpha^2)(\mu t)^{3/2}]^{-1}. \tag{10}$$

In (10), the coefficient of

$$f(n) = \alpha^{-n}[n-\alpha(n+1)] \tag{11}$$

is nonnegative, and so it suffices to study the behavior of $f(n)$. We have

$$f'(n) = \alpha^{-n}[g(\alpha) - n(1-\alpha)\ln\alpha], \tag{12}$$

where

$$g(\alpha) = 1 - \alpha + \alpha\ln\alpha. \tag{13}$$

Since $g(\alpha)$ is monotone decreasing in α and $g(1) = 0$, it follows that $g(\alpha)$ is nonnegative in $0 < \alpha < 1$, and hence, from (12), $f'(n) > 0$. Therefore $f(n)$ is monotone increasing in n, and, for large t, $\omega_{n,1}(t)$ is monotone increasing in n. Thus, $n = 0$ minimizes $\omega_{n,1}(t)$ for large t.

For the case of general N, it is shown that $\omega_{n,N}(t)$ is minimized when

$$n = \alpha/(1-\alpha), \tag{14}$$

and an approximate trade-off between N and t is developed when $n = 0$.

1.8 The Markovian Queue with Bounded Waiting Time

Gavish and Schweitzer (1977) consider a queuing system for which a customer is admitted into the system only if his waiting plus service time does not exceed an upper limit K. They start with a formulation for the $M/G/1$ first-come–first-served (FCFS) queue and give a complete analysis in the special case of $M/M/1$. Let λ be the mean arrival rate and $f(y)$ the probability density function (p.d.f.) of the customer service requirement. Each arrival is characterized by his service requirement $y \geqq 0$ and his waiting time $x \geqq 0$ for those ahead of him. If $x + y \leqq K$, he joins and remains there until he is serviced; otherwise he leaves and is lost to the system.

At time t, let $E(t)$ be the probability that the system is empty, and let $w(x, t)$ denote the p.d.f. of the virtual waiting time (v.w.t.) x, $0 < x \leqq K$. Using the method of Keilson and Kooharian (1960), one obtains the following differential equations,

$$\frac{\partial w(x,t)}{\partial t} - \frac{\partial w(x,t)}{\partial x} = \lambda f(x)E(t) - \lambda\left[\int_0^{K-x} f(u)\,du\right]w(x,t) + \lambda\int_0^x f(x-u)w(u,t)\,du, \qquad 0 < x \leqq K, \quad t \geqq 0, \tag{1}$$

$$\frac{dE(t)}{dt} = -\lambda\left[\int_0^K f(u)\,du\right]E(t) + w(0,t), \qquad t \geqq 0, \tag{2}$$

and the normalizing condition

$$1 = E(t) + \int_0^K w(x,t)\,dx, \qquad t \geqq 0. \tag{3}$$

For the equilibrium equations, let $w(x,t) \to w(x)$, $E(t) \to Q$, and set

$$w(x) = Qh(x), \qquad 0 \leqq x \leqq K. \tag{4}$$

Then, (1) and (2) yield the following integrodifferential equation for h:

$$0 = \frac{dh(x)}{dx} + \lambda f(x) - \lambda h(x)\int_0^{K-x} f(u)\,du + \lambda\int_0^x f(x-u)h(u)\,du, \qquad 0 < x \leqq K, \tag{5}$$

$$h(0) = \lambda\int_0^K f(u)\,du. \tag{6}$$

Also, from (3) we have

$$Q = \left[1 + \int_0^K h(x)\,dx\right]^{-1}. \tag{7}$$

Let us now consider the $M/M/1$ system with mean service requirement $1/\mu$, so that

$$f(y) = \mu\exp(-\mu y), \qquad y \geqq 0. \tag{8}$$

Using (8) in (5) and changing the dependent variable by

$$h(x) = \exp(-\mu x)\frac{dg(x)}{dx}, \qquad g(0) = 0, \tag{9}$$

we obtain a second-order linear differential equation, which on changing the independent variable by

$$z = \exp[-\mu(K-x)], \qquad g(x) = \tilde{g}(z), \tag{10}$$

leads to the simpler form

$$z^2\frac{d^2\tilde{g}}{dz^2} - \rho z(1-z)\frac{d\tilde{g}}{dz} + \rho\tilde{g} = -\rho, \qquad b \leqq z \leqq 1, \tag{11}$$

$$\tilde{g}(b) = 0, \qquad \left.\frac{d\tilde{g}}{dz}\right|_{z=b} = \rho(b^{-1} - 1), \tag{12}$$

where

$$\rho = \lambda/\mu, \tag{13}$$

$$b = \exp(-\mu K). \tag{14}$$

Solution of (11)–(12) is

$$\tilde{g}(z) = -1 + (z/b)^{\rho}\exp(\rho b - \rho z), \tag{15}$$

whence, using (9), we have

$$h(x) = \rho\mu b^{1-\rho}(z^{\rho-1} - z^{\rho})\exp(\rho b - \rho z), \qquad 0 \leqq x \leqq K. \tag{16}$$

It follows from differentiation that $w(x)$ is monotone decreasing in x if $\rho(1-b)^2 \leqq 1$, and has an interior maximum at $z = 1 - \rho^{-1/2}$ if $\rho(1-b)^2 > 1$. In both cases, $w(x) \to 0$ as $x \to K$, so that the v.w.t. x is unlikely to be near K.

For calculating the probability Q of server idleness, from (7) and (16) and the change of variables

$$t = \rho z = \rho b\exp(\mu x), \tag{17}$$

we have

$$\frac{1}{Q} = 1 + b^{1-\rho}\exp(\rho b)\int_{\rho b}^{\rho}\left[\left(\frac{t}{\rho}\right)^{\rho-2} - \left(\frac{t}{\rho}\right)^{\rho-1}\right]\exp(-t)\,dt, \tag{18}$$

which simplifies to

$$Q = (1-\rho)[1 - \rho b^{1-\rho}e^{\rho(b-1)} - (\rho b)^{1-\rho}e^{\rho b}\{\gamma(\rho,\rho) - \gamma(\rho,\rho b)\}]^{-1}, \quad \text{if} \quad \rho \neq 1, \tag{19}$$

where $\gamma(\cdot,\cdot)$ is the incomplete gamma function [cf. II.1(52)].

If $\rho = 1$, we have

$$Q = e^{-b}\left[e^{-1} + \int_b^1 e^{-u}u^{-1}\,du\right]^{-1}, \tag{20}$$

which after expansion of the exponential function in the integrand and term-by-term integration can be put in the form

$$Q = [\ln(1/b) + b\,{}_2F_2(1,1;2,2;-b) - {}_2F_2(1,1;2,2;-1)]^{-1}. \tag{21}$$

When $\rho = 2, 3, 4, \ldots$, (18) leads to the simpler form

$$Q = \left[b^{1-\rho} e^{-\rho(1-b)} + (b\rho)^{1-\rho}(\rho - 2)! \sum_{m=0}^{\rho-2} \frac{(\rho b)^m - \rho^m e^{-\rho(1-b)}}{m!} \right]^{-1}. \tag{22}$$

The mean v.w.t. is given by

$$E(x) = \int_0^K x w(x)\, dx = C I_1, \tag{23}$$

where, on using (4) and (16) and the substitution $z = \rho y$, we have

$$C = Q \rho b^{1-\rho} \exp(\rho b)/\mu, \tag{24}$$

$$I_1 = \int_{\rho b}^{\rho} \ln\left(\frac{y}{\rho b}\right) \left[\frac{y^{\rho-2}}{\rho^{\rho-1}} - \frac{y^{\rho-1}}{\rho^{\rho}} \right] \exp(-y)\, dy. \tag{25}$$

Now, we can write

$$I_1 = \int_{\rho b}^{\rho} u \frac{dv}{dy}\, dy, \tag{26}$$

where

$$u(y) = \ln(y/\rho b), \tag{27}$$

$$v(y) = [\rho e^{-y} y^{\rho-1} + \gamma(\rho, y)]\{(\rho - 1)\rho^{\rho}\}^{-1}. \tag{28}$$

Integration by parts then yields

$$I_1 = \mu K v(\rho) - \rho^{1-\rho}(\rho - 1)^{-1} \int_{\rho b}^{\rho} e^{-y} y^{\rho-2}\, dy - \frac{I_2}{(\rho - 1)\rho^{\rho}}, \tag{29}$$

where

$$I_2 = \int_{\rho b}^{\rho} \gamma(\rho, y) y^{-1}\, dy. \tag{30}$$

Using the relation [Abramowitz and Stegun (1964), Eq. (6.5.12)]

$$\gamma(x, y) = e^{-y} \sum_{n=0}^{\infty} \frac{y^{n+x}}{(x)_{n+1}}, \tag{31}$$

we have

$$\begin{aligned} I_2 &= \sum_{n=0}^{\infty} \int_{\rho b}^{\rho} \frac{e^{-y} y^{n+\rho-1}}{(\rho)_{n+1}}\, dy \\ &= \sum_{n=0}^{\infty} \frac{\gamma(n + \rho, y)}{(\rho)_{n+1}} \Bigg|_{y=\rho b}^{y=\rho} \\ &= \sum_{n=0}^{\infty} \sum_{m=0}^{\infty} \frac{e^{-y} y^{m+n+\rho}}{(\rho)_{n+1}(n + \rho)_{m+1}} \Bigg|_{y=\rho b}^{y=\rho}. \end{aligned} \tag{32}$$

Setting

$$(\rho)_{n+1}(n+\rho)_{m+1} = (\rho+n)(\rho)_{m+n+1}, \qquad j = m+n, \tag{33}$$

we have

$$I_2 = \sum_{j=0}^{\infty} \frac{e^{-y}y^{j+\rho}}{(\rho)_{j+1}} \left(\sum_{n=0}^{j} \frac{1}{\rho+n} \right) \Bigg|_{y=\rho b}^{\rho}, \tag{34}$$

which converges rapidly.

CHAPTER 2

Poisson Queues II

In this chapter we describe the use of special functions in Poisson queues with priorities, and those with bulk service or bulk arrivals. It includes the work on preemptive priority by Stephan (1958) and Heathcote (1959, 1960), and on alternating priorities by Yadin (1970). It also includes the work in bulk service queuing systems by Arora (1964), Ghare (1968), and Medhi (1975) as well as that on bulk arrival queues by Kabak (1968, 1970).

2.1 Two Queues with Preemptive Priority

Stephan (1958) extends the work of White and Christie (1958) on queues with preemptive priorities, taking the particular case of two priority classes. Let (i, j) denote the state of the system when there are i units of the first priority class and j units of the second priority in the system, including the unit receiving service, if any. Let the arrivals and departures be Poisson processes with parameters

(1) $$\lambda_r \quad \text{and} \quad \mu_r,$$

respectively, for the rth priority class, and let

(2) $$\rho_r = \lambda_r/\mu_r$$

be the proportion of the capacity of the system required to serve the units of the rth class. It is assumed that there is a single service facility. Let

(3) P_{ij} = steady-state probability that the system is in state (i, j).

Then, we have (White and Christie, 1958; Stephan, 1958)

(4) $$\mu_1 P_{10} - (\lambda_1 + \lambda_2) P_{00} + \mu_2 P_{01} = 0,$$

(5) $$\mu_1 P_{i+2,0} - (\lambda_1 + \lambda_2 + \mu_1) P_{i+1,0} + \lambda_1 P_{i0} = 0 \qquad (i \geqq 0),$$

(6) $$\mu_1 P_{1j} + \mu_2 P_{0,j+1} - (\lambda_1 + \lambda_2 + \mu_2) P_{0j} + \lambda_2 P_{0,j-1} = 0 \qquad (j > 0),$$

(7) $$\mu_1 P_{i+2,j} + \lambda_2 P_{i+1,j-1} - (\lambda_1 + \lambda_2 + \mu_1) P_{i+1,j} + \lambda_1 P_{ij} = 0$$
$$(i \geqq 0,\ j > 0),$$

(8) $$\sum_{i=0}^{\infty} \sum_{j=0}^{\infty} P_{ij} = 1,$$

(9) $$P_{00} = 1 - \rho_1 - \rho_2.$$

Let

(10) $$p(u, v; h) = \text{Prob}\{\text{system is in state } (i,j) \text{ at time } t_0 + h\ (h > 0) \mid \text{system was in state } (q, r) \text{ at time } t_0\},$$

where

(11) $$u = i - q, \qquad v = j - r,$$

and the transitions are subject to the following conditions:

(a) $i \geqq q > 0$, and

(b) transition from the state (q, r) to (i, j) does not include entering any state (m, n) for which $m = q - 1$.

It is shown that $p(u, v; h)$ is independent of q, r, and t_0.

Considering combinations of possible transitions during the time interval $(t_0, t_0 + h)$ and using a result (Chandrasekhar, 1943, p. 6) on random walks with absorbing barriers, we have

(12) $$p(u, v; h) = \sum_{x=0}^{\infty} \left(\frac{(\lambda_1 h)^{u+x} \exp(-\lambda_1 h)}{(u+x)!} \right) \left(\frac{(\mu_1 h)^{x} \exp(-\mu_1 h)}{x!} \right) \cdot \left(\frac{(\lambda_2 h)^{v} \exp(-\lambda_2 h)}{v!} \right) \left(\frac{u+1}{x+u+1} \right)$$
$$= \sum_{x=0}^{\infty} \frac{(u+1) \lambda_1^{u+x} \lambda_2^{v} \mu_1^{x} \exp[-(\lambda_1 + \lambda_2 + \mu_1) h]\, h^{u+v+2x}}{x!(u+1+x)!v!}.$$

From (12), we get

(13) $$\phi(u, v) = \lambda_1 \int_0^{\infty} p(u, v; h)\, dh$$
$$= \sum_{x=0}^{\infty} \frac{(u+1) \lambda_1^{u+x+1} \lambda_2^{v} \mu_1^{x}}{x!(u+x+1)!v!} \frac{(u+v+2x)!}{(\lambda_1 + \lambda_2 + \mu_1)^{u+v+2x+1}},$$

which simplifies to

(14) $$\phi(u, v) = \alpha^{u+1}\beta^{v}\frac{(u+v)!}{u!v!}\,{}_2F_1\left[\frac{u+v+1}{2}, \frac{u+v+2}{2}; u+2; z\right],$$

where

(15) $$\alpha = \lambda_1/(\lambda_1 + \lambda_2 + \mu_1),$$

(16) $$\beta = \lambda_2/(\lambda_1 + \lambda_2 + \mu_1),$$

(17) $$z = 4\lambda_1\mu_1/(\lambda_1 + \lambda_2 + \mu_1)^2.$$

Using probability arguments concerning possible transitions, Stephan (1958) establishes the result

(18) $$P_{ij} = \sum_{v=0}^{j} \phi(u, v)P_{i-u-1, j-v},$$

whence, setting $u = i - 1$, we have P_{ij} $(i > 0)$ in terms of P_{0j}. Further, P_{0j} can be evaluated by using (18) in (6) and (4), which gives

(19) $$P_{0j} = \sum_{v=0}^{j-1} c_v P_{0,j-v-1} \qquad (j > 1),$$

(20) $$P_{01} = (c_0 - 1)P_{00},$$

where

(21) $$c_0 = [\lambda_1 + \lambda_2 + \mu_2 - \mu_1\phi(0,0)]/\mu_2,$$

(22) $$c_1 = -[\lambda_2 + \mu_1\phi(0,1)]/\mu_2,$$

(23) $$c_v = -\mu_1\phi(0,v)/\mu_2 \qquad (v > 1).$$

Let us define the following marginal probability distributions of the number of units of the first and second priority, respectively:

(24) $$P_{i.} = \sum_{j=0}^{\infty} P_{ij},$$

(25) $$P_{.j} = \sum_{i=0}^{\infty} P_{ij}.$$

Then, since in preemptive priority discipline, units of any class are not affected by the presence or absence of those in lower priority classes, we have from the well-known result for the nonpriority case

(26) $$P_{i.} = \rho_1^i P_{0.},$$

(27) $$P_{0.} = 1 - \rho_1.$$

Also, from (4)–(7) and (19), it can be shown that

$$P_{.j} = \frac{P_{0,j+1}}{\rho_2} = \sum_{v=0}^{j} \frac{c_v P_{0,j-v}}{\rho_2} = \sum_{v=0}^{j-1} c_v P_{.j-v-1} + \frac{c_j(1-\rho_1-\rho_2)}{\rho_2}. \tag{28}$$

To obtain the moments of the queue length j for the lower priority class, Stephan uses the factorial function of c_g of nth order ($n > 1$), defined by

$$C_n^* = \frac{1}{n!}\sum_{g=0}^{\infty} c_g g^{[n]}, \tag{29}$$

where

$$x^{[n]} \equiv x!/(x-n)!. \tag{30}$$

Using (23) and (13) in (29), we have

$$\begin{aligned} C_n^* &= -\frac{\alpha\mu_1}{n!\mu_2}\sum_{g=n}^{\infty}\frac{g!}{(g-n)!}\sum_{x=0}^{\infty}\beta^g\frac{(2x+g)!}{g!\,x!\,(x+1)!}\left(\frac{z}{4}\right)^x \\ &= -\frac{\mu_1}{\mu_2}\alpha\beta^n\sum_{x=0}^{\infty}\frac{1}{n!\,x!\,(x+1)!}\left(\frac{z}{4}\right)^x\sum_{h=0}^{\infty}\frac{(2x+n+h)!}{h!}\beta^h, \end{aligned} \tag{31}$$

setting $g = n + h$.

Now

$$\begin{aligned} \sum_{h=0}^{\infty}\frac{(2x+n+h)!}{h!}\beta^h &= (2x+n)!\sum_{h=0}^{\infty}\frac{(2x+n+1)_h}{h!}\beta^h \\ &= (2x+n)!\,{}_1F_0(2x+n+1;-;\beta) \\ &= (2x+n)!\,(1-\beta)^{-(2x+n+1)}. \end{aligned} \tag{32}$$

Hence (31) yields

$$C_n^* = -\frac{\mu_1}{\mu_2}\alpha\beta^n(1-\beta)^{-n-1}\sum_{x=0}^{\infty}\frac{(n+2x)!}{n!\,x!\,(x+1)!}\left[\frac{z}{4(1-\beta)^2}\right]^x, \tag{33}$$

which simplifies to

$$C_n^* = -\frac{\mu_1}{\mu_2}\alpha\beta^n(1-\beta)^{-n-1}\,{}_2F_1\left[\frac{n+1}{2},\frac{n+2}{2};2;z^*\right], \tag{34}$$

where

$$z^* = 4\lambda_1\mu_1/(\lambda_1+\mu_1)^2. \tag{35}$$

Similarly, the factorial function of the first order is

$$(36)\quad C_1^* = \sum_{g=0}^{\infty} g c_g = -\rho_2 - \frac{\mu_1}{\mu_2}\alpha\beta(1-\beta)^{-2}\,{}_2F_1(1,\tfrac{3}{2};2;z^*) = \frac{-\rho_2}{(1-\rho_1)},$$

and the factorial function of order zero is

$$(37)\quad C_0^* = \sum_{g=0}^{\infty} c_g = 1,$$

where, in simplification, we use the special case of the hypergeometric function [Erdélyi *et al.* (1953), Vol. I, p. 101, Eq. (6)[1]]:

$$(38)\quad {}_2F_1(a, a+\tfrac{1}{2}; 2a+1; z) = (1-z)^{1/2}\,{}_2F_1(a+\tfrac{1}{2}, a+1; 2a+1; z)$$
$$= \{2/[1+(1-z)^{1/2}]\}^{2a}$$
$$(|z| < 1,\ (1-z)^{1/2} > 0).$$

[Cf. also II.2(26).]

Waiting Time Distribution for the Second Priority Class

Consider a unit of the second priority class that arrives at time t_0 when the system is in state (i, j). The probability for $i > 0$ that service will be started for the second queue in the interval $t_0 + h$ to $t_0 + h + dh$ is

$$(39)\quad \sum_{x=0}^{\infty} \frac{(\lambda_1 h)^x(\mu_1 h)^{x+i-1}}{x!\,(x+i-1)!}\,\frac{i}{x+i}\exp[-(\lambda_1+\mu_1)h]\mu_1\,dh$$
$$\cdot \exp[-(\lambda_1+\mu_1)\,dh].$$

Hence, the moment generating function (m.g.f.) for the portion of the waiting time h_0 is given by

$$(40)\quad G_0(i,y) = \int_0^{\infty} \exp(h_0 y) \sum_{x=0}^{\infty} \frac{\lambda_1^x \mu_1^{x+i} i h_0^{2x+i+1}}{x!\,(x+i)!}\exp[-(\lambda_1+\mu_1)h_0]\,dh_0$$
$$= \sum_{x=0}^{\infty} \frac{\lambda_1^x \mu_1^{x+i}\, i(2x+i-1)!}{x!\,(x+i)!\,(\lambda_1+\mu_1-y)^{2x+i}}$$
$$= \gamma_y^i\,{}_2F_1[i/2, (i+1)/2; i+1; Z_y]$$
$$= \gamma_y^i\,[\tfrac{1}{2}+\tfrac{1}{2}(1-Z_y)^{1/2}]^{-i} = G^i,$$

where

$$(41)\quad \gamma_y = \mu_1/(\lambda_1+\mu_1-y),$$

$$(42)\quad Z_y = 4\lambda_1\mu_1/(\lambda_1+\mu_1-y)^2,$$

$$(43)\quad G = [\lambda_1+\mu_1-y-\{(\lambda_1+\mu_1-y)^2-4\lambda_1\mu_1\}^{1/2}]/2\lambda$$

[1] Not Eq. (16), as mentioned in Stephan (1958, Appendix).

and we make use of (38) for simplification of the hypergeometric function. It will be recalled that G is the same as the α_1 given by Eq. 1.1(5), and the interpretation of this result is obvious since we are, in fact, considering the length of a busy period initiated by i units in the ordinary $M/M/1$ system and the m.g.f. is the same as the Laplace transform with $y = -s$.

After service is started for the second priority class, it may be interrupted by newly arrived elements of the first priority, and the m.g.f. of h_1, the length of interruption, is given by

$$(44)\quad G_1(y) = \int_0^\infty \exp(h_1 y) \sum_{x=0}^{\infty} \frac{(\lambda_1 h_1)^x (\mu_1 h_1)^x \exp[-(\lambda_1 + \mu_1)h_1]\,\mu_1\, dh_1}{x!(x+1)!}$$

$$= \gamma_y\, {}_2F_1(\tfrac{1}{2}, 1; 2; Z_y)$$

$$= 2\gamma_y/[1 + (1 - Z_y)^{1/2}]$$

$$= G_0(1, y) = G.$$

Stephan uses these results to find the m.g.f. of the total waiting time for the second-priority class units.

2.2 Time-Dependent Solution with Preemptive Priorities

Heathcote (1959) extends the work of Stephan (1958) described in Section 2.1 to the time-dependent case. Let

$$(1)\qquad P_{nm}(t) = \text{Prob}\{n \text{ priority and } m \text{ non-priority class units in the system at time } t\}.$$

It may be noted that in the case of just two priority classes, it is better to call them priority and nonpriority classes. It is assumed that the arrival and service processes for both the classes are Poisson, and $\lambda_1, \mu_1, \lambda_2, \mu_2$ are used to have the same meaning as in the last section. It is easy to see that the following forward differential difference equations hold for the system:

$$(2)\qquad \frac{d}{dt} P_{nm}(t) = -(\lambda_1 + \lambda_2 + \mu_1)P_{nm}(t) + \mu_1 P_{n+1,m}(t) + \lambda_1 P_{n-1,m}(t) + \lambda_2 P_{n,m-1}(t) \qquad (n > 0, m \geqq 0),$$

$$(3)\qquad \frac{d}{dt} P_{0m}(t) = -(\lambda_1 + \lambda_2 + \mu_2)P_{0m}(t) + \mu_1 P_{1m}(t) + \mu_2 P_{0,m+1}(t) + \lambda_2 P_{0,m-1}(t) \qquad (m > 0),$$

$$(4)\qquad \frac{d}{dt} P_{00}(t) = -(\lambda_1 + \lambda_2)P_{00}(t) + \mu_1 P_{10}(t) + \mu_2 P_{01}(t).$$

Let the initial state of the system be $(0, i)$, so that

$$P_{nm}(0) = \delta_{0n}\delta_{im}, \tag{5}$$

where δ_{ij} stands for the Kronecker delta. Introducing the Laplace transform

$$p_{nm}(s) = \mathscr{L}[P_{nm}(t)] = \int_0^\infty e^{-st} P_{nm}(t)\, dt, \qquad \mathrm{Re}(s) > 0, \tag{6}$$

and using (5), we may rewrite (1)–(3) as the following difference equations:

$$-(s + \lambda_1 + \lambda_2 + \mu_1)p_{nm} + \mu_1 p_{n+1,m} + \lambda_1 p_{n-1,m} + \lambda_2 p_{n,m-1} = 0 \quad (n > 0,\ m \geqq 0), \tag{7}$$

$$-(s + \lambda_1 + \lambda_2 + \mu_2)p_{0m} + \mu_1 p_{1m} + \mu_2 p_{0,m+1} + \lambda_2 p_{0,m-1} = -\delta_{hm} \quad (m > 0), \tag{8}$$

$$-(s + \lambda_1 + \lambda_2)p_{00} + \mu_1 p_{10} + \mu_2 p_{01} = -\delta_{h0}, \tag{9}$$

where the argument of $p_{nm}(s)$ has been dropped for convenience. Let us define the generating functions

$$H_n(x, s) = \sum_{m=0}^{\infty} p_{nm}(s)x^m, \tag{10}$$

$$F(z, x, s) = \sum_{n,m=0}^{\infty} p_{nm}(s)z^n x^m = \sum_{n=0}^{\infty} H_n(x, s)z^n. \tag{11}$$

Then, from (7)–(9), we have

$$F(z, x, s) = \frac{p_{00}\mu_2(1/x - 1) - x^i + [\mu_1(1/z - 1) + \mu_2(1/x - 1)]H_0(x, s)}{\lambda_1 z - (s + \lambda_1 + \lambda_2 + \mu_1 - \lambda_2 x) + \mu_1/z}. \tag{12}$$

Evaluating $H_0(x, s)$ through the use of the theory of difference equations as in Heathcote and Moyal (1959), we have

$$F(z, x, s) = \frac{p_{00}(s)\mu_2(1 - x) - x^{i+1}}{[1 - zv_1(x, s)][\mu_2 - x\{\mu_1 v_2(x, s) + \mu_2 - \mu_1\}]}, \tag{13}$$

where v_1, v_2 are the roots of

$$\mu_1 v^2 - (s + \lambda_1 + \lambda_2 + \mu_1 - \lambda_2 x)v + \lambda_1 = 0 \tag{14}$$

and are given by

$$v_1(x, s) = (1/2\mu_1)\left[s + \lambda_1 + \lambda_2 + \mu_1 - \lambda_2 x - \sqrt{(s + \lambda_1 + \lambda_2 + \mu_1 - \lambda_2 x)^2 - 4\lambda_1\mu_1}\,\right], \tag{15}$$

$$v_2(x, s) = (1/2\mu_1)\left[s + \lambda_1 + \lambda_2 + \mu_1 - \lambda_2 x + \sqrt{(s + \lambda_1 + \lambda_2 + \mu_1 - \lambda_2 x)^2 - 4\lambda_1\mu_1}\,\right]. \tag{16}$$

$p_{00}(s)$ in (13) can be determined from the fact that, for $\mathrm{Re}(s) > 0$, the zeros of the numerator and denominator of (13) must coincide within the region $|x| < 1$, $|z| < 1$. Thus, we have

$$(17) \qquad p_{00}(s) = \xi^{h+1}/[\mu_2(1-\xi)],$$

where ξ is the root, within the unit circle, of

$$(18) \qquad \mu_2 - x[\mu_1 v_2(x,s) + \mu_2 - \mu_1] = 0,$$

which, in the particular case $\mu_1 = \mu_2 = \mu$, reduces to the quadratic equation

$$(19) \qquad (\lambda_1 + \lambda_2)x^2 - (s + \lambda_1 + \lambda_2 + \mu)x + \mu = 0;$$

therefore, we have

$$(20) \qquad \xi(s) = \frac{1}{2(\lambda_1+\lambda_2)}\Big[s + \lambda_1 + \lambda_2 + \mu - \sqrt{(s+\lambda_1+\lambda_2+\mu)^2 - 4\mu(\lambda_1+\lambda_2)}\,\Big].$$

By inversion of (17), we have

$$(21) \qquad P_{00}(t) = (\mu t)^{-1} e^{-t(\lambda_1+\lambda_2+\mu)} \sum_{j=0}^{\infty} \left(\sqrt{\frac{\mu}{\lambda_1+\lambda_2}}\right)^{j+i+1} \cdot (j+i+1) I_{j+i+1}\big(2t\sqrt{(\lambda_1+\lambda_2)\mu}\big),$$

where $I_k(x)$ is the modified Bessel function of the first kind.

Also, in this case we have

$$(22) \qquad F(z,x,s) = \frac{\xi^{i+1}(1-x)(1-\xi)^{-1} - x^{i+1}}{\mu[1 - z v_1(x,s)][1 - x v_2(x,s)]},$$

which, on inversion, yields

$$(23) \qquad \begin{aligned} \phi(z,x,t) &= \mathscr{L}^{-1}[F(z,x,s)] \\ &= \sum_{k=0}^{\infty}\sum_{j=0}^{k} z^j x^{k-j} \left(\sqrt{\frac{\lambda_1}{\mu}}\right)^{2j-k} (2j-k) \\ &\quad \cdot \Big\{(1-x)\int_0^t P_{00}(t-\tau)\exp[-(\lambda_1+\lambda_2+\mu-\lambda_2 x)\tau]\,\tau^{-1} \\ &\quad \cdot I_{|2j-k|}(2\tau\sqrt{\lambda_1\mu})\,d\tau \\ &\quad - x^{i+1}\exp[-(\lambda_1+\lambda_2+\mu-\lambda_2 x)t] \\ &\quad \cdot (\mu t)^{-1} I_{|2j-k|}(2t\sqrt{\lambda_1\mu})\Big\}. \end{aligned}$$

Let $E(M|i, t)$ be the expected number of the nonpriority class units in the system at time t conditional on i at time zero; then

$$(24) \quad \mathscr{L}[E(m|i,t)] = \frac{\partial}{\partial x} F(z,x,s)\bigg|_{1,1}$$

$$= i/s + \xi^{i+1}[s(1-\xi)]^{-1} + (\lambda_2 - \mu)s^{-2} + \mu s^{-2} v_1(1,s),$$

which, on inversion, gives

$$(25) \quad E(m|i,t) = i + (\lambda_2 - \mu)t + \mu \int_0^t P_{00}(\tau)\, d\tau$$

$$+ \sqrt{\lambda_1 \mu} \int_0^t (t-\tau) \exp[-(\mu + \lambda_1)\tau]\, \tau^{-1} I_1(2\tau\sqrt{\lambda_1 \mu})\, d\tau.$$

Busy Period Distribution

To find the probability density $\gamma_i(t)$ of the time taken for the nonpriority class to be cleared through the service mechanism, given i nonpriority units in the system at time zero, Heathcote (1959) follows the absorbing barrier method described in Section 1.3 onward and shows that

$$(26) \qquad \gamma_i(t) = \mathscr{L}^{-1}[\xi^i],$$

where $\xi(s)$, for the case $\mu_1 = \mu_2 = \mu$, is given by (20). Hence, we have

$$(27) \quad \gamma_i(t) = i[\mu/(\lambda_1 + \lambda_2)]^{i/2} \exp[-(\lambda_1 + \lambda_2 + \mu)t]\, t^{-1} I_i(2t\sqrt{\mu(\lambda_1 + \lambda_2)}).$$

Several Preemptive Priority Classes

Heathcote (1960) extends the results described above (Heathcote, 1959) to the case of R priority classes with mean arrival and service rates λ_i, μ_i for class i, $i = 1, 2, \ldots, R$, and gives explicit steady-state and time-dependent solutions in the particular case $\mu_i = \mu$, $i = 1, 2, \ldots, R$. Defining

$$(28) \qquad P(t; n_1, n_2, \ldots, n_R) = \text{Prob}\{n_i \text{ members of class } i \text{ in system at time } t,\ i = 1, 2, \ldots, R\},$$

it is shown, by methods essentially similar to those outlined above, that

$$(29) \qquad p_0 = \mathscr{L}[P(t; 0, \ldots, 0)] = \beta_R^{i+1}/[\mu(1-\beta_R)] = \mu^{-1} \sum_{j=i+1}^{\infty} \beta_R^j,$$

where

$$\beta_R = (1/2S_R)[s + S_R + \mu - \sqrt{(s + S_R + \mu)^2 - 4\mu S_R}], \tag{30}$$

$$S_R = \sum_{j=1}^{R} \lambda_j, \tag{31}$$

s is the Laplace transform parameter, and it is assumed that at time 0 the system has i members, all of class R, the lowest priority class. Inversion of (29) yields

$$\begin{aligned} P_0 &= P(t;0,\ldots,0) \\ &= \frac{1}{\mu} \sum_{j=i+1}^{\infty} \exp[-t(\mu + S_R)]\, t^{-1} \left(\frac{\mu}{S_R}\right)^{i/2} iI_i(2t\sqrt{\mu S_R}). \end{aligned} \tag{32}$$

Also, it is shown that the expectation of n_R at time t is

$$\begin{aligned} E(n_R, t) = i &+ (\lambda_R - \mu)t + \mu \int_0^t P(\tau;0,\ldots,0)\, d\tau \\ &+ \sqrt{\mu S_{R-1}} \int_0^t \tau^{-1}(t - \tau) \exp[-t(\mu + S_{R-1})] \\ &\cdot I_1(2\tau\sqrt{\mu S_{R-1}})\, d\tau. \end{aligned} \tag{33}$$

2.3 Queuing with Alternating Priorities

Yadin (1970) studies a queuing system consisting of customers belonging to two *queuing units*, called unit I and unit II, which are served by a single server according to an alternating discipline in that the server switches from one unit to the other according to a certain *switch rule*, which depends on the current sizes of the lines of both units. [It may be noted that the word "unit" is used here to mean a subsystem rather than being synonymous to "customers" as customary in queuing literature.] Each system has an independent Poisson stream of arrivals, and service times are (negative) exponentially distributed.

The state of the system can be represented by a point (x, y), $x, y = 0, 1, \ldots,$ where x, y stand for the number of customers at unit I and unit II, respectively. Then, the process under study can be regarded as a random walk on the lattice in the positive quadrant of the plane. Steps in the x direction and $-x$ direction represent, respectively, arrivals and service completions at unit I with similar connotation for the y axis and unit II. The period between switches is called a *task*. There are three kinds of tasks – service of unit I, service of unit II, and the *idle period*. An idle period is clearly associated with a *two-directional random walk* (the two positive directions), while the other kinds of *tasks* are associated with a *three-directional random walk* (the two positive directions and the one negative direction corresponding to the unit being served). During the process, the transition probabilities are changed from time to time, as the tasks are changed. An *elementary process* is defined as that during which transition probabilities remain unchanged, and thus may describe a single task. A switch takes place when the random walk enters a certain set of points determined by the switch rule. Such a set will be called a *switch set* when the composite process is discussed, or an *absorbing set* when the discussion is limited to a single task. Examples of switch rules are

(a) *The zero switch rule.* One unit is served as long as its line is not empty. This rule defines absorbing set of the form

$$\{(x, y) \mid x = 0, y \geqq 0\}.$$

In other words, the absorbing set is the positive part of the y-axis.

(b) *The constant queue switch rule.* The server switches to unit II when its line reaches a specified size, say k,[2] or when the line of unit I is empty. Here the absorbing set will be of the form

$$\{(x, y) | x = 0,\ 0 \leqq y < k \text{ or } x > 0,\ y = k\}, \qquad k = 1, 2, \ldots .$$

(c) *The difference switch rule.* A switch occurs whenever the difference between the lines exceeds a certain number, or when one line becomes empty. The absorbing sets are of the form

$$\{(x, y) | x = 0,\ 0 \leqq y \leqq k \text{ or } y = x + k,\ x > 0\}, \qquad k = 1, 2, \ldots .$$

The Three-Directional Elementary Process with Absorbing y-Axis

Here we have a single task controlled by the zero switch rule. Let P_1 denote this process defined by

(i) a starting point $(x_0, 0)$,
(ii) an absorbing set S_1, which is the positive y-axis, and
(iii) the rates α (arrivals to unit I), β (service to I), and γ (arrivals to II).

These rates determine the transition probabilities of the three-directional random walk a (x direction), b ($-x$ direction), and c (y direction). Because of the Markovian nature of the process, the probability that a transition in the x direction occurs first is

$$a = \int_0^\infty \alpha \exp[-(\alpha + \beta + \gamma)t]\, dt = \frac{\alpha}{\alpha + \beta + \gamma}. \tag{1}$$

[2] This is misprinted as y in Yadin (1970).

The other two transition probabilities are obtained similarly. Clearly, we have

(2) $$\alpha/a = \beta/b = \gamma/c = \alpha + \beta + \gamma.$$

Let us define

(3) $$f_1(0, y_1|x_0, 0) = \text{conditional probability of absorption at } (0, y_1) \in S_1, \text{ given the initial state } (x_0, 0),$$

and define the generating function

$$\bar{f}_1(\theta|x, 0) = \sum_{y=0}^{\infty} f_1(0, y|x, 0)\theta^y, \qquad 0 \leqq \theta \leqq 1.$$

The method of collective marks (Van Danzig, 1947; Runnenberg, 1965), which is used here, considers θ as a probability; therefore, we restrict $0 \leqq \theta \leqq 1$. The conditional probability density of T, the time till absorption, is denoted by $g_1(t|x_0, 0)$ and its Laplace transform by

(4) $$\bar{g}_1(s|x, 0) = \int_0^{\infty} e^{-st} g_1(t|x, 0)\, dt, \qquad s \geqq 0.$$

We see that $f_1(0, 0|x, 0)$ satisfies the following equations:

(5) $$f_1(0, 0|0, 0) = 1,$$

(6) $$f_1(0, 0|x, 0) = af_1(0, 0|x + 1, 0) + bf_1(0, 0|x - 1, 0), \qquad x = 1, 2, \ldots,$$

(7) $$f_1(0, 0|x, 0) = [f_1(0, 0|1, 0)]^x.$$

The solution of (5)–(6) which satisfies (7) also is

(8) $$\begin{aligned} f_1(0, 0|x, 0) &= [(2a)^{-1}\{1 - (1 - 4ab)^{1/2}\}]^x \\ &= [(2b)\{1 + (1 - 4ab)^{1/2}\}]^x. \end{aligned}$$

Introducing the function

(9) $$M(\eta) = \frac{2}{1 + (1 - 4\eta)^{1/2}}, \qquad 0 \leqq \eta \leqq \tfrac{1}{4},$$

we can write

(10) $$f_1(0, 0|x, 0) = [bM(ab)]^x.$$

By probability arguments it is shown that

(11) $$\begin{aligned} \bar{f}_1(\theta|x, 0) &= f_1(0, 0|x, 0)\Big|_{\substack{a \to a/(1 - c\theta) \\ b \to b/(1 - c\theta)}} \\ &= [b(1 - c\theta)^{-1} M(ab(1 - c\theta)^{-2})]^x, \qquad 0 \leqq \theta \leqq 1, \end{aligned}$$

(12) $$\begin{aligned} \bar{g}_1(\gamma|x, 0) &= f_1(0, 0|x, 0) = [bM(ab)]^x \\ &= [\beta(\alpha + \beta + \gamma)^{-1} M(\alpha\beta(\alpha + \beta + \gamma)^{-2})]^x, \qquad \gamma \geqq 0. \end{aligned}$$

It may be noted that the duration of the task T is a busy period of unit I, which is an $M/M/1$ system in itself, and its distribution has been derived by many authors. An alternative approach followed by Yadin (1970) is given below.

Given $T = t$, the number of arrivals at unit II has a Poisson distribution with mean γt. Thus the distribution of the number of arrivals is a mixture of the Poisson distribution with the mixing density g_1, i.e.,

$$f_1(0, y|x, 0) = \int_0^\infty g_1(t|x, 0)\frac{(\gamma t)^y}{y!}e^{-\gamma t}\,dt. \tag{13}$$

Thus, we have

$$f_1(0, y|x, 0) = \mathscr{L}\left[\frac{(\gamma t)^y}{y!}g_1(t|x, 0)\right]. \tag{14}$$

Let Y be the random variable denoting the number of unit II arrivals during T. Let us denote by T_y the time to the yth arrival at unit II. The distribution of T_y is Erlangian with y exponential stages, i.e., E_y, and its expectation is $\tau_y = y\gamma^{-1}$. It has a limiting one point distribution as $\gamma, y \to \infty$, so that τ_y remains fixed. During the time T there are exactly y arrivals at unit II. Therefore

$$T_y \leqq T < T_{y+1}, \tag{15}$$

and, as $\gamma \to \infty$,

$$\tau_y \leqq T < \tau_{y+1} = \tau_y + \gamma^{-1}. \tag{16}$$

The density of T is given by

$$\begin{aligned} g_1(t|x, 0) &= \lim_{\gamma\to\infty} \gamma\,\text{Prob}\{t \leqq T < t + \gamma^{-1}\} = \lim_{\gamma\to\infty} \gamma\,\text{Prob}\{\tau_y = t\} \\ &= \lim_{\gamma\to\infty} \gamma\,\text{Prob}\{Y = \gamma t\} = \lim_{\gamma\to\infty} \gamma f_1(0, \gamma t|x, 0). \end{aligned} \tag{17}$$

Equations (14) and (17) may be used to obtain g_1 from f_1 (also vice versa in the case of the former). The computations of f_1 and g_1 are based on the power series expansion of f_1. We have the relation

$$f_1(0, y|x, 0) = \sum_{k=0}^{\infty} \frac{x}{2k + x + y}\binom{2k + x + y}{k, k + x, y}a^k b^{k+x} c^y \tag{18}$$

based on the consideration of all possible paths from $(x, 0)$ to absorption at $(0, y)$ and the corresponding probabilities. In particular, we have

$$f_1(0, 0|x, 0) = \sum_{k=0}^{\infty} \frac{x}{2k + x}\binom{2k + x}{k}a^k b^{k+x}. \tag{19}$$

From (10) and (19), we have the identity

$$\left[\frac{2}{1 + (1 - 4\eta)^{1/2}}\right]^x = \sum_{k=0}^{\infty} \frac{x}{2k + x}\binom{2k + x}{k}\eta^k, \tag{20}$$

$$0 \leqq \eta \leqq \tfrac{1}{4}, \quad x = 0, 1, \ldots .$$

[Note that (20) is equivalent to 2.1(38), based on a property of the hypergeometric series ${}_2F_1(a, b; c; z)$ listed in Erdélyi *et al.* (1953, Vol. I). The above, therefore, serves as a proof of this property based on probabilistic arguments. See also Takács (1962, p. 30, Eq. (19)) for another derivation.]

From (11) and (20), we have

$$\bar{f}_1(\theta|x, 0) = \sum_{k=0}^{\infty} \frac{x}{2k + x}\binom{2k + x}{k} a^k b^{k+x}(1 - c\theta)^{-(2k+x)}, \tag{21}$$

from which f_1 may be derived by the binomial expansion of $(1 - c\theta)^{-(2k+x)}$.

Also, from (12),

$$\bar{g}_1(\gamma|x, 0) = \sum_{k=0}^{\infty} \frac{x}{2k + x}\binom{2k + x}{k} \alpha^k \beta^{k+x}(\alpha + \beta + \gamma)^{-(2k+x)}, \tag{22}$$

and, noting that

$$(\alpha + \beta + \gamma)^{-(2k+x)} = \mathscr{L}[e^{-(\alpha+\beta)t} t^{2k+x-1}/(2k + x - 1)!], \tag{23}$$

we have

$$\begin{aligned} g_1(t|x, 0) &= \frac{x}{t} e^{-(\alpha+\beta)t} \sum_{k=0}^{\infty} \frac{(\alpha t)^k (\beta t)^{k+x}}{k!\,(k + x)!} \\ &= \frac{x}{t} e^{-(\alpha+\beta)t} \left(\frac{\beta}{\alpha}\right)^{x/2} I_x(2t\sqrt{\alpha\beta}). \end{aligned} \tag{24}$$

Auxiliary Functions

We introduce some functions to facilitate computation of $f_1(0, y|x, 0)$ and other probability functions. Let us define

$$M_{x,y}(\eta) = \sum_{k=0}^{\infty} \frac{x}{2k + x + y}\binom{2k + x + y}{k, k + x, y}\eta^k, \tag{25}$$

$$0 \leqq \eta < \tfrac{1}{4}; \quad x, y = 0, 1, \ldots .$$

Then (18) reduces to

$$f_1(0, y|x, 0) = b^x c^y M_{x,y}(ab). \tag{26}$$

In particular, for $y = 0$,

$$M_{x,0}(\eta) = [M(\eta)]^x, \tag{27}$$

where $M(\eta)$ is given by (9). Notice that except for $y = 0$, $M_{x,y}(\eta)$ is defined only for $0 \leqq \eta < \frac{1}{4}$. The series does not converge for $\eta = \frac{1}{4}$, $y > 0$.

Next define

$$F_{x,y}(\eta) = \sum_{k=0}^{\infty} \binom{2k + x + y}{k, k + x, y} \eta^k, \qquad 0 \leqq \eta < \tfrac{1}{4}; \quad x, y = 0, 1, \ldots, \tag{28}$$

with

$$F(\eta) = F_{0,0}(\eta) = \sum_{k=0}^{\infty} \binom{2k}{k} \eta^k = (1 - 4\eta)^{-1/2}, \qquad 0 \leqq \eta < \tfrac{1}{4}. \tag{29}$$

The functions $M(\eta)$ and $F(\eta)$ are related by

$$M(\eta) = 2F(\eta)[1 + F(\eta)]^{-1} \tag{30}$$

or

$$F(\eta) = M(\eta)[2 - M(\eta)]^{-1}. \tag{31}$$

The third auxiliary function to be defined is

$$G_{x,y}(\eta) = \sum_{k=1}^{\infty} \frac{k}{2k + x + y} \binom{2k + x + y}{k, k + x, y} \eta^k, \tag{32}$$

$$0 \leqq \eta < \tfrac{1}{4}, \quad x, y = 0, 1, \ldots,$$

with

$$G(\eta) = G_{0,0}(\eta) = \frac{1}{2} \sum_{k=1}^{\infty} \binom{2k}{k} \eta^k \tag{33}$$

$$= \tfrac{1}{2}[F(\eta) - 1] = [M(\eta) - 1][2 - M(\eta)]^{-1}.$$

[Note that the auxiliary functions defined above are expressible as hypergeometric functions. On simplification we have

$$M_{x,y}(\eta) = \binom{x + y - 1}{x - 1} {}_2F_1\left[\frac{x + y}{2}, \frac{x + y + 1}{2}; x + 1; 4\eta\right], \tag{34}$$

and, therefore,

$$M_{x,0}(\eta) = {}_2F_1[x/2, (x + 1)/2; x + 1; 4\eta], \tag{35}$$

$$M(\eta) = M_{1,0}(\eta) = {}_2F_1(\tfrac{1}{2}, 1; 2; 4\eta). \tag{36}$$

Equation (27) follows from Erdélyi *et al.* (1953, Vol. I, p. 101, Eq. (6)) [see also

II.2(26) and 2.1(38)]. Also, we can see, on simplification, that

$$(37)\quad F_{x,y}(\eta) = \binom{x+y}{x}\, {}_2F_1\left[\frac{x+y+1}{2}, \frac{x+y+2}{2}; x+1; 4\eta\right],$$

$$(38)\quad F_{0,0}(\eta) = {}_2F_1(\tfrac{1}{2}, 1; 1; 4\eta) = {}_1F_0(\tfrac{1}{2}; -; 4\eta) = (1-4\eta)^{-1/2},$$

$$(39)\quad G_{x,y}(\eta) = \binom{x+y+1}{x+1}\eta\, {}_2F_1\left[\frac{x+y+2}{2}, \frac{x+y+3}{2}; x+2; 4\eta\right],$$

$$(40)\quad G_{0,0}(\eta) = \eta\, {}_2F_1(1, \tfrac{3}{2}; 2; 4\eta),$$

and again (33) follows from (40) by use of Erdélyi *et al.*, (1953, Vol. I, p. 101, Eq. (6)) or II.2(26).]

The following relations may be easily verified:

$$(41)\quad F_{x,y}(\eta) = 2G_{x,y}(\eta) + \left(1 + \frac{y}{x}\right) M_{x,y}(\eta),$$

$$(42)\quad F_{x,y}(\eta) = \frac{1}{\eta} G_{x-1,y}(\eta), \qquad x > 0,$$

$$(43)\quad F_{x,y}(\eta) = G_{x+1,y}(\eta) + M_{x+1,y}(\eta),$$

$$(44)\quad M_{x,y}(\eta) = \frac{x}{y}[M_{x+1,y-1}(\eta) + G_{x+1,y-1}(\eta)], \qquad y > 0,$$

which yield

$$(45)\quad M_{x,y}(\eta) = \frac{x}{y} F_{x,y-1}(\eta) = \frac{x}{y\eta} G_{x-1,y-1}(\eta), \qquad x, y > 0.$$

It may be seen that (42) follows at once from (37) and (39). The remaining of the above relations easily follow from the well-known recurrence relations satisfied by ${}_2F_1(a, b; c; z)$.

Relations (41)–(43) for $y = 0$ lead to

$$(46)\quad F_{x,0}(\eta) = (1-4\eta)^{-1/2}[M(\eta)]^x = F(\eta)[M(\eta)]^x$$

and

$$(47)\quad G_{x,0}(\eta) = [2 - M(\eta)]^{-1}\eta\,[M(\eta)]^{x+2} = G(\eta)[M(\eta)]^x,$$

where in (47) we use the identity

$$(48)\quad \eta[M(\eta)]^2 = M(\eta) - 1,$$

which is obvious from (9).

From (45), (46), we have

$$(49)\quad M_{x,1}(\eta) = xF_{x,0}(\eta) = xF(\eta)[M(x)]^x,$$

which yields

$$f_1(0, 1|x, 0) = b^x c M_{x,1}(ab) = [bM(ab)]^x[cF(ab)]x. \tag{50}$$

Inversion of $\bar{g}_1$ in (24) indicates that the functions $M_{x,y}$ and $F_{x,y}$ are related to the modified Bessel function. We give below two ways of connecting these functions, viz., by Laplace transformations and by letting one parameter approach infinity. It is easy to see that

$$\left(\frac{\lambda}{s}\right)^x\left(\frac{1}{s}\right)^y M_{x,y}(\lambda^2/s^2) = \int_0^\infty e^{-st}(t^{y-1}/y!)xI_x(2\lambda t)\,dt, \qquad \mathrm{Re}(s) > 2\lambda, \tag{51}$$

$$\left(\frac{\lambda}{s}\right)^x\left(\frac{1}{s}\right)^{y+1} F_{x,y}(\lambda^2/s^2) = \int_0^\infty e^{-st}(t^y/y!)I_x(2\lambda t)\,dt, \qquad \mathrm{Re}(s) > 2\lambda, \tag{52}$$

and, in particular, for $y = 0$,

$$\left(\frac{\lambda}{s}\right)^x M_{x,0}(\lambda^2/s^2) = \left[\frac{\lambda}{s} M\left(\frac{\lambda^2}{s^2}\right)\right]^x \tag{53}$$

$$= \int_0^\infty e^{-st}t^{-1}xI_x(2\lambda t)\,dt, \qquad \mathrm{Re}(s) \geqq 2\lambda,$$

$$(\lambda/s)^x(1/s)F_{x,0}(\lambda^2/s^2) = (1/s)F(\lambda^2/s^2)[(\lambda/s)M(\lambda^2/s^2)]^x \tag{54}$$

$$= \int_0^\infty e^{-st}I_x(2\lambda t)\,dt, \qquad \mathrm{Re}(s) > 2\lambda.$$

Equations (51) and (52) follow from Erdélyi *et al.* (1953, Vol. II, p. 49, Eq. (16)), or Erdélyi *et al.* (1954, Vol. I, p. 196, Eq. (8)), both of which can be reduced to the form

$$\int_0^\infty e^{-st}t^{\mu-1}I_v(\alpha t)\,dt \tag{55}$$

$$= \frac{(\alpha/2)^v\Gamma(\mu+v)}{s^{\mu+v}\Gamma(v+1)}\,{}_2F_1\left(\frac{\mu+v}{2}, \frac{\mu+v+1}{2}; v+1; \frac{\alpha^2}{s^2}\right),$$

$$\mathrm{Re}(s \pm \alpha) > 0, \qquad \mathrm{Re}(\mu + v) > 0.$$

Also, it may be verified that

$$\lim_{y\to\infty} r^y M_{x,y}(\lambda^2) = \lim_{y\to\infty} \frac{x}{y}\lambda^{-2x}r^yI_x(2\lambda y), \tag{56}$$

$$0 < \lambda < (1-r)/2, \qquad 0 < r < 1,$$

$$\lim_{y\to\infty} r^y F_{x,y}(\lambda^2) = \lim_{y\to\infty} \lambda^{-2x}r^yI_x(2\lambda y), \tag{57}$$

$$0 < \lambda < (1-r)/2, \qquad 0 < r < 1.$$

A Four-Directional Elementary Process

Consider the extension of P_1 to the four-directional process P_2, where a fourth direction $(-y)$ has been added with associated transition rate δ. Let d be the corresponding probability, so that

$$\frac{\alpha}{a} = \frac{\beta}{b} = \frac{\gamma}{c} = \frac{\delta}{d} = \alpha + \beta + \gamma + \delta. \tag{58}$$

In this case the point of absorption may be negative, so that the absorbing set S_2 is the entire y-axis. Let $f_2(0, y_1|x_0, 0)$ be the probability of absorption at $(0, y_1)$, starting at $(x_0, 0)$. Writing difference equations for f_2 and solving them with the help of characteristic function, Yadin (1970) shows that

$$\begin{aligned} f_2(0, y|x, 0) &= \sum_{k=0}^{\infty} \sum_{j=0}^{\infty} \frac{x}{2k + x + j} \binom{2k + x + j}{k, k + x, j} a^k b^{k+x} \binom{j}{(j + y)/2} \\ &\quad \cdot c^{(j-y)/2} d^{(j+y)/2} \\ &= b^x \left(\frac{c}{d}\right)^{y/2} \sum_{j=0}^{\infty} \binom{j}{(j + y)/2} (cd)^{j/2} M_{x,j}(ab). \end{aligned} \tag{59}$$

Further manipulations yield

$$f_2(0, y|x, 0) = b^x c^{(y)_+} d^{(-y)_+} \sum_{m=0}^{\infty} \binom{2m + |y|}{m} (cd)^m M_{x, 2m+|y|}(ab), \tag{60}$$

where $(y)_+$ is the positive part of y, given by

$$(y)_+ = \max\{y, 0\}, \tag{61}$$

so that

$$(y)_+ + (-y)_+ = |y|. \tag{62}$$

The process P_1 is a special case of P_2. Another degenerate case of P_2 is P_{21}, say, where $\alpha = 0$. In this case $a = 0$, and, therefore, we can obtain the corresponding probability function f_{21} by substituting

$$M_{x,y}(0) = \frac{x}{x + y} \binom{x + y}{y} = \binom{x + y - 1}{y} \tag{63}$$

in (60), so that

$$\begin{aligned} f_{21}(0, y|x, 0) &= b^x c^{(y)_+} d^{(-y)_+} \sum_{m=0}^{\infty} \binom{2m + |y|}{m} \binom{x + 2m + |y| - 1}{2m + |y|} (cd)^m \\ &= b^x c^{(y)_+} d^{(-y)_+} F_{|y|, x-1}(cd), \end{aligned} \tag{64}$$

or, according to (45),

$$f_{21}(0, y|x, 0) = b^x c^{(y)+} d^{(-y)+} \frac{x}{|y|} M_{|y|,x}(cd). \tag{65}$$

Yadin, then, considers various other elementary processes, of which we shall briefly outline the following.

Three-Directional Elementary Process with Difference Switch Rule

This process is termed $P_{4 \cdot y}$ and is associated with the absorbing set

$$S_{4 \cdot y} = \{(x_1, y_1) \mid x_1 = (y_1 - y)_+ ; y_1 \geqq 0\} \tag{66}$$

and the transition rates α, β, γ, as before. It is shown that for the process $P_{4 \cdot 0}$ corresponding to the difference $y = 0$, the generating function of the absorption probability, given by

$$\bar{f}_{4 \cdot 0}(\theta|x_0, 0) = \sum_{y=0}^{\infty} f_{4 \cdot 0}(x, y|x_0, 0)\theta^y, \tag{67}$$

simplifies to

$$\begin{aligned} \bar{f}_{4 \cdot 0}(\theta|x, 0) &= [(b + c\theta)M(a(b + c\theta))]^x \\ &= \sum_{k=0}^{\infty} \frac{x}{2k + x} \binom{2k + x}{k} a^k (b + c\theta)^{k+x}. \end{aligned} \tag{68}$$

To obtain $\bar{g}_{4 \cdot 0}$, the Laplace transform of the density $g_{4 \cdot 0}$ of the duration of the process $P_{4 \cdot 0}$, we replace a, b, and c by $\alpha/(\alpha + \beta + \gamma + s)$, $\beta/(\alpha + \beta + \gamma + s)$, and $\gamma/(\alpha + \beta + \gamma + s)$, respectively, and set $\theta = 1$ in (68); thus[3]

$$\begin{aligned} \bar{g}_{4 \cdot 0}(s|x, 0) &= \int_0^{\infty} e^{-st} g_{4 \cdot 0}(t|x, 0)\, dt \\ &= \sum_{k=0}^{\infty} \frac{x}{2k+x} \binom{2k+x}{k} \alpha^k (\beta+\gamma)^{k+x} (\alpha+\beta+\gamma+s)^{-(2k+x)}, \end{aligned} \tag{69}$$

$\mathrm{Re}(s) \geqq 0,$

and, on inversion,

$$g_{4 \cdot 0}(t|x, 0) = \frac{x}{t} \left(\frac{\beta + \gamma}{\alpha}\right)^{x/2} \exp[-(\alpha + \beta + \gamma)t] I_x(2t\sqrt{\alpha(\beta + \gamma)}). \tag{70}$$

The probability function $f_{4 \cdot 0}$ and the density $g_{4 \cdot 0}$ are proper if and only if $\alpha \leqq \beta + \gamma$.

[3] There is an obvious misprint in Yadin (1970): instead of dt, in the integral there appears dx.

2.4 Two-Server Bulk Service Queue

Arora (1964) considers a bulk service queuing system with Poisson arrivals at a mean rate λ and two service facilities with capacities S_1 and S_2 ($S_2 \leqq S_1$). The first facility takes S_1 customers or the whole queue (whichever is less), and the second takes S_2 or the whole queue. When both channels are free, the service starts at the first. The service times for batches for the two channels are exponentially distributed with means μ_1^{-1} and μ_2^{-1}, respectively. We define the following probabilities:

(1) $E(t) = \text{Prob}\{\text{the system empty at time } t\}$,

(2) $Q(t) = \text{Prob}\{\text{the first channel busy, second empty (and no unit in queue) at } t\}$,

(3) $R(t) = \text{Prob}\{\text{the second channel busy, first empty (and no unit in queue) at } t\}$,

(4) $P_n(t) = \text{Prob}\{\text{both channels busy and } n\,(\geqq 0) \text{ units waiting in queue at time } t\}$,

and have the following difference-differential equations:

(5)
$$\frac{dE(t)}{dt} = -\lambda E(t) + \mu_1 Q(t) + \mu_2 R(t);$$

(6)
$$\frac{dQ(t)}{dt} = -(\lambda + \mu_1)Q(t) + \lambda E(t) + \mu_2 P_0(t);$$

(7)
$$\frac{dR(t)}{dt} = -(\lambda + \mu_2)R(t) + \mu_1 P_0(t);$$

(8)
$$\frac{dP_0(t)}{dt} = -(\lambda + \mu_1 + \mu_2)P_0(t) + \lambda[Q(t) + R(t)] + \mu_1 \sum_{m=1}^{S_1} P_m(t) + \mu_2 \sum_{k=1}^{S_2} P_k(t);$$

(9)
$$\frac{dP_n(t)}{dt} = -(\lambda + \mu_1 + \mu_2)P_n(t) + \lambda P_{n-1}(t) + \mu_1 P_{n+S_1}(t) + \mu_2 P_{n+S_2}(t) \qquad (n \geqq 1).$$

Let us define the generating function

$$F(x, t) = \sum_{n=0}^{\infty} P_n(t)x^n. \tag{10}$$

From (8) and (9), we have

$$\begin{aligned}\frac{\partial F(x, t)}{\partial t} = &- [\lambda + \mu_1 + \mu_2 - \lambda x - \mu_1 x^{-S_1} - \mu_2 x^{-S_2}]F(x, t) \\ &+ \mu_1 \sum_{m=0}^{S_1 - 1} (1 - x^{m-S_1})P_m(t) + \mu_2 \sum_{k=0}^{S_2 - 1} (1 - x^{k-S_2})P_k(t) \\ &- (\mu_1 + \mu_2)P_0(t) + \lambda[Q(t) + R(t)].\end{aligned} \tag{11}$$

Taking Laplace transforms of (5)–(7) and (11), and using small letters to denote L.T.s, we have

$$- (s + \lambda)e(s) + 1 + \mu_1 q(s) + \mu_2 r(s) = 0, \tag{12}$$

$$- (s + \lambda + \mu_1)q(s) + \lambda e(s) + \mu_2 p_0(s) = 0, \tag{13}$$

$$- (s + \lambda + \mu_2)r(s) + \mu_1 p_0(s) = 0, \tag{14}$$

$$\begin{aligned}f(x, s) = x^{S_1}\Bigg[&\mu_1 \sum_{m=0}^{S_1 - 1} (1 - x^{m-S_1})p_m(s) + \mu_2 \sum_{k=0}^{S_2 - 1} (1 - x^{k-S_2})p_k(s) \\ &- (\mu_1 + \mu_2)p_0(s) + \lambda\{q(s) + r(s)\}\Bigg] \\ &\cdot [- \lambda x^{S_1 + 1} + (s + \lambda + \mu_1 + \mu_2)x^{S_1} - \mu_2 x^{S_1 - S_2} - \mu_1]^{-1},\end{aligned} \tag{15}$$

where we have assumed that the system starts with empty state, so that

$$E(0) = 1, \tag{16}$$

$$Q(0) = R(0) = P_n(0) = 0, \qquad n \geqq 0. \tag{17}$$

It can be shown by Rouché's theorem that, for $S_2 \leqq S_1$, the equation

$$\lambda x^{S_1 + 1} - (s + \lambda + \mu_1 + \mu_2)x^{S_1} + \mu_2 x^{S_1 - S_2} + \mu_1 = 0 \tag{18}$$

has S_1 roots inside and one outside $|x| = 1$. Let the roots inside $|x| = 1$ be denoted by $x_i(s)$, $i = 1, 2, \ldots, S_1$, and the one outside by $x_0(s)$. The numerator of (15) should vanish at $x_i(s)$, $i = 1, 2, \ldots, S_1$, thus giving rise to S_1 equations, which along with (12), (13), and (14) are sufficient to determine the $S_1 + 3$ unknowns, viz., $p_n(s)$, $n = 0, 1, \ldots, S_1 - 1$, $e(s)$, $q(s)$, and $r(s)$. Now, since the degree of the numerator of (15) is less than that of the denominator by one, we can write

$$f(x, s) = A(s)/[x_0(s) - x], \tag{19}$$

and, therefore,

(20) $$p_n(s) = A(s)/[x_0(s)]^{n+1},$$

where $A(s)$ is an unknown to be determined. Setting $x = 1$ in (15) and (19), we have

(21) $$A(s)/(x_0 - 1) = f(1, s) = [-(\mu_1 + \mu_2)p_0(s) + \lambda\{q(s) + r(s)\}]/s,$$

which gives $A(s)$ in terms of $p_0(s)$, $q(s)$, and $r(s)$, which are determined as above.

The Busy Period Distribution

For the two-server queuing system, we define two types of busy periods: (a) the duration of time during which at least one server remains busy, and (b) the time during which both the servers are busy. Let us first consider

Case (a) – At Least One Server Busy This type of busy period starts with the arrival of a customer at the empty service facility and lasts until it next becomes empty for the first time. For simplicity, consider the case when $S_1 = S_2 = S$ and $\mu_1 = \mu_2 = \mu$. We introduce an absorbing barrier at the empty state, and let the system start with $n = 1$. Let

(22) $$U(t) = P(t) + Q(t).$$

Then we have

(23) $$\frac{dE(t)}{dt} = \mu U(t),$$

(24) $$\frac{dU(t)}{dt} = -(\lambda + \mu)U(t) + 2\mu P_0(t),$$

(25) $$f(x, s) = x^S\left[2\mu \sum_{m=0}^{S-1} (1 - x^{m-S})p_m(s) - 2\mu p_0(s) + \lambda u(s)\right] \cdot [-\{\lambda x^{S+1} - (s + \lambda + 2\mu)x^S + 2\mu\}]^{-1}.$$

Clearly, the probability density of busy period durations is given by

(26) $$\Gamma_1(t) = dE(t)/dt,$$

and with the help of (22)–(26), it can be shown that its L.T. is given by

(27) $$\gamma_1(s) = se(s) = \mu u(s) = [\mu(s + 2\mu)x_0 - 2\mu^2]/[\{s(s + \lambda + \mu) + 2\mu(s + \mu)\}x_0 - 2\mu(s + \mu)].$$

This can be expressed in the form

$$(28)\qquad \gamma_1(s) = \frac{\mu(s+2\mu)x_0 - 2\mu^2}{[s(s+\lambda+\mu) + 2\mu(s+\mu)]x_0} \cdot \sum_{m=0}^{\infty}\left[\frac{2\mu(s+\mu)}{\{s(s+\lambda+\mu)+2\mu(s+\mu)\}x_0}\right]^m$$

$$= \sum_{m=0}^{\infty} \frac{\mu(2\mu)^m(s+2\mu)(s+\mu)^m x_0^{-m}}{[s(s+\lambda+\mu)+2\mu(s+\mu)]^{m+1}} - \sum_{m=0}^{\infty} \frac{\mu(2\mu)^{m+1}(s+\mu)^m x_0^{-m-1}}{[s(s+\lambda+\mu)+2\mu(s+\mu)]^{m+1}},$$

where x_0 is the root, with modulus greater than one, of the equation

$$(29)\qquad \lambda x^{S+1} - (s+\lambda+2\mu)x^S + 2\mu = 0.$$

To determine x_0, using Lagrange's theorem (Whittaker and Watson, 1927, pp. 132–133; Takács, 1962, p. 235), on writing (29) as

$$(30)\qquad x = (s+\lambda+2\mu)/\lambda - (2\mu/\lambda)(1/x^S),$$

we have

$$(31)\qquad x_0^{-v} = \left[a^{-v} + v\sum_{j=1}^{\infty}\{(-1)^{j+1}/j!\}(2\mu/\lambda)^j \cdot \frac{d^{j-1}}{da^{j-1}}(a^{-(Sj+v+1)})\right]_{a=(s+\lambda+2\mu)/\lambda}$$

$$= v\sum_{j=0}^{\infty}\frac{(2\mu/\lambda)^j}{j!}\frac{\{(S+1)j+v-1\}!}{(Sj+v)!} \cdot [(s+\lambda+2\mu)/\lambda]^{-[(S+1)j+v]}.$$

On inversion, with the help of Erdélyi *et al.* (1954, Vol. I, p. 144, Eq. (3)), we have

$$(32)\qquad \mathscr{L}^{-1}(x_0^{-v}) = \frac{v}{t}\sum_{j=0}^{\infty}\frac{(2\mu/\lambda)^j e^{-(\lambda+2\mu)t}(\lambda t)^{(S+1)j+v}}{j!(Sj+v)!}$$

$$= \left(\frac{v}{t}\right)\left(\frac{1}{b^v}\right)e^{-(\lambda+2\mu)t}I_v^S(2\lambda bt),$$

where [cf. II.5(22)]

$$(33)\qquad I_v^S(z) = \sum_{n=0}^{\infty}\frac{(z/2)^{(S+1)n+v}}{n!\,(nS+v)!},$$

$$(34)\qquad b^{S+1} = 2\mu/\lambda.$$

Using (32), Arora (1964) has inverted (28) in terms of convolution integrals involving $I_\nu^S(z)$, given by (33), and the roots α_1, α_2 of

$$s^2 + s(\lambda + 3\mu) + 2\mu^2 = 0. \tag{35}$$

The expression is quite complicated and we shall not reproduce it here. The mean and variance of the distribution have been found by differentiating (27), as usual.

In the particular case $S = 1$, i.e., the $M/M/2$ with customers served singly, (27) with the help of (29) simplifies to

$$\gamma_1(s) = \sum_{m=1}^{\infty} \left(\frac{\mu}{\lambda}\right)^m x_0^{-m}, \tag{36}$$

whence, on inversion,

$$\Gamma_1(t) = e^{-(\lambda+2\mu)t} t^{-1} \sum_{m=1}^{\infty} m\left(\frac{2\lambda}{\mu}\right)^{-m/2} I_m(2\sqrt{2\lambda\mu}\, t). \tag{37}$$

And, in this case, the mean and variance simplify to

$$\text{mean} = 1/[\mu(1 - \rho)], \tag{38}$$

$$\text{variance} = 1/[\mu^2(1 - \rho)^3]. \tag{39}$$

Equations (36) and (37) have also been obtained by Chaudhry and Templeton (1973).

Case (b) – Both Channels Busy Proceeding as in Case (a), it is shown that the Laplace transform of the busy period density function in this case is given by

$$\gamma_2(s) = \frac{2\mu(x_0 - 1)}{[(s + 2\mu)x_0 - 2\mu]} = \sum_{i=0}^{\infty} \frac{(2\mu)^{i+1}}{(s + 2\mu)^{i+1}} (x_0^{-i} - x_0^{-(i+1)}), \tag{40}$$

which, on inversion, gives

$$\Gamma_2(t) = \sum_{i=0}^{\infty} (2\mu)^{i+1} \int_0^t e^{-(2\mu t + \lambda\tau)} \frac{(t - \tau)^i}{i!} \cdot \left\{\frac{i}{b^i} I_i^S(2\lambda b\tau) - \frac{i+1}{b^{i+1}} I_{i+1}^S(2\lambda b\tau)\right\} d\tau. \tag{41}$$

For $S = 1$, we have

$$\gamma_2(s) = (2\mu/\lambda)x_0^{-1}, \tag{42}$$

$$\Gamma_2(t) = \sqrt{2\mu/\lambda}\, t^{-1} e^{-(\lambda+2\mu)t} I_1(2t\sqrt{2\lambda\mu}), \tag{43}$$

$$\text{mean} = 1/[2\mu(1 - \rho)], \tag{44}$$

$$\text{variance} = (1 + \rho)/[4\mu^2(1 - \rho)^3]. \tag{45}$$

Equations (42) and (43) have also been derived by Chaudhry and Templeton (1973).

2.5 Multichannel Queuing System with Bulk Service

Ghare (1968) generalizes Arora's (1964) results by considering an $M/M/c$ queuing system where the c channels are identical, each with capacity b. The mean arrival and service (for each channel) rates are λ and μ, respectively. Let us define

(1) $$P_{m,n}(t) = \text{Prob}\{m \text{ channels busy and } n \text{ units } \textit{waiting} \text{ in queue at time } t\}.$$

Obviously, this is nonzero only when $0 \leqq m \leqq c$, $0 \leqq n < \infty$, and $(c-m)n = 0$. Then we have

(2) $$\frac{d}{dt}P_{c,n}(t) = -(\lambda + c\mu)P_{c,n}(t) + \lambda P_{c,n-1}(t) + c\mu P_{c,n+b}(t) \qquad (n > 0),$$

(3) $$\frac{d}{dt}P_{c,0}(t) = -(\lambda + c\mu)P_{c,0}(t) + \lambda P_{c-1,0}(t) + c\mu \sum_{k=1}^{b} P_{c,k}(t),$$

(4) $$\frac{d}{dt}P_{m,0}(t) = -(\lambda + m\mu)P_{m,0}(t) + \lambda P_{m-1,0}(t) + (m+1)\mu P_{m+1,0}(t) \qquad (1 \leqq m < c),$$

(5) $$\frac{d}{dt}P_{0,0}(t) = -\lambda P_{0,0}(t) + \mu P_{1,0}(t).$$

Define the generating function

(6) $$F(x,t) = \sum_{k=0}^{\infty} P_{c,k}(t)x^k.$$

Let us assume the initial condition

(7) $$P_{c,0}(0) = 1,$$

i.e., all channels are busy and no unit is waiting to be serviced; then

(8) $$F(x,0) = 1.$$

From (2) and (3) we have

(9) $$\frac{d}{dt}F(x,t) = -(\lambda + c\mu - \lambda x - c\mu x^{-b})F(x,t) + c\mu \sum_{k=0}^{b-1} (1 - x^{k-b})P_{c,k}(t) - c\mu P_{c,0}(t) + \lambda P_{c-1,0}(t).$$

Taking the Laplace transform of (9), and simplifying, we have

$$(10)\quad f(x,s)=\left[1+c\mu\sum_{k=0}^{b-1}(1-x^{k-b})p_{c,k}(s)-c\mu p_{c,0}(s)+\lambda p_{c-1,0}(s)\right]\cdot[s+\lambda+c\mu-\lambda x-\mu x^{-b}]^{-1}.$$

By Rouché's theorem it can be shown that the denominator of (10) has b zeros inside and 1 zero outside $|x|=1$. Let the zero outside $|x|=1$ be denoted by x_0. Then, as in 2.4(19), we can write

$$(11)\quad f(x,s)=A/(x_0-x),$$

where A is an unknown function to be determined, and we have

$$(12)\quad p_{c,n}(s)=A/x_0^{n+1},\qquad n\geqq 0.$$

Let us write the L.T. of $P_{m,0}(t)$ simply as p_m; then, from (4) and (5), we have

$$(13)\quad (m+1)\mu p_{m+1}=(\lambda+s+m\mu)p_m-\lambda p_{m-1},$$

$$(14)\quad \mu p_1=(\lambda+s)p_0.$$

The general solution of the set of Eqs. (13) and (14) can be written [see Jackson and Henderson (1966)]

$$(15)\quad p_m=p_0\sum_{j=0}^{m}\frac{(\lambda/\mu)^{m-j}\Gamma(j+s/\mu)}{\Gamma(s/\mu)j!(m-j)!}=p_0\Phi_m(s)\qquad(0\leqq m<c),$$

where

$$(16)\quad \Phi_m(s)=(\lambda/\mu)^m(1/m!)\,{}_2F_0(-m,s/\mu;-;-\mu/\lambda).$$

Writing (10) for $x=1$ and substituting from (11), (12), and (15), we have

$$(17)\quad f(1,s)=A/(x_0-1)=[1-c\mu A/x_0+\lambda p_0\Phi_{c-1}(s)]/s,$$

whence

$$(18)\quad A=x_0(x_0-1)[1+\lambda p_0\Phi_{c-1}(s)][sx_0+c\mu(x_0-1)]^{-1}.$$

Also, since the total probability is one, we have

$$(19)\quad \frac{A}{x_0-1}+\sum_{0}^{c-1}p_m=\frac{1}{s}.$$

Substituting from (18) and (15) in (19), and solving for p_0, we get

$$(20)\quad p_0=c\mu(x_0-1)\left[s\{\lambda x_0\Phi_{c-1}(s)+(sx_0+c\mu(x_0-1))\}\sum_{0}^{c-1}\Phi_m(s)\right]^{-1}.$$

Equations (12), (15), and (20) give the Laplace transforms of all the state probabilities.

The steady-state probabilities are found, using the Tauberian theorem as usual, in the form

(21) $$P_{c,n} = P_{0,0}\{(\lambda/\mu)^c/c!\}k^{-n},$$

(22) $$P_{m,0} = P_{0,0}\{(\lambda/\mu)^m/m!\},$$

(23) $$P_{0,0} = \left[\frac{(\lambda/\mu)^c}{c!(1-1/k)} + \sum_0^{c-1}\left(\frac{(\lambda/\mu)^m}{m!}\right)\right]^{-1},$$

where k is the greatest root of[4]

(24) $$\lambda/(c\mu) = (k^b - 1)/[k^b(k-1)].$$

2.6 Poisson Queue with General Bulk Service Rule

Medhi (1975) considers the queuing system $M/M_{a,b}/1$ with a general service rule (Neuts, 1967a). Under this rule service is rendered in batches of minimum size a and maximum size b, so that every time the server becomes free and finds at least a units waiting in the queue, he takes into service b units or the whole queue, whichever is less, and otherwise he remains idle until the queue size increases to a. Mean arrival rate is λ, and mean service rate for the batches is μ. The service time distribution is assumed to be independent of the batch size. Let us say that the system is in state $(0, q)$ when the server is idle and there are $(q \mid 0 \leqq q \leqq a-1)$ units waiting in the queue, and in state $(1, n)$ when the server is busy and n units are waiting in the queue. Let

(1) $P_{0q}(t) = \text{Prob}\{\text{system is in state } (0, q) \text{ at time } t\}$, $\quad 0 \leqq q \leqq a-1$,

(2) $P_{1n}(t) = \text{Prob}\{\text{system is in state } (1, n) \text{ at time } t\}$, $\quad n \geqq 0$.

We have

(3) $$\frac{d}{dt}P_{1n}(t) = -(\lambda+\mu)P_{1n}(t) + \lambda P_{1,n-1}(t) + \mu P_{1,n+b}(t) \qquad (n \geqq 1),$$

(4) $$\frac{d}{dt}P_{10}(t) = -(\lambda+\mu)P_{10}(t) + \lambda P_{0,a-1}(t) + \mu\sum_{k=a}^{b} P_{1k}(t),$$

(5) $$\frac{d}{dt}P_{00}(t) = -\lambda P_{00}(t) + \mu P_{10}(t),$$

(6) $$\frac{d}{dt}P_{0q}(t) = -\lambda P_{0q}(t) + \lambda P_{0,q-1}(t) + \mu P_{1q}(t) \qquad (1 \leqq q \leqq a-1),$$

where (6) will be redundant if $a = 1$.

[4] In Ghare's (1968) paper the equation corresponding to (24) seems to be in error and the word "greatest" is missing. Also, there is an obvious misprint in the equation corresponding to (23).

The steady-state probabilities are given (Borthakur, 1971) by

(7) $P_{1n} = pd^{-n}P_{00}, \qquad n \geqq 0,$

(8) $P_{0q} = d^{-q}(d^q + d^{q-1} + \cdots + 1)P_{00}$

$= (d - d^{-q})P_{00}/(d-1), \qquad 1 \leqq q \leqq a-1,$

(9) $P_{00} = d^{a-1}(d-1)/[(a+p)d^a - (d^{a-1} + d^{a-2} + \cdots + 1)],$

where

(10) $$p = \lambda/\mu,$$

(11) $$d = \lim_{s \to 0} z_0(s),$$

$z_0(s)$ being the root, lying outside the unit circle $|z| = 1$, of the equation

(12) $$s + \lambda + \mu - \lambda z - \mu z^{-b} = 0.$$

It may be noted [cf. 2.5(24)] that d is a solution of the equation

(13) $$\lambda/\mu = p = d^{-b}(d^b - 1)/(d-1).$$

Assuming the system to be in steady state, we proceed to find the waiting time distribution.

Waiting Time Distribution

Let $w(t)$ be the probability density function (p.d.f.) of the waiting time *in the queue* for an arriving unit, and let

(14) $f(\alpha, k; t) \equiv$ the p.d.f. of gamma distribution with parameters α, k

$= \alpha^k t^{k-1} e^{-\alpha t}/(k-1)!, \qquad t > 0, \quad k = 1, 2, \ldots,$

(15) $$\Gamma_x(\alpha, k) = \int_0^x f(\alpha, k; t)\, dt = 1 - \sum_{r=0}^{k-1} \frac{e^{-\alpha x}(\alpha x)^r}{r!}.$$

It may be noted that

(16) $$\Gamma_x(\alpha, k) = \gamma(k, \alpha x)/(k-1)!,$$

where $\gamma(k, \alpha x)$ is the incomplete gamma function.

Then, from the usual probability arguments, by considering all possible states of the system that the arriving customer will find, we have

(17) $$w(t) = \sum_{q=0}^{a-2} P_{0q} f(\lambda, a-1-q; t) + \sum_{k=0}^{\infty} \sum_{r=a-1}^{b-1} P_{1,r+kb} f(\mu, k+1; t)$$

$$+ \sum_{k=0}^{\infty} \sum_{r=0}^{a-2} P_{1,r+kb} h_z(t), \qquad 0 < t < \infty,$$

where $h_z(t)$ is the p.d.f. of the random variable (r.v.) Z defined as

$$(18)\qquad Z = \max\{\text{gamma variate with parameters } \lambda, a-1-r; \text{ gamma variate with parameters } \mu, k+1\}$$

and is given by

$$(19)\quad h_z(t) = f(\lambda, a-1-r; t)\Gamma_t(\mu, k+1) + \Gamma_t(\lambda, a-1-r)f(\mu, k+1; t).$$

Let us write

$$(20)\qquad \sum_{k=0}^{m-1} \frac{z^k}{k!} \equiv e(m, z), \qquad m \geqq 1,$$

$$(21)\qquad e^{-z}e(m, z) \equiv E(m, z).$$

Then, using (7), (8), (20), and (21), we can simplify (17) as follows:

$$(22)\quad w(t) = [\lambda P_{00}d/(d-1)](1-d^{-b})\exp[-\mu(1-d^{-b})t] + E(a-1, \lambda t)[1-\exp\{-\mu(1-d^{-b})t\}], \qquad 0 < t < \infty.$$

Also, the probability of not having to wait is given by

$$(23)\qquad \text{Prob}\{T = 0\} = P_{0,a-1}.$$

Equation (22) can be used to find the moments of the waiting time distribution. Expressions for the mean and variance are obtained by Medhi (1975). The particular case $a = 1$ corresponds to the usual bulk service rule.

2.7 Blocking and Delays in Bulk Arrival Queues

Kabak (1968) considers the queuing system $M^{(n)}/M/c$, in which there are n arrivals at each arrival epoch, the arrival epochs occur in a Poisson fashion, and service is exponential, the mean arrival and service rates being λ and 1, respectively. He considers the two types of this system, viz., (1) the loss system, $M^{(n)}/M/c$: (c), in which no items are allowed to wait, and (2) the delay system, $M^{(n)}/M/c$: (∞), in which any number of items may wait. Also, it is assumed that all arrivals of previous arrival epochs are served before any arrivals of the latest epoch begin service. This is known as *gating*; within each batch the arrivals may be thought of as either being first come, first served, or, equivalently, served at random. Thus the rule first come, first served is applied to batches rather than individual items.

From usual probability considerations, we obtain the steady-state probabilities as given below.

For the loss system

(1) $$P_i = \frac{\lambda}{i} \sum_{j=0}^{i-1} P_j \qquad (1 \leqq i \leqq c;\ i \leqq n),^5$$

(2) $$P_i = \frac{\lambda}{i} \sum_{j=i-n}^{i-1} P_j \qquad (1 \leqq i \leqq c;\ i \geqq n),^5$$

(3) $$\sum_{i=0}^{c} P_i = 1.$$

Equations (1) and (2) can be written in a compact form as

(4) $$P_i = \frac{\lambda}{i} \sum_{j=\max(0,i-n)}^{i-1} P_j.$$

For the delay system

(5) $$P_i = \frac{\lambda}{i} \sum_{j=\max(0,i-n)}^{i-1} P_j \qquad (1 \leqq i \leqq c-1),$$

(6) $$P_i = \frac{\lambda}{c} \sum_{j=\max(0,i-n)}^{i-1} P_j \qquad (i \geqq c),$$

(7) $$\sum_{i=0}^{\infty} P_i = 1.$$

Again, we can combine (5) and (6) as

(8) $$P_i = \frac{\lambda}{\min(i, c)} \sum_{j=\max(0,i-n)}^{i-1} P_j.$$

The two types of system are studied for the determination of (a) the *probability of blocking*, i.e., the probability that an arrival will not be served immediately, and (b) the *delay distribution*, i.e., the probability distribution of the delay that arrivals will suffer, in the delay system. We consider below the determination of the delay distribution.

Delay Distribution

The probability that the system is in state $j + n$ immediately after an arrival epoch is P_j. When $j + n \leqq c$, no arrivals are delayed. If $j + n = c + k$, $k = 1, 2, \ldots$, then k arrivals are delayed if $k \leqq n$, and all n arrivals are delayed if $k \geqq n$. When $k = n + i$ $(i = 0, 1, 2, \ldots)$, the n delayed arrivals must wait for

[5] Kabak (1968) has $0 \leqq i \leqq c$ here.

$i + 1, i + 2, i + 3, \ldots$ departures before each of the n arrivals begins its service. When we have $1 \leqq k < n < c$, or $n - c \leqq k \leqq n$ $(> c)$, the k arrivals are delayed and must wait for $1, 2, 3, \ldots, k$ departures before all delayed arrivals are served.

When $n \leqq c$, the probability that an arrival will have to wait for d departures is given by

$$P_{n \leqq c}(d) = n^{-1} \sum_{i=d}^{d+c-1} P_i, \tag{9}$$

or, equivalently,

$$P_{n \leqq c}(d) = n^{-1} \sum_{i=d+c-n}^{d+c-1} P_i. \tag{10}$$

And, when $n \geqq c$,

$$P_{n \geqq c}(d) = n^{-1} \sum_{i=0}^{d+c-1} P_i \qquad (d \leqq n - c), \tag{11}$$

$$P_{n \geqq c}(d) = n^{-1} \sum_{i=d+c-n}^{d+c-1} P_i \qquad (d \geqq n - c). \tag{12}$$

Equations (10)–(12) can be combined in the form

$$P(d) = n^{-1} \sum_{i=\max(0,d+c-n)}^{d+c-1} P_i.$$

Now, the time occupied by d departures is the gamma distribution with d degrees of freedom, viz.,

$$g(d) = c^d t^{d-1} \exp(-ct)/\Gamma(d). \tag{13}$$

From (12) and (13), we have

$$\text{Prob}\{\text{delay} > t\} = n^{-1} \sum_{d=1}^{\infty} \sum_{i=\max(0,d+c-n)}^{d+c-1} P_i \int_t^{\infty} g(d)\, dt. \tag{14}$$

It may be noted that

$$\int_t^{\infty} g(d)\, dt = \frac{\Gamma(d, ct)}{\Gamma(d)}, \tag{15}$$

where $\Gamma(\alpha, x)$ stands for the incomplete gamma function defined by

$$\Gamma(\alpha, x) = \int_x^{\infty} e^{-z} z^{\alpha-1}\, dz. \tag{16}$$

Variable Number of Arrivals

Kabak (1970) considers a generalization of the above by studying the system $M^{(x)}/M/c$, where the number of arrivals at each arrival epoch is a random variable x with probability function $f(x)$, and the interarrival times and service times at each of the c channels are exponentially distributed with parameters λ and μ, respectively. From the recurrence relations satisfied by the steady-state probabilities P_i, it is shown that

(17) $$P_i = P_0 G(i), \qquad i = 0, 1, 2, \ldots, u,$$

(18) $$P_0 = \left[\sum_{i=0}^{u} G(i) \right]^{-1},$$

where

(19) $$G(i) = \frac{\lambda}{\mu(i)} \sum_{j=1}^{i} F(j) G(i-j), \qquad i = 1, 2, \ldots, u,$$

(20) $$F(j) = \sum_{x=j}^{\infty} f(x),$$

(21) $$\mu(i) = \begin{cases} i\mu, & 1 \leqq i \leqq c \\ c\mu, & i \geqq c, \end{cases}$$

and u is the upper limit on the allowable number of items in the system, so that

(22) $$u = \begin{cases} c & \text{for the loss system} \\ \infty & \text{for the delay system.} \end{cases}$$

Proceeding as in Kabak (1968), it is shown that

(23) $$P(d) = \sum_{i=0}^{d+c-1} P_i \sum_{x=c-i+d}^{\infty} \frac{f(x)}{x}, \qquad d = 1, 2, 3, \ldots,$$

(24) $$\text{Prob}\{\text{delay} > t\} = \sum_{d=1}^{\infty} \sum_{i=0}^{d+c-1} P_i \sum_{x=c-i+d}^{\infty} \frac{f(x)}{x} \int_t^{\infty} g(d)\, dt,$$

where

(25) $$g(d) = (c\mu)^d t^{d-1} \exp(-c\mu t)/\Gamma(d), \qquad t \geqq 0, \quad d = 1, 2, 3, \ldots,$$

so that

(26) $$\int_t^{\infty} g(d)\, dt = \frac{\Gamma(d, c\mu t)}{\Gamma(d)},$$

where $\Gamma(\alpha, x)$ is given by (16).

CHAPTER 3

Queues with Variable Parameters

In this chapter we deal with the studies in queuing theory in which special functions have been used and the interarrival and service times both follow the negative exponential distribution, the parameters of either or both of these distributions being variable. We begin with the work of Clarke (1953, 1956) and Dobbie (1961) on queues with time-dependent parameters, and proceed to the work on queues with state-dependent parameters by Ledermann and Reuter (1954), Saaty (1961b), Iglehart (1965), Reynolds (1968), and Hadidi (1969, 1974, 1976). As a special case of queues with state-dependent parameters, we include the work on queues with balking and reneging by Ancker and Gafarian (1963b,c).

3.1 Queues with Time-Dependent Parameters

Clarke (1953, 1956) considers a queuing system with Poisson arrivals and exponential service times, the parameters of the arrival and service distributions being, respectively, $\lambda(t)$ and $\mu(t)$, which are functions of the time t. Letting

$$(1) \qquad P_n(t) = P_{in}(t) = \text{Prob}\{N(t) = n \mid N(0) = i\} \qquad (n, i = 0, 1, 2, \ldots),$$

where

$$(2) \qquad N(t) = \text{number in the system at time } t,$$

we have the following forward set of Kolmogorov equations (Clarke, 1956):

$$(3) \qquad \frac{d}{dt}P_n(t) = -\{\lambda(t) + \mu(t)\}P_n(t) + \lambda(t)P_{n-1}(t) + \mu(t)P_{n+1}(t) \quad (n > 0)$$

$$(4) \qquad \frac{d}{dt}P_0(t) = -\lambda(t)P_0(t) + \mu(t)P_1(t).$$

Also, the initial conditions are

(5) $$P_{in}(0) = \delta_{in},$$

the Kronecker delta. On rescaling the time axis by introducing a new variable

(6) $$\tau = \int_0^t \mu(s)\,ds$$

in place of t, Eqs. (3)–(5) become

(7) $$\frac{d}{d\tau}P_n = -[\rho(\tau)+1]P_n + \rho(\tau)P_{n-1} + P_{n+1} \qquad (n>0),$$

(8) $$\frac{d}{d\tau}P_0 = -\rho(\tau)P_0 + P_1,$$

(9) $$P_{in}(0) = \delta_{in},$$

where P_n is now regarded as function of τ, and

(10) $$\rho(\tau) = \lambda(t)/\mu(t),$$

the *instantaneous relative traffic intensity* of the process. We assume that $\mu(t)$ has only discrete zero to make the rescaling (6) possible. Let us write

(11) $$R(\tau) = \int_0^t \lambda(s)\,ds \Big/ \left[\int_0^t \mu(s)\,ds\right] = \frac{1}{\tau}\int_0^\tau \rho(\sigma)\,d\sigma,$$

which might be termed the *smoothed relative traffic intensity* of the process.

Introducing a new system of dependent variables

(12) $$Q_n(\tau) = P_n \exp[\tau\{1 + R(\tau\}],$$

we can write Eqs. (7)–(9) as

(13) $$\frac{d}{d\tau}Q_n(\tau) = \rho(\tau)Q_{n-1}(\tau) + Q_{n+1}(\tau) \qquad (n>0),$$

(14) $$\frac{d}{d\tau}Q_0(\tau) = Q_0(\tau) + Q_1(\tau),$$

(15) $$Q_{in}(0) = \delta_{in}.$$

Introducing the generating function

(16) $$Q(z,\tau) = \sum_{n=0}^{\infty} \frac{Q_{in}(\tau)(z-\tau)^n}{n!},$$

we can write (13) in the form of the "telegrapher's equation"

$$\frac{\partial^2 Q}{\partial\tau\,\partial z} = \rho(\tau)Q, \tag{17}$$

with the two boundary conditions given by (14) and (15), which transform into

$$\left.\frac{\partial Q(z,\tau)}{\partial\tau}\right|_{z=\tau} = Q(\tau,\tau), \tag{18}$$

$$Q(z,0) = z^i/i!. \tag{19}$$

We shall first solve for $Q(z,\tau)$ in terms of the (unknown) function

$$f(\tau) = \frac{\partial Q(0,\tau)}{\partial\tau}, \tag{20}$$

using the classical Riemann method with boundary conditions (19) and (20) [see, for example, Courant and Hilbert (1950, p. 316)]. The condition (18) is then used to derive an integral equation for $f(\tau)$. The Riemann function associated with (17) is

$$I_0[2\{(R(\tau)\tau - R(\sigma)\sigma)(z-\zeta)\}^{1/2}], \tag{21}$$

where $I_n(x)$ is the modified Bessel function. Application of standard methods and integration formulas for Bessel functions [Watson (1944, p. 373)] gives the solution

$$Q(z,\tau) = A_i(0,\tau,z) + \int_0^\tau A_0(\sigma,\tau,z)f(\sigma)\,d\sigma, \tag{22}$$

where

$$A_n(\sigma,\tau,z) = z^{n/2}[R(\tau)\tau - R(\sigma)\sigma]^{-n/2}I_n[2\{(R(\tau)\tau - R(\sigma)\sigma)z\}^{1/2}] \tag{23}$$

$$(n = 0, \pm 1, \pm 2, \ldots).$$

From (22) and (23), we have

$$P_{in} = \exp[-\tau\{1+R(\tau)\}]\left.\frac{\partial^n Q(z,\tau)}{\partial z^n}\right|_{z=\tau} \tag{24}$$

$$= \exp[-\tau\{1+R(\tau)\}]\left\{A_{i-n}(0,\tau,\tau) + \int_0^\tau A_{-n}(\sigma,\tau,\tau)f(\sigma)\,d\sigma\right\}.$$

From (18) and (22), we can obtain the following Volterra-type integral equation for $f(\tau)$:

$$f(\tau) = B_i(0,\tau) + \int_0^\tau B_0(\sigma,\tau)f(\sigma)\,d\sigma, \tag{25}$$

where

$$B_n(\sigma, \tau) = A_n(\sigma, \tau, \tau) - \rho(\tau)A_{n+1}(\sigma, \tau, \tau). \tag{26}$$

In the special case of *constant traffic* intensity, when we have $\rho(\tau) \equiv \rho$, and equivalently, $R(\tau) \equiv \rho$, we can obtain a power series solution of (25) in the form

$$f(\tau) = \frac{\tau^i}{i!} + (1 - \rho) \sum_{k=i+1}^{\infty} \frac{\tau^k}{k!}, \tag{27}$$

and it can be shown that

$$P_{in} = e^{-\tau(1+\rho)}\Bigg[A_{i-n}(0, \tau, \tau) + \rho^n A_{n+i+1}(0, \tau, \tau) + (1 - \rho)\rho^n \sum_{k=n+i+2}^{\infty} A_k(0, \tau, \tau)\Bigg]. \tag{28}$$

When $\rho < 1$, by proceeding to limits as $\tau \to \infty$ in (28), we have the well-known equilibrium distribution

$$P_n = (1 - \rho)\rho^n. \tag{29}$$

For $\rho \geqq 1$, no such limiting distribution exists and we study the mean $M(\tau)$ and variance $\sigma^2(\tau)$ of the distribution. Using (7) and (8), these are given by

$$\frac{d}{d\tau} M(\tau) = \rho(\tau) - 1 + P_0, \tag{30}$$

$$\frac{d}{d\tau} \sigma^2(\tau) = 2[\rho(\tau) + \{\rho(\tau) - 1\}M(\tau)] - [1 + 2M(\tau)]\frac{dM(\tau)}{d\tau}, \tag{31}$$

whence

$$M(\tau) = [R(\tau) - 1]\tau + \int_0^\tau P_0(\sigma)\, d\sigma + i, \tag{32}$$

$$\sigma^2(\tau) = 2\int_0^\tau [\rho(\sigma) + \{\rho(\sigma) - 1\}M(\sigma)]\, d\sigma - M(\tau) - [M(\tau)]^2 + i + i^2, \tag{33}$$

where, in writing Eqs. (30)–(33), we temporarily drop the restriction $\rho(\tau) \equiv$ constant. If $\rho(\tau)$ is bounded for all $\tau \geqq 0$, then we see from (32) and (33) that both $M(\tau)$ and $\sigma(\tau) = O(\tau)$ as $\tau \to \infty$.

In the case when ρ is constant and greater than 1, from (32) and (28), on using standard integration formulas for Bessel functions [Watson (1944, p. 386)], we have

$$M(\tau) = (\rho - 1)\tau + \frac{1}{\rho^i(\rho - 1)} + i + o(1). \tag{34}$$

From (33) and (34), we have

$$\sigma(\tau) = \sqrt{\tau(\rho + 1)} + o(\sqrt{\tau}). \tag{35}$$

If we assume $\rho = 1$, (28) gives

$$P_0(\tau) = e^{-2\tau}[I_i(2\tau) + I_{i+1}(2\tau)] = \frac{1}{\sqrt{\pi\tau}}\left[1 + O\left(\frac{1}{\tau}\right)\right], \tag{36}$$

on using an asymptotic property of $I_n(z)$; therefore,

$$M(\tau) = \int_0^\tau P_0(\sigma)\,d\sigma + i = 2\sqrt{\frac{\tau}{\pi}} + O(1), \tag{37}$$

$$\sigma(\tau) = \sqrt{2\tau(1 - 2/\pi)} + O(1). \tag{38}$$

3.2 The Double-Ended Queue with Time-Dependent Rates

Dobbie (1961) considers the double-ended queuing problem described in Section 1.3 with the further generalization that the mean arrival rates of taxis and passengers are, respectively, $\alpha(t)$ and $\beta(t)$, functions of the time t. Let

$$P_n(t) = \begin{cases} \text{Prob}\{n \text{ taxis waiting at time } t\} & (n \geqq 0) \\ \text{Prob}\{-n \text{ passengers waiting at time } t\} & (n < 0) \end{cases} \tag{1}$$

be defined for $n = 0, \pm 1, \pm 2, \ldots$ [cf. 1.3(13)]. Then, we have

$$\frac{d}{dt} P_n(t) = [P_{n-1}(t) - P_n(t)]\alpha(t) - [P_n(t) - P_{n+1}(t)]\beta(t) \tag{2}$$

$$(n = 0, \pm 1, \pm 2, \ldots).$$

Introducing the Laurent generating function

$$F(z, t) = \sum_{n=-\infty}^{\infty} P_n(t) z^n, \tag{3}$$

we can write (2) in the form

$$\frac{\partial F(z, t)}{\partial t} = [(z - 1)\alpha(t) + (z^{-1} - 1)\beta(t)]F(z, t). \tag{4}$$

Let i be the initial state of the system, so that we have the boundary condition

$$F(z, 0) = z^i. \tag{5}$$

The solution of (4) with the boundary condition (5) is

(6) $$F(z,t) = z^i \exp[(z-1)A(t) + (z^{-1}-1)B(t)]$$
$$= z^i \exp[zA(t) + z^{-1}B(t)]\exp[-A(t) - B(t)],$$

where

(7) $$A(t) = \int_0^t \alpha(\tau)\,d\tau,$$

(8) $$B(t) = \int_0^t \beta(\tau)\,d\tau.$$

Now, $P_n(t)$ is the coefficient of z^n in $F(z,t)$; so we require the coefficient of z^{n-i} in the expansion of

(9) $$\exp(zA + z^{-1}B) = \sum_{j=0}^{\infty}\left(\frac{z^j A^j}{j!}\right)\sum_{k=0}^{\infty}\left(\frac{z^{-k}B^k}{k!}\right).$$

If $n - i \geqq 0$, the coefficient of z^{n-i} is

(10) $$\sum_{k=0}^{\infty}\frac{A^{k+n-i}B^k}{(k+n-i)!k!} = \left(\frac{A}{B}\right)^{(n-i)/2} I_{n-i}(2\sqrt{AB}).$$

If $n - i \leqq 0$, then the coefficient is

(11) $$\sum_{j=0}^{\infty}\frac{A^j B^{j+i-n}}{j!(j+i-n)!} = \left(\frac{A}{B}\right)^{(n-i)/2} I_{i-n}(2\sqrt{AB}).$$

Since we have

(12) $$I_{-\nu}(x) = I_\nu(x) \qquad \text{for integral } \nu,$$

we can write, from (10) and (11),

(13) $$P_n(t) = [A(t)/B(t)]^{(n-i)/2} I_{n-i}[2\sqrt{A(t)B(t)}]$$
$$\cdot \exp[-A(t) - B(t)] \qquad (n = 0, \pm 1, \pm 2, \ldots).$$

3.3 Spectral Theory

In their famous and oft-quoted paper, Ledermann and Reuter (1954) use spectral theory to find the time-dependent solution of the birth-and-death process, or more generally, of the queuing process with state-dependent parameters. Let

(1) $$P_{ij}(t) = \text{Prob}\{\text{system state at time } t \text{ is } j \mid \text{state was } i \text{ at } t = 0\}.$$

Then the process is represented by the following infinite set of "forward" differential equations:

$$\frac{d}{dt}P_{ij}(t) = \lambda_{j-1}P_{i,j-1}(t) - (\lambda_j + \mu_j)P_{ij}(t) + \mu_{j+1}P_{i,j+1}(t) \qquad (j = 1, 2, \ldots), \tag{2}$$

$$\frac{d}{dt}P_{i0}(t) = -\lambda_0 P_{i0}(t) + \mu_1 P_{i1}(t), \tag{3}$$

with the initial conditions

$$P_{ij}(0) = \delta_{ij}. \tag{4}$$

In matrix notation we can write Eqs. (2)–(4) as

$$\frac{d}{dt}P(t) = P(t)A, \tag{5}$$

$$P(0) = I, \tag{6}$$

where

$$P(t) = [P_{ij}(t)] \qquad (i, j = 0, 1, 2, \ldots), \tag{7}$$

I is the infinite identity matrix, and A is the infinite matrix given by

$$A = \begin{bmatrix} -\lambda_0 & \lambda_0 & 0 & 0 & \cdots & & \\ \mu_1 & -(\lambda_1+\mu_1) & \lambda_1 & 0 & \cdots & & \\ 0 & \mu_2 & -(\lambda_2+\mu_2) & \lambda_2 & \cdots & & \\ \vdots & \vdots & \vdots & \ddots & & & \\ & & & & \mu_n & -(\lambda_n+\mu_n) & \lambda_n \\ & & & & & & \ddots \end{bmatrix}. \tag{8}$$

Let us define as the nth section of A the $(n + 1) \times (n + 1)$ matrix

$$A^{(n)} = \begin{bmatrix} -\lambda_0 & \lambda_0 & 0 & 0 & 0 & \cdots & \\ \mu_1 & -(\lambda_1+\mu_1) & \lambda_1 & 0 & 0 & \cdots & \\ 0 & \mu_2 & -(\lambda_2+\mu_2) & \lambda_2 & 0 & \cdots & \\ \vdots & \vdots & \vdots & \vdots & \vdots & & \\ & & & & \mu_{n-1} & -(\lambda_{n-1}+\mu_{n-1}) & \lambda_{n-1} \\ & & & & 0 & \mu_n & -(\lambda_n+\mu_n) \end{bmatrix}, \tag{9}$$

derived from A by deleting all elements except those in the first $n + 1$ rows and columns. The system of differential equations

$$\frac{d}{dt}X(t) = X(t)A^{(n)} \qquad [X(0) = I^{(n)}] \tag{10}$$

has the unique solution

(11) $$X(t) = F^{(n)}(t) = \exp(tA^{(n)}).$$

In order to examine the limit, as $n \to \infty$, we write the matrix $F^{(n)}(t)$ in its spectral form; thus,

(12) $$F^{(n)}(t) = \sum_{r=0}^{n} \exp(t\alpha_r^{(n)}) D_r^{(n)},$$

where $\alpha_r^{(n)}$ $(r = 0, 1, \ldots, n)$ are the eigenvalues of $A^{(n)}$ and $D_r^{(n)}$ are the corresponding idempotent matrices. Then, if

(13) $$\lim_{n \to \infty} F^{(n)}(t) = F(t),$$

the infinite matrix $F(t)$ satisfies (5) and the initial conditions. By definition $\alpha_r^{(n)}$ are the roots of the equation

(14) $$\phi_n(\xi) = |\xi I^{(n)} - A^{(n)}| = 0,$$

where $|A|$ stands for the determinant of A; so we can write

(15) $$\phi_n(\xi) = \prod_{r=0}^{n} (\xi - \alpha_r^{(n)}).$$

It is easy to see that

(16) $$\phi_n(0) = \lambda_0 \lambda_1 \cdots \lambda_n \qquad (n = 1, 2, \ldots).$$

When $\lambda_0 = 0$, we have $\phi_n(0) = 0$; therefore, in this case we can write

(17) $$\phi_n(\xi) = \xi \bar{\phi}_n(\xi).$$

It is shown that the eigenvalues $\alpha_r^{(n)}$ of $A^{(n)}$ are distinct and negative or zero, say

(18) $$0 \geqq \alpha_0^{(n)} > \alpha_1^{(n)} > \cdots > \alpha_n^{(n)},$$

with

(19) $$\alpha_0^{(n)} = 0 \quad \text{if and only if} \quad \lambda_0 = 0.$$

Let

(20) $$x_r^{(n)}, \; y_r^{(n)\prime} \qquad (r = 0, 1, \ldots, n),$$

be, respectively, the left and right eigenvectors corresponding to the $\alpha_r^{(n)}$. Define

(21) $$\theta_r^{(n)} = x_r^{(n)} y_r^{(n)\prime} \qquad (r = 0, 1, \ldots, n).$$

Then we have

(22) $$D_r^{(n)} = y_r^{(n)\prime} x_r^{(n)} / \theta_r^{(n)} = (d_r^{(n)}(i, j)),$$

and, from (12) and (22), the (i, j)th element of $F^{(n)}(t)$ is

$$f_{ij}^{(n)}(t) = \sum_{r=0}^{n} \exp(t\alpha_r^{(n)})\, d_r^{(n)}(i, j) \qquad (i, j = 0, 1, \ldots, n). \tag{23}$$

For the section $A^{(n)}$, we define a *spectral function* $\rho^{(n)}(x)$ as follows: $\rho^{(n)}(x)$ is to be a nondecreasing step function, defined for $-\infty < x < \infty$, with discontinuities $1/\theta_r^{(n)}$ at

$$x = \alpha_r^{(n)} \qquad (r = 0, 1, \ldots, n) \tag{24}$$

and

$$\rho^{(n)}(0) = 0. \tag{25}$$

Explicitly, $\rho^{(n)}(x)$ is given by

$$\rho^{(n)}(x) = \begin{cases} 0, & x \geqq \alpha_0^{(n)} \\ -\left(\dfrac{1}{\theta_0^{(n)}} + \dfrac{1}{\theta_1^{(n)}} + \cdots + \dfrac{1}{\theta_r^{(n)}}\right), & \alpha_{r+1}^{(n)} \leqq x < \alpha_r^{(n)} \ (r = 0, \ldots, n-1) \\ -\left(\dfrac{1}{\theta_0^{(n)}} + \cdots + \dfrac{1}{\theta_n^{(n)}}\right), & x < \alpha_n^{(n)}. \end{cases} \tag{26}$$

Now, it can be shown that

$$d_r^{(n)}(i, j) = l_{i-1} m_j \phi_{j-1}(\alpha_r^{(n)}) \phi_{j-1}(\alpha_r^{(n)}) / \theta_r^{(n)}, \tag{27}$$

where

$$l_{-1} = \lambda_0, \qquad l_0 = 1, \qquad l_\nu = (\lambda_1 \lambda_2 \cdots \lambda_\nu)^{-1} \qquad (\nu \geqq 1), \tag{28}$$

$$m_0 = 1, \qquad m_\nu = (\mu_1 \mu_2 \cdots \mu_\nu)^{-1} \qquad (\nu \geqq 1). \tag{29}$$

So, from (23), (26), and (27), we have

$$f_{ij}^{(n)}(t) = l_{i-1} m_j \int \phi_{i-1}(x) \phi_{j-1}(x) e^{tx}\, d\rho^{(n)}(x) \qquad (0 \leqq i, j \leqq n), \tag{30}$$

where the integral is taken from $-\infty$ to ∞, i.e., effectively from $-\infty$ to 0, $\rho^{(n)}(x)$ being constant for $x \geqq 0$.

By a limiting procedure, which is justified, we have

$$f_{ij}(t) = l_{i-1} m_j \int \phi_{i-1}(x) \phi_{j-1}(x) e^{tx}\, d\rho(x) \qquad (i, j \geqq 0) \tag{31}$$

and, for $t = 0$,

$$f_{ij}(0) = \delta_{ij} = l_{i-1} m_j \int \phi_{i-1}(x) \phi_{j-1}(x)\, d\rho(x). \tag{32}$$

It can be shown that

(33) $$\theta_0^{(n)} = 1 \qquad \text{if} \quad \lambda_0 = 0,$$

and that

(34) $$\theta_r^{(n)} = \sum_{v=0}^{n} l_{v-1} m_v \{\phi_{v-1}(\alpha_r^{(n)})\}^2$$

in all other cases, i.e., for $0 \leqq r \leqq n$ if $\lambda_0 > 0$ and for $1 \leqq r \leqq n$ if $\lambda_0 = 0$.

Case $\lambda_0 = 0$ In this case we use (17), and define a function $\bar{\rho}^{(n)}(x)$ to have discontinuities $1/\bar{\theta}_r^{(n)}$ at $x = \alpha_r^{(n)}$, $r = 1, 2, \ldots, n$ ($r = 0$ is excluded), and to have $\bar{\rho}^{(n)}(0) = 0$, where

(35) $$\bar{\theta}_r^{(n)} = \theta_r^{(n)}/|\alpha_r^{(n)}| \qquad (r = 1, \ldots, n).$$

Then, by proceeding to limits as before, we have

(36) $$f_{ij}(t) = -\, l_{i-1} m_j \int x \bar{\phi}_{i-1}(x) \bar{\phi}_{j-1}(x) e^{tx}\, d\bar{\rho}(x) \qquad (i \geqq 1, j \geqq 1),$$

(37) $$f_{0j}(t) = \delta_{0j} \qquad (j \geqq 0),$$

(38) $$f_{i0}(t) = \zeta_i - l_{i-1} \int \bar{\phi}_{i-1}(x) e^{tx}\, d\bar{\rho}(x) \qquad (i \geqq 1),$$

where

(39) $$\zeta_0 = 1, \qquad \zeta_i = 1 - \left(\sum_{v=0}^{i-1} w_v^{-1}\right) \Big/ \left(\sum_{v=0}^{\infty} w_v^{-1}\right) \qquad (i \geqq 1),$$

(40) $$w_0 = 1, \qquad w_n = \frac{\lambda_1 \lambda_2 \cdots \lambda_n}{\mu_1 \mu_2 \cdots \mu_n} = \frac{m_n}{l_n} \qquad (n \geqq 1).$$

Examples

We take three particular cases of the above general model, which have applications in queuing theory.

Case I—Constant Coefficients ($\lambda_0 = 0$) Let

(41) $$\lambda_0 = 0, \qquad \lambda_n = \lambda > 0, \qquad \mu_n = \mu > 0 \qquad (n \geqq 1).$$

Then, we have

(42) $$\bar{\phi}_0(x) = 1,$$

(43) $$\bar{\phi}_1(x) = x + \lambda + \mu,$$

(44) $$\bar{\phi}_n(x) - (x + \lambda + \mu)\bar{\phi}_{n-1}(x) + \lambda\mu\bar{\phi}_{n-2}(x) = 0 \qquad (n \geqq 2).$$

If we set

$$x = -(\lambda + \mu) + 2\sqrt{\lambda\mu}\cos\omega, \tag{45}$$

$$\bar{\phi}_n(x) = (\sqrt{\lambda\mu})^n \Phi_n(\omega), \tag{46}$$

we find that

$$\Phi_n(\omega) = \frac{\sin(n+1)\omega}{\sin\omega}. \tag{47}$$

Hence $\Phi_n(\omega)$ vanishes at

$$\omega = \omega_r^{(n)} \qquad (r = 1, \ldots, n), \tag{48}$$

where

$$\omega_r^{(n)} = r\pi/(n+1), \tag{49}$$

so that

$$\alpha_r^{(n)} = -(\lambda + \mu) + 2\sqrt{\lambda\mu}\cos\omega_r^{(n)}. \tag{50}$$

Therefore

$$\bar{\theta}_r^{(n)} = -\frac{\alpha_r^{(n)}}{\mu}\sum_{v=1}^{n}\left[\frac{\sin v\omega_r^{(n)}}{\sin\omega_r^{(n)}}\right]^2, \tag{51}$$

and, on evaluating the trigonometric sum involved,

$$1/\bar{\theta}_r^{(n)} = -\frac{2\mu}{n+1}\frac{\sin^2\omega_r^{(n)}}{\alpha_r^{(n)}} = -\frac{2\mu}{\pi}\frac{\sin^2\omega_r^{(n)}}{\alpha_r^{(n)}}\Delta\omega, \tag{52}$$

where we write

$$\Delta\omega = \frac{\pi}{n+1} \quad (= \omega_{r+1}^{(n)} - \omega_r^{(n)}), \tag{53}$$

and by a limiting process

$$d\bar{\rho}(x) = -\frac{2\mu}{\pi}\frac{\sin^2\omega}{x}\,d\omega. \tag{54}$$

Hence

$$f_{ij}(t) = \frac{2}{\pi}(\mu/\lambda)^{(i-j)/2}\exp[-(\lambda+\mu)t] \cdot \int_0^\pi \sin i\omega \sin j\omega \exp(2t\sqrt{\lambda\mu}\cos\omega)\,d\omega \qquad (i \geqq 1, j \geqq 1), \tag{55}$$

$$f_{0j}(t) = \delta_{0j} \qquad (j \geqq 0), \tag{56}$$

$$(57)\quad f_{i0}(t) = \zeta_i - \frac{2}{\pi}(\mu/\lambda)^{(i-1)/2}\exp[-(\lambda+\mu)t] \cdot \int_0^\pi \sin i\omega \exp(2t\sqrt{\lambda\mu}\cos\omega)\frac{\mu\sin\omega\, d\omega}{\lambda+\mu-2\sqrt{\lambda\mu}\cos\omega} \qquad (i \geqq 1),$$

with

$$(58)\qquad \zeta_i = \begin{cases} 1 & \text{if} \quad \lambda \leqq \mu \\ (\mu/\lambda)^i & \text{if} \quad \lambda > \mu. \end{cases}$$

The integrals occurring above can be expressed in terms of the modified Bessel functions $I_m(z)$, given by

$$(59)\qquad I_m(z) = \frac{1}{\pi}\int_0^\pi \cos m\omega \exp(z\cos\omega)\, d\omega.$$

Therefore, for $i \geqq 1, j \geqq 1$, we have

$$(60)\quad f_{ij}(t) = (\mu/\lambda)^{(i-j)/2}\exp[-(\lambda+\mu)t]\{I_{i-j}(2\sqrt{\lambda\mu}\, t) - I_{i+j}(2\sqrt{\lambda\mu}\, t)\}.$$

For $j = 1$, this simplifies to

$$(61)\qquad f_{i1}(t) = (i/\mu)(\mu/\lambda)^{i/2}\exp[-(\lambda+\mu)t]\, t^{-1} I_i(2\sqrt{\lambda\mu}\, t),$$

and since

$$(62)\qquad f'_{i0}(t) = \mu f_{i1}(t), \qquad f_{i0}(0) = 0 \qquad (i \geqq 1),$$

it follows that

$$(63)\qquad f_{i0}(t) = i\left(\frac{\mu}{\lambda}\right)^{i/2}\int_0^t \exp[-(\lambda+\mu)\tau]\, \tau^{-1} I_i(2\sqrt{\lambda\mu}\, \tau)\, d\tau.$$

It can be easily seen that (63) gives the distribution function of the duration of the busy period initiated by i customers in the $M/M/1$ system (Takács, 1962, p. 38). The corresponding result for the case $i = 1$ was obtained by Kendall (1951). It is obvious that (63) holds for all positive λ, μ and not only for $\lambda \leqq \mu$ as implied in Kendall (1951). Confirming Kendall's conjecture, we see that when $\lambda > \mu$, we have an infinitely long busy period with a positive probability equal to

$$(64)\qquad 1 - f_{10}(\infty) = 1 - \zeta_1 = 1 - (\mu/\lambda).$$[1]

Case II — Constant Coefficients ($\lambda_0 > 0$) Suppose now that

$$(65)\qquad \lambda_n = \lambda > 0 \quad (n \geqq 0); \qquad \mu_n = \mu > 0 \quad (n \geqq 1).$$

[1] There is an obvious misprint in Ledermann and Reuter (1954).

Then we have

$$\phi_{-1}(x) = 1, \tag{66}$$

$$\phi_0(x) = x + \lambda, \tag{67}$$

$$\phi_n(x) - (x + \lambda + \mu)\phi_{n-1}(x) + \lambda\mu\phi_{n-2}(x) = 0 \qquad (n \geqq 2). \tag{68}$$

As before, we have

$$\begin{aligned}\phi_n(x) &= (\sqrt{\lambda\mu})^n F_n(\omega) \\ &= \frac{(\lambda\mu)^{n/2}}{\sin\omega}[\sqrt{\lambda\mu}\sin(n+2)\omega - \mu\sin(n+1)\omega],\end{aligned} \tag{69}$$

where x is given by (45). Using analysis similar to the above, it is shown that

$$\begin{aligned}f'_{ij}(t) = (\mu/\lambda)^{(i-j)/2}\exp[-(\lambda+\mu)t]\{&-(\lambda+\mu)I_{i-j} + \sqrt{\lambda\mu}\,I_{i-j-1} \\ &+ \sqrt{\lambda\mu}\,I_{i+j+1} + \lambda I_{i+j+2} - 2\sqrt{\lambda\mu}\,I_{i+j+1} + \mu I_{i+j}\},\end{aligned} \tag{70}$$

where $I_m \equiv I_m(2\sqrt{\lambda\mu}\,t)$. Expressions for $f_{ij}(t)$ can be obtained by integration, using $f_{ij}(0) = \delta_{ij}$. Equation (70) gives the time-dependent solution of $M/M/1$. In case $\lambda = \mu$, the expression for $f_{ij}(t)$ simplifies to

$$f_{ij}(t) = e^{-2\lambda t}[I_{i-j}(2\lambda t) + I_{i+j+1}(2\lambda t)]. \tag{71}$$

Case III – Linear Coefficients Now let

$$\lambda_n = n\lambda, \qquad \mu_n = n\mu \qquad (n \geqq 0), \quad \lambda, \mu > 0. \tag{72}$$

The above analysis can be completed only in the case $\lambda = \mu$. In this case, we have

$$\bar{\phi}_0(x) = 1, \tag{73}$$

$$\bar{\phi}_1(x) = x + 2\lambda, \tag{74}$$

$$\bar{\phi}_n(x) - (x + 2n\lambda)\bar{\phi}_{n-1}(x) + \lambda^2 n(n-1)\bar{\phi}_{n-2}(x) = 0, \tag{75}$$

from which it is easy to deduce that

$$\bar{\phi}_n(x) = \lambda^n n!\,L_n^{(1)}(-x/\lambda), \tag{76}$$

where $L_n^{(\alpha)}(z)$ is the Laguerre polynomial, given by

$$L_n^{(\alpha)}(z) = \frac{e^z z^{-\alpha}}{n!}\frac{d^n}{dz^n}(e^{-z}z^{n+\alpha}). \tag{77}$$

The solution in the general case, given by (72), is quoted from Kendall (1949) in terms of ξ_t, η_t defined by

$$\frac{\xi_t}{\mu} = \frac{\eta_t}{\lambda} = \frac{e^{(\lambda-\mu)t} - 1}{\lambda e^{(\lambda-\mu)t} - \mu} \qquad (\lambda \neq \mu), \tag{78}$$

$$\xi_t = \eta_t = \lambda t/(1 + \lambda t) \qquad (\lambda = \mu). \tag{79}$$

Kendall shows that $f_{ij}(t)$ (for $i \geqq 1$) is the coefficient of z^j in the generating function

(80) $$[\xi_t + (1 - \xi_t)(1 - \eta_t)z/(1 - \eta_t z)]^i.$$

3.4 Stochastic Processes with Absorbing Barriers

Saaty (1961b) studies Cases I and III discussed in Section 3.3 with the modification that E_0 and E_N, the states of the system corresponding to system sizes 0 and N, respectively, are absorbing states. Thus, he has

(1) $$\lambda_0 = \mu_0 = 0,$$

(2) $$\lambda_n = \mu_n = 0 \qquad \text{for} \quad n \geqq N.$$

Also, two alternative assumptions are made on λ_n and μ_n for $0 < n < N$:

For case (1):

(3) $$\lambda_n = n\lambda, \qquad \mu_n = n\mu.$$

For case (2):

(4) $$\lambda_n = \lambda, \qquad \mu_n = \mu.$$

Case (1), which has been dealt with in greater detail, has applications in epidemiology and is used to study the growth of bacteria from an initial dosage to an infectious size at which symptoms of the disease will appear.

Case (2) serves to determine the distribution of busy period in $M/M/1$ and has also been discussed by Heathcote and Moyal (1959) [see Section 1.1].

The assumptions cited in (1)–(4) are used in analyzing the general birth–death model given by the following "forward" equations, in the notation of Section 3.3:

(5) $$\frac{d}{dt}P_{in}(t) = -(\lambda_n + \mu_n)P_{in}(t) + \lambda_{n-1}P_{i,n-1}(t) + \mu_{n+1}P_{i,n+1}(t) \qquad (n > 0, \quad i = 0, 1, 2, \ldots),$$

(6) $$\frac{d}{dt}P_{i0}(t) = -\lambda_0 P_{i0}(t) + \mu_1 P_{i1}(t) \qquad (i = 0, 1, 2, \ldots),$$

with initial conditions

(7) $$P_{in}(0) = \delta_{in}.$$

The associated "backward" equations are

$$\frac{d}{dt}P_{in}(t) = -(\lambda_i + \mu_i)P_{in}(t) + \lambda_i P_{i+1,n}(t) + \mu_i P_{i-1,n}(t) \qquad (n \geqq 0, \quad i > 0), \tag{8}$$

$$\frac{d}{dt}P_{0n} = -\lambda_0 P_{0n}(t) + \lambda_0 P_{1n}(t) \qquad (n \geqq 0). \tag{9}$$

Considering Case (1), let $Q_n(t)$ be the probability density of the time till absorption at E_N starting from E_n; then

$$Q_n(t) = \frac{d}{dt}P_{nN}(t). \tag{10}$$

From (8) and (9), setting $i = n$ and $n = N$, we have

$$\frac{d}{dt}Q_n(t) = -n(\lambda + \mu)Q_n(t) + n\mu Q_{n-1}(t) + n\lambda Q_{n+1}(t), \tag{11}$$

$$Q_0(t) = Q_N(t) = 0. \tag{12}$$

Similarly, from (5),

$$\frac{d}{dt}P_{nN}(t) = \lambda(N-1)P_{n,N-1}(t), \tag{13}$$

whence, on using (7), we have, for initial conditions,

$$Q_n(0) = \lambda(N-1)\delta_{n,N-1}. \tag{14}$$

Using the generating function

$$Q(z,t) = \sum_{n=1}^{N-1} Q_n(t)z^{n-1}, \tag{15}$$

so that, from initial conditions,

$$Q(z,0) = \lambda(N-1)z^{N-2}, \tag{16}$$

we have, from (11), the Lagrange equation

$$-\frac{\partial Q}{\partial t} + [\mu z^2 - (\lambda + \mu)z + \lambda]\frac{\partial Q}{\partial z} = (\lambda + \mu - 2\mu z)Q + \mu Q_{N-1}(t)z^{N-1}, \tag{17}$$

on solving which, and then writing the coefficient of z^{n-1} with the help of

Leibniz's theorem, as in Section 1.2, we have

$$(18)\quad Q_n(t) = -\frac{(-1)^{n-1}\mu\alpha^2}{(n-1)!}$$

$$\cdot\int_0^t \Bigg[Q_{N-1}(\tau)e^{-\alpha(t-\tau)}\frac{\{\lambda(1-e^{-\alpha(t-\tau)})\}^{N-n}}{\{\lambda-\mu e^{-\alpha(t-\tau)}\}^{N+1}}[\mu-\lambda e^{-\alpha(t-\tau)}]^{n-1}$$

$$\cdot\sum_{i=0}^{n-1}\binom{n-1}{i}\frac{\Gamma(N+i+1)}{\Gamma(N-n+i+1)}$$

$$\cdot\left\{\frac{-\beta\lambda^2(1-e^{-\alpha(t-\tau)})^2}{(\mu-\lambda e^{-\alpha(t-\tau)})(\lambda-\mu e^{-\alpha(t-\tau)})}\right\}^i\Bigg]d\tau$$

$$+\frac{(-1)^{n-1}\lambda\alpha^2}{(n-1)!}e^{-\alpha t}\frac{\{\lambda(1-e^{-\alpha t})\}^{N-n-1}}{\{\lambda-\mu e^{-\alpha t}\}^N}\{\mu-\lambda e^{-\alpha t}\}^{n-1}$$

$$\cdot\sum_{i=0}^{n-1}\binom{n-1}{i}\frac{\Gamma(N+i)}{\Gamma(N-n+i)}\left\{\frac{-\beta\lambda^2(1-e^{-\alpha t})}{(\mu-\lambda e^{-\alpha t})(\lambda-\mu e^{-\alpha t})}\right\}^i,$$

where

$$(19)\qquad \alpha=\lambda-\mu, \qquad \beta=\mu/\lambda.$$

For $n = N - 1$ we have a Volterra-type integral equation in $Q_{N-1}(t)$. Alternatively, we can take Laplace transforms, for which we note that the integral occurring in (18) is a convolution. [See, for example, Srivastava and Buschman (1977).] Also, the second term on the right-hand side of (18) can be simplified to

$$(20)\quad \frac{\alpha^2e^{-\alpha t}}{\mu(n-1)!}(-1)^{n-1}\sum_{i=0}^{n-1}\binom{n-1}{i}\frac{\Gamma(N+i)}{\Gamma(N-n+i)}(-1)^i\beta^n$$

$$\cdot\sum_{j=0}^{n-i-1}\binom{n-i-1}{j}(-1/\beta)^je^{-j\alpha t}(1-e^{-\alpha t})^{N-n+2i-1}(1-\beta e^{-\alpha t})^{-N-i},$$

and its Laplace transform (L.T.) can be written, on using Erdélyi *et al.* (1954, Vol. I, p. 145, Eq. (14)),

$$(21)\quad \int_0^\infty e^{-st}(1-e^{-t})^{v-1}(1-ze^{-t})^{-\mu}\,dt = B(s,v)\,{}_2F_1(\mu,s;s+v;z),$$

$$\mathrm{Re}(v)>0, \qquad |\arg(1-z)|<\pi, \qquad \mathrm{Re}(s)>0,$$

where $B(\alpha, \beta)$ is the beta function. The L.T. of the first term on the right-hand side of (18) can be written by using the convolution theorem of Laplace transforms and the relation (21). The analysis becomes too complicated and is not pursued further.

Explicit Solution to Case (1) for Large N

$Q_n(t)$ can be determined easily for large N without the use of Laplace transforms. By direct integration of the forward equations (11), it can be shown that

(22) $$\int_0^\infty Q_n(t)\,dt = \frac{1-\beta^n}{1-\beta^N}.$$

Consequently, for $\mu < \lambda$ (which is obviously the case of interest) and with $n = N - 1$, we have

(23) $$\lim_{N\to\infty} \int_0^\infty Q_{N-1}(t) = 1.$$

Also, from initial conditions,

(24) $$\lim_{N\to\infty} Q_{N-1}(0) \to \infty.$$

Hence, $Q_{N-1}(t)$ behaves like a Dirac delta function, and so we have

(25) $$\int_0^t Q_{N-1}(\tau) f(t-\tau)\,d\tau = f(t).$$

Using (25) in (18), we have

(26) $$\frac{Q_n(t)}{\lambda} = (-1)^{n-1}\frac{\alpha^2}{(n-1)!}e^{-\alpha t}\frac{A^{N-n-1}}{C^{N+1}}B^{n-1}\lambda \cdot \sum_{i=0}^{n-1}\binom{n-1}{i}\left(-\frac{\beta A^2}{BC}\right)^i \frac{(N+i-1)!}{(N+i-n)!} \cdot \{(N+i)(1-\beta) - n(1-\beta e^{-\alpha t})\},$$

where

(27) $$A = \lambda(1-e^{-\alpha t}),$$
$$B = \mu - \lambda e^{-\alpha t},$$
$$C = \lambda - \mu e^{-\alpha t}.$$

On further simplification, we come across terms of the form

(28) $$N^p e^{-p\alpha t} \qquad (p > 0 \text{ an integer}),$$

and using the following relation that can be proved from $Q_1(t)$,

(29) $$\alpha t \sim \log N,$$

(26) finally yields

$$\frac{Q_n(t)}{\lambda} = (1-\beta)^3 \frac{(1-e^{-\alpha t})^{N-n-1}}{(1-\beta e^{-\alpha t})^{N+n}} \cdot \sum_{k=0}^{n-1} \binom{n}{k+1} \beta^{n-k-1}(1-\beta)^{2k} \frac{(Ne^{-\alpha t})^{k+1}}{k!} = (1-\beta)\beta^n \exp\left(\frac{-\beta x}{1-x}\right) \frac{x}{n!} \frac{dL_n(-x)}{dx}, \tag{30}$$

where

$$x = [(1-\beta)^2/\beta]Ne^{-\alpha t} \tag{31}$$

and $L_n(z)$ is the Laguerre polynomial given by

$$L_n(z) = L_n^{(0)}(z) = \frac{e^z}{n!} \frac{d^n}{dz^n}(z^n e^{-z}), \tag{32}$$

[cf. 3.3(77)].

Explicit Solution to Case (2) for Large N

In this case, using (4), we can write[2]

$$\frac{d}{dt} Q_n(t) = -(\lambda+\mu)Q_n(t) + \mu Q_{n-1}(t) + \lambda Q_{n+1}(t), \tag{33}$$

$$Q_0(t) = Q_N(t) = 0, \tag{34}$$

$$Q_n(0) = \lambda \delta_{n,N-1}. \tag{35}$$

Using the generating function given in (15), we have

$$z \frac{\partial Q}{\partial t} = \{\mu z^2 - (\lambda+\mu)z + \lambda\}Q(z,t) - \mu Q_{N-1} z^{N-1} - \lambda Q_1. \tag{36}$$

Using Laplace transforms and Rouché's theorem as usual, we have, on inversion of transforms,

$$Q_n(t) = e^{-(\lambda+\mu)t} \left(\frac{\lambda}{\mu}\right)^{(N-n)/2} (\lambda\mu)^{1/2} \sum_{k=0}^{\infty} [I_{(2k+1)N-n} - I_{(2k+1)N-n+1} - I_{(2k+1)N+n-1} + I_{(2k+1)N+n+1}], \tag{37}$$

where I_ν stands for the modified Bessel function of the first kind with the suppressed argument to be understood as $2\sqrt{\lambda\mu}\, t$.

[2] There is an obvious misprint in Saaty (1961b, Eq. (6.1)).

3.5 Queuing with Balking and Reneging

Ancker and Gafarian (1963b) consider the $M/M/1$ queuing system with further assumptions about the balking and reneging behavior of the customers. The mean arrival and service rates are λ and μ, respectively, and the queue discipline is first come, first served. An arriving customer balks with probability n/N and, therefore, joins the system with probability

(1) $$e_n = 1 - n/N \qquad (n = 0, 1, 2, \ldots, N),$$

where n is the number in the system and N is the maximum number allowed in the system. After joining the queue, each customer waits for service a certain time, which is a random variable with density function

(2) $$d(t) = \alpha e^{-\alpha t},$$

and if service does not begin by then, he departs and is lost to the system. Denoting by $P_n(t)$ the transient probability of n in the system as usual, we have the following equations:

(3) $$P_0'(t) = -\lambda P_0(t) + \mu P_1(t),$$

(4) $$P_n'(t) = -[\lambda(1-n/N)+\mu+(n-1)\alpha]P_n(t)+\lambda[1-(n-1)/N]P_{n-1}(t)$$
$$+ (\mu + n\alpha)P_{n+1}(t) \qquad (n = 1, 2, \ldots, N-1),$$

(5) $$P_N'(t) = -[\mu + (N-1)\alpha]P_N(t) + (\lambda/N)P_{N-1}(t).$$

If P_n is the steady-state probability of n in the system and

(6) $$\gamma = \mu/\alpha,$$

(7) $$\delta = \lambda/(N\alpha),$$

the steady-state equations are

(8) $$0 = -N\delta P_0 + \gamma P_1,$$

(9) $$0 = -[(N-n)\delta + \gamma + n - 1]P_n + (N-n+1)\delta P_{n-1}$$
$$+ (\gamma + n)P_{n+1} \qquad (n = 1, 2, \ldots, N-1),$$

(10) $$0 = -[\gamma + N - 1]P_N + \delta P_{N-1}.$$

From Eqs. (8)–(10) we have

(11) $$P_{n+1} = [(N-n)\delta/(\gamma + n)]P_n \qquad (n = 0, 1, 2, \ldots, N-1)$$

and by repeated substitution

(12) $$P_n = \left(\delta^n \prod_{j=0}^{n-1} (N-j) \Big/ \prod_{k=0}^{n-1} (\gamma + k)\right) P_0$$
$$= [N!\Gamma(\gamma)\delta^n/\{(N-n)!\Gamma(\gamma+n)\}]P_0 \qquad (n = 1, 2, \ldots, N).$$

P_0 is computed from the normalizing condition

$$\sum_{n=0}^{N} P_n = 1, \tag{13}$$

which gives

$$P_0 = \left(1 + \sum_{n=1}^{N} \frac{N!\Gamma(\gamma)\delta^n}{(N-n)!\Gamma(\gamma+n)}\right)^{-1}. \tag{14}$$

In order to evaluate the finite sum occurring in (14), let us write

$$f(x) = \sum_{n=1}^{N} \frac{N!\Gamma(\gamma)x^n}{(N-n)!\Gamma(\gamma+n)} \qquad (x > 0). \tag{15}$$

Then, $f(x)$ satisfies the differential equation

$$f'(x) + [(\gamma - 1 - Nx)/\{x(1+x)\}]f(x) = N/(1+x). \tag{16}$$

The solution of (16) is

$$f(x) = [(1+x)^{\gamma-1+N}/x^{\gamma-1}]\{NB_y(\gamma, N) + C\}, \tag{17}$$

where C is an arbitrary constant,

$$B_y(\gamma, N) = \int_0^y \xi^{\gamma-1}(1-\xi)^{N-1}\, d\xi \tag{18}$$

is the incomplete beta function [cf. Section II.1], and

$$y = x/(1+x). \tag{19}$$

The arbitrary constant C is zero, since we have, from (15),

$$\lim_{x\to\infty} \{f(x)/x\} = N/\gamma. \tag{20}$$

Hence, from (17) and (19), we find that[3]

$$f(x) = NB_y(\gamma, N)/\{y^{\gamma-1}(1-y)^N\}. \tag{21}$$

[3] Incidentally, the sum in (14) and (15) can be written at once as a hypergeometric function, and we thus have the alternative (but essentially equivalent) expression for $f(x)$

$$f(x) = {}_2F_1(-N, 1; \gamma; -x) - 1, \tag{21*}$$

so that (cf. Section II.2)

$$\begin{aligned} P_0 &= \{{}_2F_1(-N, 1; \gamma; -\delta)\}^{-1} = (1-z)^N/\{{}_2F_1(-N, \gamma-1; \gamma; z)\} \\ &= z^{\gamma-1}(1-z)^N/\{(\gamma-1)B_z(\gamma-1, N+1)\}, \end{aligned} \tag{22*}$$

in view of the relation II.1(60), z being defined by (23).

Thus, (14) becomes

$$P_0 = [1 + NB_z(\gamma, N)/\{z^{\gamma-1}(1-z)^N\}]^{-1}, \tag{22}$$

where, for convenience,

$$z = \delta/(1+\delta). \tag{23}$$

The mean number in the queue is

$$\begin{aligned} L_q &= \sum_{n=1}^{N-1} nP_{n+1} = P_0\{xf'(x) - f(x)\}_{x=\delta} \\ &= [1/(1+\delta)][N\delta - (\delta+\gamma)(1-P_0)]. \end{aligned} \tag{24}$$

The mean number in the system is

$$L = \sum_{n=1}^{N} nP_n = L_q + 1 - P_0 = [1/(1+\delta)][N\delta - (\gamma - 1)(1 - P_0)]. \tag{25}$$

The probability that there are R or more in the system is

$$\begin{aligned} Q_R &= \sum_{n=R}^{N} P_n \\ &= P_0\left(\frac{N!\Gamma(\gamma)B_z(\gamma+R-1, N-R+1)}{(N-R)!\Gamma(\gamma+R-1)z^{\gamma-1}(1-z)^N}\right). \end{aligned} \tag{26}$$

Let us define the following events,

$$\begin{aligned} B &= \text{balking}, \\ W &= \text{waiting in queue by those who join}, \\ R &= \text{reneging}, \\ A &= \text{acquiring service}, \end{aligned} \tag{27}$$

and let $P(B)$ stand for "probability of balking," etc. Then

$$P(B) = \sum_{n=1}^{N} P_n(1 - e_n) = \frac{L}{N}, \tag{28}$$

$$\begin{aligned} P(W) &= \sum_{n=1}^{N-1} P_n e_n = 1 - P_0 - P(B) \\ &= P_0(N-1)B_z(\gamma, N-1)/[z^{\gamma-1}(1-z)^{N-1}].^4 \end{aligned} \tag{29}$$

It can be shown [see Ancker and Gafarian (1962)] that for reneging systems of this type, the probability that a new arrival will survive to be serviced, given n

[4] Ancker and Gafarian (1963b) have an obvious misprint in this result.

in the system on arrival and given that it joins, is

(30) $$\beta_n = \gamma/(\gamma + n).$$

Hence

(31) $$P(R) = \sum_{n=1}^{N-1} P_n(1 - \beta_n)e_n = 1 - \left(\frac{L}{N}\right) - \frac{\gamma(1 - P_0)}{N\delta},$$

(32) $$P(A) = \sum_{n=0}^{N-1} P_n\beta_n e_n = \frac{\gamma(1 - P_0)}{N\delta}.$$

Also,

(33) $$P(A, W) = \sum_{n=1}^{N-1} P_n\beta_n e_n = \frac{P_0(N - 1)B_z(\gamma + 1, N - 1)}{z^{\gamma}(1 - z)^{N-1}},$$

(34) $$P(A \mid W) = \frac{P(A, W)}{P(W)} = \frac{B_z(\gamma + 1, N - 1)}{zB_z(\gamma, N - 1)}.$$

Waiting Times

We consider those items that join the queue and have a positive waiting time, thus leaving aside items that balk or those which have a zero waiting time. Define

(35) T_a = time spent in queue by an item acquiring service,

T_r = time spent in queue by an item that reneges,

T_q = time spent in queue by an item that joins (both those that renege and that acquire service).

From these definitions

(36) $$P\{t \leqq T_a \leqq t + dt\} = P\{t \leqq T_q \leqq t + dt \mid (A, W)\}$$
$$= P\{t \leqq T_q \leqq t + dt, (A, W)\}/P(A, W),$$

so that the probability density $f_a(t)$ of T_a is given by

(37) $$f_a(t) = \sum_{n=1}^{N-1} P_n e_n \beta_n f_n * f_{n-1} * \cdots * f_j * \cdots * f_1/P(A, W),$$

where $f_j(t)$ is the density function for the time to reduce by one a queue with j members in it, and is given by

(38) $$f_j(t) = (\mu + j\alpha)e^{-(\mu + j\alpha)t}.$$

$f_a(t)$ can be found from (37) by using characteristic functions and inverting with the help of contour integration as is done in Ancker and Gafarian (1963a). Alternatively, we can use known results on Laplace transforms. Thus, if $\Phi_a(s)$ and $\phi_j(s)$ are the Laplace transforms of $f_a(t)$ and $f_j(t)$, respectively, we have from (37)

$$P(A, W)\Phi_a(s) = \sum_{n=1}^{N-1} P_n e_n \beta_n \prod_{j=1}^{n} \phi_j(s), \tag{39}$$

where, from (38),

$$\phi_j(s) = \frac{\mu + j\alpha}{\mu + j\alpha + s} = \frac{\gamma + j}{\gamma + j + (s/\alpha)} \qquad \text{by} \quad (6). \tag{40}$$

Substituting for P_n, e_n, β_n, and $\phi_j(s)$ in (39) and simplifying, we have

$$[P(A, W)/P_0]\Phi_a(s) = [(N-1)\delta/(\gamma + 1 + s/\alpha)] \cdot {}_2F_1(-N+2, 1; \gamma + 2 + s/\alpha; -\delta). \tag{41}$$

Now, using a particular case of Erdélyi *et al.* (1954, Vol. I, p. 220, Eq. (22)), we have

$$(1/s)\, {}_2F_1(a, 1; s+1; z) = \mathscr{L}\{{}_1F_0[a; -; z(1 - e^{-t})]\}, \qquad \operatorname{Re}(s) > 0. \tag{42}$$

Also, if

$$f(s) = \mathscr{L}[F(t)], \tag{43}$$

it can be easily verified that

$$f((s+a)/b) = \mathscr{L}[be^{-at}F(bt)]. \tag{44}$$

Inverting (41) with the help of (42) and (44), we have

$$\begin{aligned} [P(A, W)/P_0]f_a(t) &= (N-1)\alpha\delta e^{-(\mu+\alpha)t} \cdot {}_1F_0[-N+2; -; -\delta(1 - e^{-\alpha t})] \\ &= \lambda(1 - 1/N)e^{-(\mu+\alpha)t}[1 + \delta(1 - e^{-\alpha t})]^{N-2}, \end{aligned} \tag{45}$$

by (7), and hence

$$f_a(t) = \lambda z^{\gamma}(1 - z)^{N-1} e^{-(\mu+\alpha)t} \cdot [1 + \delta(1 - e^{-\alpha t})]^{N-2}/[NB_z(\gamma + 1, N - 1)]. \tag{46}$$

It is easy to show that the mean of $f_a(t)$ is

$$W_a = \int_0^\infty t f_a(t)\, dt = [\alpha B_z(\gamma + 1, N - 1)]^{-1} \cdot \sum_{j=0}^{N-2} \left[(-1)^j \binom{N-2}{j} z^{\gamma+1+j} \Big/ (\gamma + 1 + j)^2 \right]. \tag{47}$$

It can be shown [see Ancker and Gafarian (1963a)] that for such reneging systems

(48) $$f_q(t) = P(A|W)[f_a(t) - \alpha e^{-\alpha t} g(t)] + \alpha e^{-\alpha t},$$

(49) $$f_r(t) = \alpha e^{-\alpha t}[1 - P(A|W) g(t)]/[1 - P(A|W)],$$

where

(50) $$g(t) = \int_0^t e^{\alpha\tau} f_a(\tau)\, d\tau.$$

From the above, it can be shown that

(51) $$\begin{aligned} g(t) &= \left(\frac{P_0}{P(A|W)}\right)\left(\frac{(N-1)\lambda}{N}\right)\int_0^t e^{-\mu\tau}[1 + \delta(1 - e^{-\alpha\tau})]^{N-2}\, d\tau \\ &= \left(\frac{P_0}{P(A|W)}\right)\left(\frac{N-1}{z^{\gamma-1}(1-z)^{N-1}}\right) \\ &\quad \cdot [B_z(\gamma, N-1) - B_{z\exp(-\alpha t)}(\gamma, N-1)]. \end{aligned}$$

Ancker and Gafarian (1963b) consider a modification of the model described above wherein the reneging assumption is dropped. This modified model can, with suitable interpretation of the system parameters, be looked upon as the "machine interference" model, where the input to the system is from a finite source of N machines [see, for example, Cox and Smith (1961, Chapter IV)]. Defining a new parameter

(52) $$\rho = \lambda/(N\mu),$$

the steady-state results for the model are

(53) $$P_{n+1} = (N-n)\rho P_n \qquad (n = 0, 1, 2, \ldots, N-1),$$

(54) $$\begin{aligned} P_n &= P_0 \rho^n \prod_{j=0}^{n-1} (N - j) \\ &= [N!/(N-n)!]\rho^n P_0 \qquad (n = 1, 2, \ldots, N), \end{aligned}$$

(55) $$P_0 = e^{-1/\rho}[N!\rho^N\{1 - I(1/(\rho\sqrt{N+1}), N)\}]^{-1},$$

where $I(u, p)$ is the well-tabled incomplete gamma function ratio (Pearson, 1957), defined by

(56) $$I(u, p) = \int_0^{u\sqrt{p+1}} \frac{e^{-v} v^p\, dv}{\Gamma(p+1)}.$$

From the above, various results corresponding to those for the general model can be obtained as before.

Model II

Ancker and Gafarian (1963c) consider a different model, which they call model II, the foregoing model and its modification being called models I and IA, respectively. Here we make the same assumptions about reneging behavior as in model I, but the balking behavior is different in that we assume that an arriving customer joins the system with probability

$$e_n = \begin{cases} \beta/n & (0 \leqq \beta \leqq 1; n = 1, 2, \ldots) \\ 1 & (n = 0), \end{cases} \tag{57}$$

where n is the number in the system and β is a measure of customer willingness to join the queue. Obviously, in the present model there is no upper limit on n. With the same notation as above, we have

$$P_0'(t) = -\lambda P_0(t) + \mu P_1(t), \tag{58}$$

$$P_1'(t) = -(\lambda\beta + \mu)P_1(t) + \lambda P_0(t) + (\mu + \alpha)P_2(t), \tag{59}$$

$$\begin{aligned} P_n'(t) = &-[\lambda\beta/n + \mu + (n-1)\alpha]P_n(t) \\ &+ [\lambda\beta/(n-1)]P_{n-1}(t) + (\mu + n\alpha)P_{n+1}(t) \qquad (n = 2, 3, \ldots). \end{aligned} \tag{60}$$

The steady-state equations are

$$0 = -(\varepsilon/\beta)P_0 + \gamma P_1, \tag{61}$$

$$0 = -(\varepsilon + \gamma)P_1 + (\varepsilon/\beta)P_0 + (\gamma + 1)P_2, \tag{62}$$

$$\begin{aligned} 0 = &-[(\varepsilon/n) + \gamma + n - 1]P_n + [\varepsilon/(n-1)]P_{n-1} \\ &+ (\gamma + n)P_{n+1} \qquad (n = 2, 3, \ldots), \end{aligned} \tag{63}$$

where γ is as defined in (6), and

$$\varepsilon = \lambda\beta/\alpha. \tag{64}$$

From the steady-state equations, by induction we have

$$P_n = P_0\varepsilon^n \Big/ \left[\beta(n-1)! \prod_{j=0}^{n-1} (\gamma + j)\right], \tag{65}$$

whence, on using the normalizing condition,

$$P_0 = \left[1 + \left(\frac{\varepsilon}{\beta}\right) \sum_{n=0}^{\infty} \frac{\varepsilon^n}{n!(\gamma)_{n+1}}\right]^{-1} = [1 + \varepsilon^{1-\gamma/2}\Gamma(\gamma)I_\gamma(2\sqrt{\varepsilon})/\beta]^{-1}, \tag{66}$$

where $I_\gamma(x)$ is the modified Bessel function of the first kind and order γ. The mean number in the queue is

$$L_q = \sum_{n=1}^{\infty} nP_{n+1} = \left(\frac{P_0}{\beta}\right)\left(\frac{\Gamma(\gamma)}{(\sqrt{\varepsilon})^{\gamma-3}}\right) I_{\gamma+1}(2\sqrt{\varepsilon}), \tag{67}$$

and the mean number in the system is

$$L = L_q + 1 - P_0. \tag{68}$$

With the notation of (27), we have

$$P(B) = \sum_{n=1}^{\infty} P_n(1 - e_n) = 1 - \frac{P_0 \Gamma(\gamma) I_{\gamma-1}(2\sqrt{\varepsilon})}{(\sqrt{\varepsilon})^{\gamma-1}}, \tag{69}$$

$$\begin{aligned} P(W) &= \sum_{n=1}^{\infty} P_n e_n = 1 - P(B) - P_0 \\ &= P_0[\Gamma(\gamma) I_{\gamma-1}(2\sqrt{\varepsilon})/(\sqrt{\varepsilon})^{\gamma-1} - 1], \end{aligned} \tag{70}$$

$$P(R) = \sum_{n=1}^{\infty} P_n(1 - \beta_n)e_n = P_0\left(\frac{\Gamma(\gamma)}{(\sqrt{\varepsilon})^{\gamma-1}}\right) I_{\gamma+1}(2\sqrt{\varepsilon}) = \left(\frac{\beta}{\varepsilon}\right) L_q, \tag{71}$$

$$P(A) = \sum_{n=0}^{\infty} P_n e_n \beta_n = \frac{P_0 \Gamma(\gamma + 1) I_\gamma(2\sqrt{\varepsilon})}{(\sqrt{\varepsilon})^\gamma} = \left(\frac{\beta\gamma}{\varepsilon}\right)(1 - P_0), \tag{72}$$

$$\begin{aligned} P(A, W) &= \sum_{n=1}^{\infty} P_n e_n \beta_n = P_0\left(\frac{\Gamma(\gamma + 1) I_\gamma(2\sqrt{\varepsilon})}{(\sqrt{\varepsilon})^\gamma} - 1\right) \\ &= \left(\frac{\beta\gamma}{\varepsilon}\right)(1 - P_0) - P_0, \end{aligned} \tag{73}$$

$$\begin{aligned} P(A|W) &= P(A, W)/P(W) \\ &= [\Gamma(\gamma+1) I_\gamma(2\sqrt{\varepsilon}) - (\sqrt{\varepsilon})^\gamma]/[\sqrt{\varepsilon}\, \Gamma(\gamma) I_{\gamma-1}(2\sqrt{\varepsilon}) - (\sqrt{\varepsilon})^\gamma]. \end{aligned} \tag{74}$$

Waiting Times

Following the notation given in (35), the model II analog of (39) is

$$\begin{aligned} \left(\frac{P(A, W)}{P_0}\right) \Phi_a(s) &= \sum_{n=1}^{\infty} \left(\frac{(\alpha\varepsilon)^n}{n!}\right) \prod_{j=1}^{n} \left(\frac{1}{\mu + j\alpha + s}\right) \\ &= \left(\frac{\alpha\varepsilon}{\mu + s + \alpha}\right) {}_1F_2\left[1; 2, \frac{\mu + s}{\alpha} + 2; \varepsilon\right]. \end{aligned} \tag{75}$$

Again, from a particular case of Erdélyi *et al.* (1954, Vol. I, p. 220, Eq. (22)), we have

$$(1/s)\, {}_1F_2[1; 2, s + 1; z] = \mathscr{L}\{{}_0F_1[-; 2; z(1 - e^{-t})]\}, \qquad \mathrm{Re}(s) > 0. \tag{76}$$

Inverting (75) with the help of (76) and (44), we have

(77) $$[P(A, W)/P_0]f_a(t) = \alpha\varepsilon e^{-(\mu+\alpha)t}\,{}_0F_1[-;2;\varepsilon(1-e^{-\alpha t})]$$
$$= \lambda\beta e^{-(\mu+\alpha)t} I_1[2\sqrt{\varepsilon(1-e^{-\alpha t})}]/\sqrt{\varepsilon(1-e^{-\alpha t})},$$

where $I_1(x)$ is the modified Bessel function of the first kind and order one. The mean of $f_a(t)$ is

(78) $$W_a = -\frac{\partial}{\partial s}\Phi_a(s)\bigg|_{s=0}$$
$$= \frac{P_0}{\alpha P(A, W)}\sum_{n=1}^{\infty}\left(\frac{\varepsilon^n}{n!(\gamma+1)_n}\right)\sum_{k=1}^{n}\left(\frac{1}{\gamma+k}\right).$$

Also, $f_q(t)$ and $f_r(t)$ are given by (48) and (49), respectively, where $g(t)$, defined by (50), is

(79) $$g(t) = \frac{P_0\lambda\beta}{P(A, W)}\int_0^t e^{-\mu\tau}\frac{I_1[2\sqrt{\varepsilon(1-e^{-\alpha\tau})}]}{\sqrt{\varepsilon(1-e^{-\alpha\tau})}}\,d\tau$$
$$= \left(\frac{P_0}{P(A, W)}\right)\left[\left(\frac{\Gamma(\gamma)}{(\sqrt{\varepsilon})^{\gamma-1}}\right)I_{\gamma-1}(2\sqrt{\varepsilon}) - 1\right.$$
$$\left. - \sum_{n=0}^{\infty}\left(\frac{\varepsilon^{n+1}}{n!(n+1)!}\right)B_{\exp(-\alpha t)}(\gamma, n+1)\right].$$

Model IIA

Here we allow balking only, so that no reneging is permitted. The model is similar, but not identical, to the one considered by Cox and Smith (1961, p. 44). The results for this model can be derived from those for model II by making $\alpha \to 0$. It can be shown that

(80) $$P_n = P_0\rho^n/[\beta(n-1)!],$$

(81) $$P_0 = \beta/(\beta + \rho e^{\rho}),$$

where

(82) $$\rho = \lambda\beta/\mu.$$

Also, we have

(83) $$Q_R = [\rho e^{\rho}/(\beta + \rho e^{\rho})]I[\rho/\sqrt{R-1}, R-2],$$

where $I(u, p)$ is given by (56).

3.6 Limiting Behavior of the Many-Server Queue

Iglehart (1965) studies the limiting behavior of the many-server queue as the number of servers becomes large by considering it as a birth–death process. Such a process $X(t)$ may be defined as a continuous time Markov chain, whose state space is the nonnegative integers and whose transition probabilities

$$P_{ij}(t) = \mathrm{Prob}\{X(t+\tau) = j \mid X(t) = i\} \qquad (i, j = 0, 1, 2, \ldots) \tag{1}$$

are given by the matrix differential equation

$$P'(t) = AP(t), \tag{2}$$

where $P(t)$ is the infinite matrix

$$P(t) = [P_{ij}(t)] \qquad (i, j = 0, 1, 2, \ldots), \tag{3}$$

and

$$A = \begin{bmatrix} -(\lambda_0 + \mu_0) & \lambda_0 & 0 & 0 & \cdots \\ \mu_1 & -(\lambda_1 + \mu_1) & \lambda_1 & 0 & \cdots \\ 0 & \mu_2 & -(\lambda_2 + \mu_2) & \lambda_2 & \cdots \\ \vdots & \vdots & \vdots & \vdots & \end{bmatrix}. \tag{4}$$

Following Karlin and McGregor (1957a), we have

$$P_{ij}(t) = \pi_j \int_0^\infty e^{-xt} Q_i(x) Q_j(x)\, d\Psi(x), \tag{5}$$

where the $\{Q_i(x)\}$ are polynomials in x defined by the recurrence relations

$$Q_0(x) \equiv 1, \tag{6}$$

$$-xQ_0(x) = -(\lambda_0 + \mu_0)Q_0(x) + \lambda_0 Q_1(x), \tag{7}$$

$$-xQ_n(x) = \mu_n Q_{n-1}(x) - (\lambda_n + \mu_n)Q_n(x) + \lambda_n Q_{n+1}(x) \qquad (n \geqq 1), \tag{8}$$

the sequence $\{\pi_j\}$ is defined by

$$\pi_0 = 1, \tag{9}$$

$$\pi_n = (\lambda_0 \lambda_1 \cdots \lambda_{n-1})/(\mu_1 \mu_2 \cdots \mu_n) \qquad (n \geqq 1), \tag{10}$$

and Ψ is a positive regular measure on $[0, \infty)$. We also know that the polynomials $\{Q_i(x)\}$ are orthogonal with respect to the measure Ψ, since

$$\int_0^\infty Q_i(x) Q_j(x)\, d\Psi(x) = \delta_{ij} \pi_j^{-1} \qquad (i, j = 0, 1, \ldots). \tag{11}$$

For the many-server queue $M/M/n$ with mean arrival rate λ and mean service rate per customer μ, the number $X_n(t)$ of the number of customers in the system at time t is a birth and death process with

(12) $$\lambda_i \equiv \lambda \qquad (i = 0, 1, \ldots),$$

(13) $$\mu_i = \begin{cases} i\mu & (0 \leqq i \leqq n) \\ n\mu & (i \geqq n + 1). \end{cases}$$

For this process, it follows from the general theory of Karlin and McGregor (1957a) that the parameters $\{\lambda_i\}$ and $\{\mu_i\}$ uniquely determine $P_{ij}(t)$ and that the sample paths of $X_n(t)$ are jump functions with probability one. Further, it can be shown that

(14) $$Q_k(x) = c_k(x/\mu; \lambda/\mu) \qquad (k \leqq n),$$

where [cf. II.3(68)]

(15) $$c_k(x; a) = \{x!(-a)^{-k}/(x-k)!\}\, {}_1F_1(-k; x-k+1; a)$$

are the Poisson–Charlier polynomials (Erdélyi *et al.*, 1953, Vol. II, p. 226, Eq. (5)).

Further, let $X_n(t)$ be the birth and death process with parameters

(16) $$\lambda_i^{(n)} = n\rho,$$

(17) $$\mu_i^{(n)} = \begin{cases} i, & i \leqq n \\ n, & i > n \quad (i = 0, 1, \ldots; n = 1, 2, \ldots), \end{cases}$$

where

(18) $$0 < \rho < 1.$$

Clearly, the process $X_n(t)$ represents the number in the system at time t in an n-server queue with fixed traffic intensity ρ. Define a sequence $\{Y_n(t)\}$ by

(19) $$Y_n(t) = [X_n(t) - n\rho]/(n\rho)^{1/2}, \qquad n = 1, 2, \ldots .$$

Iglehart (1965) proves that the sequence $\{Y_n(t)\}$ is weakly convergent in the Markov sense to the Ornstein–Uhlenbeck (O–U) diffusion process, where an O–U process is uniquely determined by the backward equation

(20) $$\frac{\partial u}{\partial t} = \frac{\partial^2 u}{\partial x^2} - x\frac{\partial u}{\partial x} \qquad (-\infty < x < \infty).$$

The elementary solution of (20) is given by [cf. Karlin and McGregor (1960, p. 170)]

(21) $$p(t; x, y) = \sum_{n=0}^{\infty} e^{-nt} H_n\left(\frac{x}{\sqrt{2}}\right) H_n\left(\frac{y}{\sqrt{2}}\right) w_n,$$

where

(22) $$w_n = (\pi^{1/2} 2^{n+1} n!)^{-1}$$

and $H_n(x)$ are the Hermite polynomials, given by [cf. Erdélyi *et al.* (1953, Vol. II, p. 194, Eq. (19))]

$$\sum_{n=0}^{\infty} \frac{H_n(x)t^n}{n!} = \exp(2xt - t^2) \qquad \text{(all finite } x, t). \tag{23}$$

[See also II.8(20) and II.8(21).] Furthermore, it can be shown that

$$p(t; x, y)e^{-y^2/2} = \frac{1}{\sqrt{2\pi(1 - e^{-2t})}} \exp\left[-\frac{(xe^{-t} - y)^2}{2(1 - e^{-2t})}\right]. \tag{24}$$

3.7 A Multiserver Queue with Discouragement

Reynolds (1968) considers a multichannel queuing model in which the arrival rate is a decreasing function of the queue length. Similar single-channel models incorporating discouragement for arrivals have been studied by Ancker and Gafarian (1963b,c), which have been discussed in Section 3.5, as well as by Morse (1958), and Cox and Smith (1961), and Reynolds' paper can be taken as the multiserver extension of the model studied by Cox and Smith.

Consider an $M/M/c$ queue (Reynolds, 1968) with state-dependent parameters λ_n, μ_n given by

$$\lambda_n = \begin{cases} \lambda & (n < c) \\ \lambda/(n - c + 2) & (n \geqq c), \end{cases} \tag{1}$$

$$\mu_n = \begin{cases} n\mu & (n \leqq c) \\ c\mu & (n \geqq c), \end{cases} \tag{2}$$

where n is the number in the system. Let

$$S = \sum_{n=0}^{\infty} \prod_{i=1}^{n} \left(\frac{\lambda_{i-1}}{\mu_i}\right). \tag{3}$$

It is known (Cox and Miller, 1965) that if S is convergent, a steady-state distribution exists with

$$P_n = S^{-1} \prod_{i=1}^{n} \left(\frac{\lambda_{i-1}}{\mu_i}\right), \qquad n = 1, 2, \ldots,\text{[5]} \tag{4}$$

$$P_0 = S^{-1}. \tag{5}$$

[5] There is an obvious error in Reynolds (1968).

Define the generating function

(6)
$$Q(x) = \sum_{n=0}^{\infty} P_n x^n.$$

Then, from (1) and (2), we have

(7)
$$Q(x) = P_0\left[\sum_{n=0}^{c-2} \frac{(\rho x)^n}{n!} + \frac{(\rho x)^{c-1}\exp(\rho x/c)}{(c-1)!}\right],$$

where

(8)
$$\rho = \lambda/\mu.$$

The value of P_0 is given by the fact that

(9)
$$Q(1) = 1.$$

For $c = 1$, we have

(10)
$$Q(x) = \exp[\rho(x-1)].$$

For $c \geqq 2$, we can write

(11)
$$Q(x) = f(x, c, \rho)/f(1, c, \rho),$$

where

(12)
$$f(x, c, \rho) = \exp(\rho x)[1 - G(c-1, \rho x)] + (\rho x)^{c-1}\exp(\rho x/c)/(c-1)!,$$

(13)
$$P_0 = [f(1, c, \rho)]^{-1},$$

(14)
$$G(a, b) = \int_0^b \frac{e^{-t}t^{a-1}\,dt}{(a-1)!} = \frac{\gamma(a, b)}{\Gamma(a)},$$

$\gamma(a, b)$ being the incomplete gamma function, and we follow the convention that

(15)
$$G(a, b) = 1 \qquad \text{for} \quad a \leqq 0.$$

If P_w is the probability that a customer has to wait for service to begin, we have

(16)
$$P_w = 1 - \sum_{n=0}^{c-1} P_n.$$

From (7) it follows that

(17)
$$P_w = P_0\rho^{c-1}[\exp(\rho/c) - 1]/(c-1)!.$$

The mean number in the system is

(18)
$$L = Q'(1) = P_0[\rho e^{\rho}\{1 - G(c-2, \rho)\} + \rho^{c-1}\{c(c-1) + \rho\}\exp(\rho/c)/c!].$$

For the special case $c = 1$, we have

(19) $$L = \rho.$$

For $c \geqq 2$, it follows that

(20) $$L = \rho - (c-1)\rho^{c+1}/\{(2c)c!\} + o(\rho^{c+1}).$$

Thus for sufficiently small ρ and $c \geqq 2, d > 0$,

(21) $$L_{c+d} > L_c,$$

where the subscript indicates the number of channels.

Also, for large ρ,

(22) $$L \sim (c-1) + \rho/c.$$

The mean number waiting in the queue is

(23) $$\begin{aligned} L_q &= \sum_{n=c+1}^{\infty} (n-c)P_n \\ &= \frac{d}{dx}\left[\left(Q(x) - \sum_{n=0}^{c-1} P_n x^n\right)\Big/ x^c\right]_{x=1} \\ &= P_0[\rho^{c-1}/(c-1)! + \rho^{c-1}(\rho - c)\exp(\rho/c)/c!]. \end{aligned}$$

For small ρ, we have

(24) $$L_q = \rho^{c+1}/\{(2c)\, c!\} + o(\rho^{c+1}).$$

The inequality analogous to (21) is thus reversed, giving

(25) $$(L_q)_{c+d} < (L_q)_c.$$

Also, for large ρ,

(26) $$L_q \sim \rho/c - 1.$$

The quantity λ is a measure of the potential customer demand, since it denotes the mean arrival rate under the ideal case when c is infinite. For a finite c, the mean arrival rate $\bar{\lambda}$ falls below λ, and its determination is useful for finding an optimal value of c in a particular situation. We have

(27) $$\bar{\lambda} = \sum_{n=0}^{\infty} P_n \lambda_n,$$

so that

(28) $$\begin{aligned} \frac{\bar{\lambda}}{\lambda} &= \sum_{n=0}^{c-1} P_n + \sum_{n=c}^{\infty} \frac{P_n}{n-c+2} \\ &= P_0\left(e^{\rho}\{1 - G(c-1, \rho)\} + \frac{c\rho^{c-2}\{\exp(\rho/c) - 1\}}{(c-1)!}\right), \end{aligned}$$

whence it follows that the proportional loss in customers is

$$(29) \qquad 1 - \bar{\lambda}/\lambda = P_0[\rho^{c-2}\{(\rho - c)\exp(\rho/c) + c\}/(c-1)!].$$

Reynolds (1968) finds the waiting time distribution through the use of characteristic functions and contour integration. As before, one may alternatively use the Laplace transforms. Thus if $\phi(s)$ is the Laplace transform of the density function of the unconditional distribution of waiting time, i.e., the time a customer has to wait for service to begin, and $\pi_n(s)$ is the L.T. of the conditional density of the waiting time given n customers in the system, then

$$(30) \qquad \pi_n(s) = \begin{cases} 1 & (n < c) \\ [c\mu/(c\mu + s)]^{n-c+1} & (n \geqq c), \end{cases}$$

and, therefore,

$$(31) \qquad \phi(s) = \sum_{n=0}^{\infty} P_n \pi_n(s)$$

$$= P_0\left(e^{\rho}\{1 - G(c-1, \rho)\} + \frac{\rho^{c-1}\exp\{\rho\mu/(c\mu + s)\}}{(c-1)!}\right).$$

From (31), we can find the mean waiting time in the queue to be

$$(32) \qquad W_{\mathrm{q}} = -\phi'(0) = P_0\rho^c \exp(\rho/c)/\{(c\mu)c!\}.$$

For inversion of $\phi(s)$, we can write it in the form

$$(33) \quad \phi(s) = P_0\left[\sum_{n=0}^{c-1}\left(\frac{\rho^n}{n!}\right) + \sum_{n=c}^{\infty}\left(\frac{\rho^n \mu^{n-c+1}}{(c-1)!(n-c+1)!(c\mu + s)^{n-c+1}}\right)\right].$$

The first term on the right-hand side corresponds to the discrete component of the distribution and gives the probability that the waiting time is zero. The second term is the L.T. of the density function, say $p(t)$, of the continuous component. The second term can be written in the form

$$(34) \qquad \{\rho^{c-1}/(c-1)!\}\{\rho\mu/(c\mu + s)\}\, {}_1F_1[1; 2; \rho\mu/(c\mu + s)].$$

From a particular case of Erdélyi *et al.* (1954, Vol. I, p. 293, Eq. (2)), with $k = 0$, $\mu = \frac{1}{2}$, we have

$$(35) \qquad \mathscr{L}^{-1}[(\alpha/s)\, {}_1F_1(1; 2; \alpha/s)] = \sqrt{\alpha/t}\, I_1(2\sqrt{\alpha t}).$$

Using (35) for inversion we have

$$(36) \qquad p(t) = \{\rho^{c-1}/(c-1)!\}e^{-c\mu t}\sqrt{\rho\mu/t}\, I_1(2\sqrt{\rho\mu t}),$$

whence we see that the distribution function of the waiting time is

$$(37) \quad F(t) = P_0\left[e^{\rho}\{1 - G(c-1, \rho)\} + \left(\frac{\sqrt{\lambda}\,\rho^{c-1}}{(c-1)!}\right)\int_0^t e^{-c\mu\tau}\tau^{-1/2} I_1(2\sqrt{\lambda\tau})\, d\tau\right],$$

which can be expressed as

(38) $$F(t) = P_0\left[e^{\rho}\{1 - G(c-1,\rho)\} + \frac{1}{2}\left(\frac{\rho^{c-1}}{(c-1)!}\right) \cdot \sum_{j=0}^{\infty} \frac{(4\lambda t)^{j+1}}{\{(j+1)!\}^2}\, {}_1F_1(j+1; j+2; -c\mu t)\right].$$

3.8 A Potentially Infinite Capacity Queuing System

Hadidi (1969) considers the $M/M/1$ queuing system with a state-dependent service rate, given by

(1) $$\mu_n = (n+1)\mu,$$

when there are n customers in the *queue*. The mean arrival rate is taken as unity. The process is thus formally analogous to $M/M/\infty$, except that customers may have to wait for service in the former system, whereas in the latter waiting has no relevance [see Hadidi and Conolly (1969a)].

Let P_n be the steady-state probability that there are n in the *queue* (excluding the one in service) at the epoch that a service interval commences, and let P_{ij} be the transition probabilities from one such service epoch to the next. For the calculation of P_{00} we require that no arrival or one arrival occurs during the service time of a customer who leaves an empty queue behind him. Thus

(2) $$P_{00} = \int_0^{\infty} e^{-t}\mu_0 e^{-\mu_0 t}\,dt + \int_{t=0}^{\infty}\int_{\tau=0}^{t} e^{-\tau}\,d\tau\, e^{-(t-\tau)}e^{-\mu_0\tau}\mu_1 e^{-\mu_1(t-\tau)}\,dt$$
$$= \frac{\mu_0}{\mu_0+1} + \frac{\mu_1}{(1+\mu_0)(1+\mu_1)}.$$

Using (1), we have, from (2),

(3) $$P_{00} = \frac{1}{\rho+1} + \frac{2\rho}{(\rho+1)(\rho+2)},$$

where

(4) $$\rho = 1/\mu.$$

[Note that $\lambda = 1$ here.]

Similarly, we have

$$P_{ij} = \frac{\rho^{j-i+1}(j+2)}{(\rho+i+1)(\rho+i+2)\cdots(\rho+j+2)} \qquad (j \geqq i-1). \tag{5}$$

Now, we have

$$P_n = \sum_{i=0}^{n+1} P_i P_{in} \qquad (n \geqq 1), \tag{6}$$

from which we can show by induction that

$$P_n = \frac{P_0 \rho^{n+1}}{(1+\rho)(n+1)!} \qquad (n \geqq 1), \tag{7}$$

so that

$$P_0 = (1+\rho)e^{-\rho}. \tag{8}$$

Let $b_k(t)$ $(k \geqq 1)$ be the probability density function of a single service interval which commences with k customers in the *system*. During the service interval there could be 0, 1, 2,... arrivals. Let $b_k^{(j)}(t)$ be the contribution to $b_k(t)$ corresponding to j arrivals. Then

$$b_k^{(0)}(t) = \mu k e^{-(1+\mu k)t}, \tag{9}$$

$$b_k^{(1)}(t) = \mu(k+1)\int_0^t \exp[-(1+\mu k)\tau] \cdot \exp[-\{1+\mu(k+1)\}(t-\tau)]\,d\tau, \tag{10}$$

etc., whence, in general,

$$b_k^{(j)}(t) = \mu(k+j)\exp[-(1+\mu k)t]\{\rho\Lambda(t)\}^j/j!, \tag{11}$$

where

$$\Lambda(t) = 1 - e^{-\mu t}, \tag{12}$$

and, summing over j, we have

$$b_k(t) = [k\mu + \Lambda(t)]\exp[-(1+k\mu)t + \rho\Lambda(t)]. \tag{13}$$

Hence, the unconditional p.d.f. of the service interval is given by

$$b(t) = \sum_{n \geqq 0} P_n b_{n+1}(t), \tag{14}$$

which simplifies to

$$b(t) = e^{-t} + [\mu e^{-\mu t} - (1-e^{-\mu t})^2]\exp[-t - \rho e^{-\mu t}], \tag{15}$$

and its Laplace transform

$$b^*(s) = \int_0^\infty e^{-st} b(t)\,dt \tag{16}$$

is given by

$$b^*(s) = (1+s)^{-1} + se^{-\rho} - s^2 \rho^{-(\rho s + \rho - 1)} \gamma(\rho s + \rho, \rho), \tag{17}$$

where $\gamma(\alpha, z)$ is the incomplete gamma function. From (17), we have for the moments of the service time distribution

$$E(t) = 1 - e^{-\rho}, \tag{18}$$

$$E(t^2) = 2[1 - \rho^{1-\rho}\gamma(\rho, \rho)]. \tag{19}$$

Waiting Time

For the determination of the waiting time distribution, we define an auxiliary function $b_{kr}(\omega)$, which is the p.d.f. of exactly r services having combined length ω (> 0), given that the original one starts with k customers in the system. As regards the waiting time problem, only values of k such that $k \geqq r$ are of interest. With this definition,

$$b_{k1}(\omega) = b_k(\omega), \tag{20}$$

where $b_k(\omega)$ is as defined earlier in this section. Further,

$$b_{kr}(\omega) = \sum_{j=0}^{\infty} \int_0^\omega b_{k1}^{(j)}(t) b_{k+j-1,r-1}(\omega - t)\,dt, \tag{21}$$

with $b_{k1}^{(j)}(t)$ as introduced earlier. Thus

$$b_{kr}(\omega) = \mu \int_0^\omega \exp[-(1+\mu k)t] \sum_{j=0}^{\infty} \left(\frac{k+j}{j!}\right) \cdot \{\rho \Lambda(t)\}^j b_{k+j-1,r-1}(\omega - t)\,dt, \tag{22}$$

where $\Lambda(t)$ is given by (12). To solve the above integral-difference equation, we introduce the functions

$$h_{ij}(\omega) \qquad (i = 1, 2, \ldots, j+1;\, j = 1, 2, \ldots), \tag{23}$$

which are defined iteratively by

$$h_{11}(\omega) = 1, \tag{24}$$

$$h_{21}(\omega) = \rho\Lambda(\omega), \tag{25}$$

$$h_{ij}(\omega) = \int_0^\omega \exp[\mu(\omega - t)] \sum_{n=1}^{\min(i,j)} \binom{j-n+1}{i-n} z^{i-n} h_{n,j-1}(\omega - t)\,dt, \tag{26}$$

$$(i = 1, 2, \ldots, j+1;\, j = 2, 3, \ldots),$$

where

$$(27) \qquad z = \rho \Lambda(t) \exp[-\mu(\omega - t)].$$

Then, it can be proved by induction that

$$(28) \quad b_{kr}(\omega) = \mu^r \exp[\rho\Lambda(\omega) - (1 + \mu k)\omega] \sum_{i=0}^{r} \left(\frac{k!}{(k+i-r)!} \right) h_{i+1,r}(\omega).$$

Also, solving (26) for various values of i, j and using induction, it is shown that

$$(29) \qquad h_{i+1,j+1}(\omega) = [\rho^i \{\psi(\omega)\}^j / j!] \cdot \sum_{k=0}^{\min(i,j)} (-1)^{i-k} \binom{j}{k} \sigma_{j-i+2,i-k}(\omega) \left(\frac{\omega}{\psi(\omega)} \right)^k$$

$$[i \geqq 0, j \geqq \max(0, i-1)],$$

where

$$(30) \qquad \psi(\omega) = e^{\mu\omega} \Lambda(\omega)/\mu,$$

$$(31) \qquad \sigma_{ij}(\omega) = \binom{i+j-1}{j} e^{-j\mu\omega} - \binom{i+j-2}{j-1} e^{-(j-1)\mu\omega} \qquad (i, j \geqq 1),$$

$$(32) \qquad \sigma_{i0}(\omega) = 1 \qquad \text{for all} \quad i.$$

Using (29) in (28), we have, after some algebraic simplification, the following solution of (22):

$$(33) \quad b_{kr}(\omega) = \mu[\mu\psi(\omega)]^{r-1} \exp[\rho\Lambda(\omega) - (1 + \mu k)\omega] \cdot \sum_{i=0}^{r-1} \binom{k}{r-i-1} \left(\frac{\{T(\omega)\}^i}{i!} \right) \{\rho\Lambda(\omega) + k + i + 1 - r\},$$

where

$$(34) \qquad T(\omega) = \rho e^{-\mu\omega} [\{\mu\omega/\Lambda(\omega)\} - 1].$$

Now, if p_{ij} is the transition probability from one *arrival* epoch to the next, it can be shown (Hadidi and Conolly, 1969a) that

$$(35) \qquad p_{ij} = (i+1)!\,\Gamma(\rho + j)/[\Gamma(\rho + i + 2) j!],$$

whence the steady-state probability at an arrival epoch is given by

$$(36) \qquad \bar{P}_r = \lim_{n \to \infty} P_r^{(n)} = e^{-\rho} \rho^r / r!.$$

The steady-state p.d.f. of the waiting time (including service) is, therefore, given by

$$(37) \qquad f(\omega) = \sum_{r \geqq 0} \bar{P}_r b_{r+1,r+1}(\omega),$$

which, with the help of (33) and (36), yields

$$(38)\qquad f(\omega) = \mu \exp[-\rho\sigma_{21}(\omega) - (1+\mu)\omega][\{1 + 2\rho\Lambda(\omega)\}I_0(2\phi^{1/2}) + \{\phi^{1/2} + (\rho\Lambda(\omega))^2\phi^{-1/2}\}I_1(2\phi^{1/2})],$$

where

$$(39)\qquad \phi = \rho\Lambda(\omega)T(\omega),$$

and $I_\nu(\cdot)$ stands for the Bessel function frequently encountered. From (38), we have the average waiting time

$$(40)\qquad E(\omega) = \rho.$$

Values of the variance of ω for various values of ρ have been tabulated in Hadidi (1969).

3.9 Busy Periods of Queues with State-Dependent Parameters

Hadidi (1974) studies the busy period distributions of two queuing systems with state-dependent parameters. The first of these is the system called $M/M_n/1$, which has been discussed in Section 3.8 and is characterized by

$$(1)\qquad \lambda_n = \lambda \quad (n \geqq 0), \qquad \mu_n = n\mu \quad (n \geqq 1),$$

and the second, called $M_n/M/1$, has

$$(2)\qquad \lambda_n = \lambda/(n+1) \quad (n \geqq 0), \qquad \mu_n = \mu \quad (n \geqq 1),$$

where n is the number in the *system*.

Let us define

$$(3)\qquad \gamma_{kr}(t) \quad (1 \leqq k \leqq r)$$

to be the conditional probability density that r consecutive uninterrupted services, the last one of which leaves no customers in the system, have a combined duration t, given that the first service starts with k customers in the system. It is evident that $\gamma_{kr}(t)$ is the joint p.d.f. of the duration of a busy period starting with k in the system during which r customers are served. Further, let $b_k^{(j)}(t)$ be defined, as in the last section, to be the transition probability over a single service interval for a transition from k to $k + j - 1$ in the system, where obviously $j \geqq 0$. Then, we have

$$\gamma_{kr}(t) = \sum_{j=\max(0,2-k)}^{r-k} \int_0^t b_k^{(j)}(\tau)\gamma_{k+j-1,r-1}(t-\tau)\,d\tau; \tag{4}$$

$$r \geqq k \qquad \text{for} \quad k \geqq 2,$$

$$r \geqq k+1 \qquad \text{for} \quad k = 1.$$

The lower limit of summation is due to the fact that for $k = 1$ and $r \geqq 2$, there should be no gap in the service of the r customers. Clearly, we have

$$\gamma_{11}(t) = \mu \exp[-(\lambda+\mu)t] \qquad \text{for} \quad M/M_n/1, \tag{5}$$

$$\gamma_{11}(t) = \mu \exp[-(\tfrac{1}{2}\lambda+\mu)t] \qquad \text{for} \quad M_n/M/1. \tag{6}$$

Also, for $M/M_n/1$,

$$\gamma_k^{(0)}(t) = k\mu \exp[-(\lambda+k\mu)t], \tag{7}$$

$$\gamma_k^{(1)}(t) = (k+1)\mu \int_0^t \lambda \exp[-(\lambda+k\mu)\tau] \exp[-\{\lambda+(k+1)\mu\}(t-\tau)]\,d\tau \tag{8}$$

[cf. 3.8(9), (10)]. Let us use the asterisk to denote Laplace transforms so that

$$f^*(s) = \int_0^\infty e^{-st} f(t)\,dt. \tag{9}$$

Then, from (4) we have

$$\gamma_{kr}^*(s) = \mu \sum_{j=\max(0,2-k)}^{r-k} \frac{a_{kj}\lambda^j}{f_k \cdots f_{k+j}} \gamma_{k+j-1,r-1}^*(s), \tag{10}$$

where

$$a_{kj} = k + j, \tag{11}$$

$$f_k = \lambda + k\mu + s \qquad \text{for} \quad M/M_n/1$$

and

$$a_{kj} = k + j + 1, \tag{12}$$

$$f_k = \lambda + (k+1)(\mu + s) \qquad \text{for} \quad M_n/M/1.$$

Introducing the generating function

$$\Gamma_k(x,t) = \sum_{j \geqq 0} x^{k+j} \gamma_{k,k+j}(t) \qquad (|x| \leqq 1), \tag{13}$$

we obtain, from (10), the difference equation

$$\Gamma_k^*(x,s) = \mu x \sum_{j \geqq 0} \frac{a_{kj}\lambda^j}{f_k \cdots f_{k+j}} \Gamma_{k+j-1}^*(x,s), \tag{14}$$

which is valid for $k \geqq 1$, provided we define

(15) $$\Gamma_0^*(x, s) \equiv 1.$$

Writing (14) out, we see that

(16) $$\lambda \Gamma_k^*(x, s) = f_{k-1}\Gamma_{k-1}^*(x, s) - a_{k-1,0}\mu x \Gamma_{k-2}^*(x, s) \qquad (k \geqq 2).$$

$\Gamma_k^*(1, s)$ is the Laplace transform of the probability density of a busy period starting with k customers irrespective of the number of customers served therein, and it is evident that

(17) $$\Gamma_k^*(1, 0) = 1 \qquad \text{for all} \quad k.$$

It may be mentioned that Conolly (1974), following the technique of Conolly (1959), has obtained the following infinite continued fraction representation for $\Gamma_1^*(x, s)$ for the model with general λ_n and μ_n:

(18) $$\Gamma_1^*(x, s) = \frac{\mu_1 x}{Z_1 -}\,\frac{\lambda_1 \mu_2 x}{Z_2 -}\,\frac{\lambda_2 \mu_3 x}{Z_3 -} \cdots,$$

where

(19) $$Z_n = s + \lambda_n + \mu_n.$$

Hadidi solves (16) separately for the two systems considered here.

For the system $M/M_n/1$ Introducing the generating function

(20) $$L(x, s, y) = \sum_{k \geqq 1} y^k \Gamma_k^*(x, s) \qquad |y| < 1,$$

we have from (16), using (11), that

(21) $$\frac{\partial L}{\partial y} + \frac{(\lambda + s)y - \mu x y^2 - \lambda}{\mu y^2(1 - xy)} L = \frac{xy - \rho \Gamma_1^*}{y(1 - xy)},$$

where, as usual,

(22) $$\rho = \lambda/\mu.$$

Now, (21) is a linear equation of the first order and can be solved by standard methods. The integrating factor is

(23) $$y^\alpha (1 - xy)^{1-\alpha} e^{\rho/y},$$

where

(24) $$\alpha = (\lambda + s)/\mu - \rho x.$$

Using the fact that

(25) $$L(1, 0, y) = y/(1 - y),$$

we have the solution of (21) as

$$L = (1 - xy)^{\alpha - 1} \sum_{i \geqq 0} (-xy)^i \binom{-\alpha}{i} \cdot [\Gamma_1^* y \,{}_2F_0(1, \alpha+i+1; -; y/\rho) - (x/\rho) y^2 \,{}_2F_0(1, \alpha+i+2; -; y/\rho)]. \tag{26}$$

Taking the coefficient of y^k in (26), we have

$$\Gamma_k^* = \sum_{i=0}^{k-1} \binom{\alpha - 1}{i} (-x)^i (A_{k-i}\Gamma_1^* - B_{k-i}x), \tag{27}$$

where

$$\rho^{i-1} A_i(x, s) = \sum_{j=0}^{i-1} (-\rho x)^j \binom{-\alpha}{j} \frac{\Gamma(\alpha + i)}{\Gamma(\alpha + j + 1)}, \qquad i \geqq 1, \tag{28}$$

$$\rho^{i-1} B_i(x, s) = \sum_{j=0}^{i-2} (-\rho x)^j \binom{-\alpha}{j} \frac{\Gamma(\alpha + i)}{\Gamma(\alpha + j + 2)}, \qquad i \geqq 2, \tag{29}$$

$$B_1(x, s) = 0. \tag{30}$$

Hadidi and Conolly (1969b) have shown that

$$\Gamma_1^*(1, s) = (\lambda + s) p_0^*(s) / [1 + \lambda p_0^*(s)], \tag{31}$$

where

$$p_0^*(s) = \mu e^{-\rho} \left[\frac{1}{s(s + \mu)} + \sum_{r=1}^{\infty} \frac{\rho^r}{r!\,(s + r\mu)\{s + (r + 1)\mu\}} \right]. \tag{32}$$

$\Gamma_k^*(1, s)$ is determined by (27), (31), and (32).

For the system $M_n/M/1$ Using (12) and (16), and $L(x, s, y)$ given by (20), we now have

$$\frac{\partial L}{\partial y} + \frac{(\lambda + \mu + s)y - 2\mu x y^2 - \lambda}{y^2(\mu + s - \mu x y)} L = \frac{2\mu x y - \lambda \Gamma_1^*}{y(\mu + s - \mu x y)}. \tag{33}$$

Solving (33), we have

$$L = (1 - \beta y)^{\alpha - 1} \sum_{i \geqq 0} (-\beta y)^i \binom{-\alpha}{i} \left[\Gamma_1^* y \,{}_2F_0\left(1, \alpha + i + 2; -; \frac{y}{\eta}\right) - \left(\frac{2xy^2}{\rho}\right) {}_2F_0\left(1, \alpha + i + 3; -; \frac{y}{\eta}\right) \right], \tag{34}$$

where now

$$\beta = \mu x/(\mu + s), \tag{35}$$

$$\eta = \lambda/(\mu + s), \tag{36}$$

$$\alpha = \eta(1 - \beta). \tag{37}$$

Taking coefficients as before, we have

$$\Gamma_k^* = \sum_{i=0}^{k-1} \binom{\alpha - 1}{i} (-\beta)^i \left[A'_{k-1}\Gamma_1^* - \left(\frac{2x}{\rho}\right) B'_{k-1} \right], \tag{38}$$

where

$$\eta^{i-1} A'_i(x, s) = \sum_{j=0}^{i-1} \binom{-\alpha}{j} (-\beta\eta)^j \frac{\Gamma(\alpha + i + 1)}{\Gamma(\alpha + j + 2)}, \qquad i \geqq 1, \tag{39}$$

$$\eta^{i-2} B'_i(x, s) = \sum_{j=0}^{i-2} \binom{-\alpha}{j} (-\beta\eta)^j \frac{\Gamma(\alpha + i + 1)}{\Gamma(\alpha + j + 3)}, \qquad i \geqq 2, \tag{40}$$

$$B'_1(x, s) = 0. \tag{41}$$

$\Gamma_1^*(1, s)$ is again given by (31), where now we have

$$p_0^*(s) = \frac{1}{s} - \left(\frac{\lambda + s}{s(\mu + s)}\right) \exp\left(-\frac{\lambda\mu}{(\mu + s)^2}\right) \cdot \sum_{k=0}^{\infty} \frac{(\lambda\mu)^k}{k!(\mu + s)^{2k}\{\lambda s/(s + \mu)^2 + k + 1\}} \tag{42}$$

[see Natvig (1973)]. As before, (38), (31), and (42) determine $\Gamma_k^*(1, s)$. Hadidi (1974) evaluates $E_k(t)$, the expected length of a busy period starting with k customers, for various values of k by derivation of $\Gamma_k^*(1, s)$ with respect to s at $s = 0$.

Hadidi (1976) considers a service policy of allotting to each customer a random length of time, as required and available, governed by the density function $b(t)$, given by 3.8(15), and shows that, for this model,[6] the probability of an empty queue is

$$P_0 = e^{-\rho}, \tag{43}$$

and the mean and the variance of waiting time in the steady state are

$$E(w) = (1 - \alpha)/(1 - \rho), \tag{44}$$

$$\operatorname{Var}(w) = \frac{1 + \alpha(\alpha - 2)}{4(1 - \rho)^2} - \frac{2(\alpha\rho \ln \rho - 1 - e^{-\rho})}{1 - \rho}, \tag{45}$$

[6] Hadidi's equivalent model is $M/G/1$, where $b(t)$ has the form given by 3.8(15).

where

(46) $$\alpha = \rho^{1-\rho}\gamma(\rho, \rho).$$

Hadidi (1976) also considers the case where momentary arrival rate depends on the state of the system, and shows that the "effective" interarrival time distribution in this case has a mean given by

(47) $$E(t) = \frac{\rho e^{\rho}}{e^{\rho} - 1}.$$

CHAPTER 4

Queues with Poisson Arrivals or Service

Here we describe the work on $M/G/1$ and $G/M/1$ queues. We start with Gaver (1954) describing the effect of various service time distributions in queuing systems and proceed to Luchak (1956, 1957) and Hasofer (1964, 1965) dealing with queuing processes with time-dependent Poisson input and general service. We include the work on double queues by Kashyap (1965b, 1966, 1967) and that on occupation time or busy period in $M/G/1$ queues by Takács (1964, 1976), Kashyap (1970a), Gaur and Kashyap (1975), and Heyman (1974). We describe the queuing system with Pareto-distributed service times discussed by Harris (1968b) and with service in phases with correlated service times studied by Alagaraja (1970) and some theorems on queues with balking discussed by Dick (1970). Finally, we discuss an approximate analysis of $M/G/1$ and $G/M/1$ queues given by Kashyap and Srivastava (1974), and results on a sequence of $M/G/1$ queues in heavy traffic given by Ott (1977).

4.1 The Effect of Service Distributions on Waiting Times

Consider a single-server queue with Poisson arrivals, where service is demanded and rendered in phases (Gaver, 1954), each customer requiring j phases with probability c_j $(j = 1, 2, \ldots)$. Let λ be the mean arrival rate, and let the time spent in each phase be exponentially distributed with mean $1/\mu$. Thus we use Erlang's method (Brockmeyer *et al.*, 1948) of exponential phases of service to simulate nonexponential service time distributions. The traffic intensity is given by

(1) $$\rho = \frac{\lambda}{\mu} \sum_{j=1}^{\infty} jc_j.$$

Defining

$$P_n(t) = \text{Prob}\{n \textit{ phases} \text{ in system at time } t\}, \tag{2}$$

we have the following differential difference equations:

$$P_0'(t) = -\lambda P_0(t) + \mu P_1(t); \tag{3}$$

$$P_n'(t) = -(\lambda + \mu)P_n(t) + \mu P_{n+1}(t) + \lambda \sum_{j=1}^{n} P_{n-j}(t)c_j \qquad (n = 1, 2, \ldots). \tag{4}$$

Using the generating function

$$G(z, t) = \sum_{n=0}^{\infty} P_n(t)z^n, \tag{5}$$

we have, from (3) and (4),

$$\frac{\partial G}{\partial t} + \left(\lambda + \mu - \frac{\mu}{z} - \lambda \sum_{j=1}^{\infty} c_j z^j\right)G(z, t) = \mu\left(1 - \frac{1}{z}\right)P_0(t). \tag{6}$$

Let $W(x, t)$ be the probability density of the time a customer arriving at time t waits in queue before service; then[1]

$$W(x, t) = P_0(t)\,\delta(x) + \sum_{n=1}^{\infty} \frac{P_n(t)e^{-\mu x}\mu^n x^{n-1}}{(n-1)!}. \tag{7}$$

Let $w(s, t)$ be the L.T. of $W(x, t)$, given by

$$w(s, t) = \int_0^{\infty} e^{-sx} W(x, t)\,dx. \tag{8}$$

Then (7) gives

$$w(s, t) = P_0(t) + \sum_{n=1}^{\infty} P_n(t)\left(\frac{\mu}{\mu + s}\right)^n = G\left(\frac{\mu}{\mu + s}, t\right). \tag{9}$$

Also, let $B(v)$ be the service time density function with L.T. $b(s)$; then we have

$$B(v) = \sum_{j=1}^{\infty} \frac{c_j e^{-\mu v}\mu^j v^{j-1}}{(j-1)!}, \tag{10}$$

whence

$$b(s) = \sum_{j=1}^{\infty} c_j\left(\frac{\mu}{\mu + s}\right)^j. \tag{11}$$

[1] We prefer the following form to Gaver's (1954, Eq. (4)), in which the lower limit of summation is $n = 0$, since this form has the facility of easy interpretation of the $n = 0$ term.

Setting $z = \mu/(\mu + s)$ in (6), and using (9) and (11), we obtain

$$\frac{\partial w(s,t)}{\partial t} + [\lambda\{1 - b(s)\} - s]w(s,t) = -sP_0(t). \tag{12}$$

For the steady state, we set $\partial w/\partial t = 0$ and obtain

$$w(s) = P_0/[1 - (\lambda/s)\{1 - b(s)\}]. \tag{13}$$

Letting $s \to 0$, we have

$$w(0) = P_0/(1 - \rho) = 1, \tag{14}$$

so that

$$P_0 = 1 - \rho. \tag{15}$$

Thus, for the steady state to exist we should have $\rho < 1$. Gaver considers the following particular cases.

Case I—k-Erlang or Pearson Type III Service Times In this case each customer is assumed to require the same number, k, of phases, so that

$$c_j = \delta_{jk} \qquad (k = 1, 2, \ldots), \tag{16}$$

the Kronecker delta. Therefore

$$b(s) = [\mu/(\mu + s)]^k. \tag{17}$$

For different values of k, we have the Erlang family of distributions. For $k = 1$, we evidently have the exponential service case. If we take limits such that $k \to \infty$, $\mu \to \infty$ in such a way that $k/\mu = T$, then

$$b(s) \to e^{-sT}, \tag{18}$$

which is the Laplace transform of a Dirac delta function located at T, i.e.,

$$B(v) = \delta(v - T), \tag{19}$$

so that we have the deterministic case of constant service time equal to T.

From (13), using (15) and (17), we have

$$w(s) = \frac{1 - \rho}{1 - (\lambda/s)[1 - \mu^k/(\mu + s)^k]}, \tag{20}$$

where

$$\rho = k\lambda/\mu. \tag{21}$$

Case II Here we assume the number of phases to have the geometric distribution, given by

$$c_j = (1 - a)a^{j-1} \qquad (j = 1, 2, \ldots), \tag{22}$$

which, from (10), corresponds to a negative exponential service time distribution with variable parameter:

(23) $$B(v) = (1 - a)\mu e^{-(1-a)\mu v}.$$

Case III Let the number of phases be distributed according to the modified Poisson distribution

(24) $$c_j = e^{-a}a^j/[(1 - e^{-a})j!] \qquad (j = 1, 2, \ldots).$$

Then, from (10), we have for the service time density

(25) $$B(v) = [e^{-(a+\mu v)}/(1 - e^{-a})]\sqrt{a\mu/v}\, I_1(2\sqrt{a\mu v}).$$

Using Erdélyi *et al.* (1954, Vol. I, p. 197, Eq. (16)), we have

(26) $$b(s) = [\exp\{a\mu/(\mu + s)\} - 1]/(e^a - 1)$$

and, therefore,

(27) $$w(s) = \frac{1 - \rho}{1 - (\lambda/s)[1 - \{\exp(a\mu/(\mu + s)) - 1\}/(e^a - 1)]}.$$

Gaver uses (20) and (27) to find the mean and variance of waiting times. He also draws the graphs of the service time density and average waiting times for Cases I and III and points out close resemblance for large values of k and a, respectively. He observes that the distribution given by (25) can fit a greater number of situations than the Erlang or Pearson type III family will handle and that the former has the additional merit of being integrable in terms of the well-tabulated Bessel functions.

4.2 Time-Dependent Arrival Rate and Service in Phases

Luchak (1956) considers a generalization of the model studied by Gaver (1954) and discussed in Section 4.1 by postulating a time-dependent mean arrival rate $\lambda(t)$ and a time-dependent parameter $\mu(t)$ of the exponential phases that comprise service. As before (Section 4.1), c_j is the probability that a service consists of j phases ($j = 1, 2, \ldots$). Considering first the case where μ is constant, we have the following equations for the system:

(1) $$\frac{dP_0(t)}{dt} = -\lambda(t)P_0(t) + \mu P_1(t);$$

(2) $$\frac{dP_n(t)}{dt} = -[\lambda(t) + \mu]P_n(t) + \mu P_{n+1}(t) + \lambda(t)\sum_{j=1}^{n} c_j P_{n-j}(t) \qquad (n \geqq 1).$$

Let us write [cf. Section 3.1]

$$\rho(t) = \lambda(t)/\mu, \tag{3}$$

$$\mu\, dt = d\tau, \tag{4}$$

$$P_n(\tau) = q_n(\tau)\exp\left[-\int_0^\tau \{1 + \rho(t)\}\, dt\right]; \tag{5}$$

then we have, from (1) and (2),

$$\frac{dq_0(\tau)}{d\tau} = q_1(\tau) + q_0(\tau), \tag{6}$$

$$\frac{dq_n(\tau)}{d\tau} = q_{n+1}(\tau) + \rho(\tau)\sum_{j=1}^{n} c_j q_{n-j}(\tau) \qquad (n \geqq 1). \tag{7}$$

We can write (6) and (7) in the vector-matrix form

$$\frac{d\mathbf{q}(\tau)}{d\tau} = \mathbf{q}(\tau)\mathbf{A}(\tau), \tag{8}$$

where

$$\mathbf{q}(\tau) = [q_0(\tau) \quad q_1(\tau) \quad q_2(\tau) \quad \cdots] \tag{9}$$

is an infinite row vector and

$$\mathbf{A}(\tau) = \begin{bmatrix} 1 & c_1\rho(\tau) & c_2\rho(\tau) & c_3\rho(\tau) & \cdots \\ 1 & 0 & c_1\rho(\tau) & c_2\rho(\tau) & \cdots \\ 0 & 1 & 0 & c_1\rho(\tau) & \cdots \\ 0 & 0 & 1 & 0 & \cdots \\ \vdots & \vdots & \vdots & \vdots & \end{bmatrix} \tag{10}$$

is an infinite matrix. To solve (8) with initial conditions

$$P_n(0) = q_n(0) = \omega_n \quad (n = 0, 1, 2, \ldots), \qquad 0 \leqq \omega_n \leqq 1, \qquad \sum_{n=0}^{\infty} \omega_n = 1, \tag{11}$$

we write

$$q_n(\tau) = \sum_{i=0}^{\infty} \omega_i q_{in}(\tau) \tag{12}$$

and solve

$$\frac{d\mathbf{Q}(\tau)}{d\tau} = \mathbf{Q}(\tau)\mathbf{A}(\tau) \tag{13}$$

with

$$\mathbf{Q}(0) = \mathbf{I}, \tag{14}$$

where $\mathbf{Q}(\tau)$ is the matrix $[q_{in}(\tau)]$ and $\mathbf{I}$ is the infinite unit matrix [see Ledermann and Reuter (1954)]. From (12), we note that it is necessary to obtain only the first column of $\mathbf{Q}(\tau)$ in order to get $q_0(\tau)$ and hence $P_0(\tau)$. Then $P_n(\tau)$ $(n = 1, 2, \ldots)$ can be found from (1) and (2).

Determination of c_j and μ

As in the last section, the service time density is

(15) $$B(v) = \sum_{j=1}^{\infty} \frac{c_j e^{-\mu v} \mu^j v^{j-1}}{(j-1)!},$$

whence, by Taylor's formula,

(16) $$\mu^j c_j = \left[\left(\frac{d}{dt}\right)^{j-1} \{B(v)e^{\mu v}\}\right]_{v=0}.$$

The c_j must satisfy the conditions

(17) $$\sum_{j=1}^{\infty} c_j = 1,$$

(18) $$0 \leqq c_j \leqq 1.$$

For general $B(v)$, (17) is automatically satisfied, while the c_j calculated from (16) may not satisfy (18) for any chosen value of μ, and Luchak points out that this is the case with the normal distribution

(19) $$B(v) = (2a/\sqrt{2\pi}) \exp(-a^2 v^2).$$

Such distributions, therefore, cannot be simulated by the above method. The best method of finding c_j and μ for simulating a given distribution will be to equate the first few moments M_n of the given distribution to those of (15), thus getting

(20) $$M_n = \mu^{-n} \sum_{j=1}^{\infty} \frac{(n+j-1)!c_j}{(j-1)!} \qquad (n = 1, 2, \ldots),$$

which may be solved for an appropriate solution subject to $\mu \geqq 0$ and (17), (18). Many distributions can be satisfactorily fitted by the Erlang family

(21) $$c_j = \delta_{jk},$$

where k and μ are found from the first two equations of (20), i.e., by equating only the mean and the variance.

Solution of Equations

We define infinite matrices $\boldsymbol{\alpha}$ and $\boldsymbol{\beta}$ as follows:

(22)
$$\boldsymbol{\alpha} = \begin{bmatrix} 1 & 0 & 0 & 0 & \cdots \\ 1 & 0 & 0 & 0 & \cdots \\ 0 & 1 & 0 & 0 & \cdots \\ 0 & 0 & 1 & 0 & \cdots \\ 0 & 0 & 0 & 1 & \cdots \\ \vdots & \vdots & \vdots & \vdots & \end{bmatrix},$$

(23)
$$\boldsymbol{\beta} = \begin{bmatrix} 0 & 1 & 0 & 0 & 0 & \cdots \\ 0 & 0 & 1 & 0 & 0 & \cdots \\ 0 & 0 & 0 & 1 & 0 & \cdots \\ 0 & 0 & 0 & 0 & 1 & \cdots \\ \vdots & \vdots & \vdots & \vdots & \vdots & \end{bmatrix}.$$

It can be verified by multiplication that $\boldsymbol{\alpha}$ and $\boldsymbol{\beta}$ have the following properties:

(a) $\boldsymbol{\beta\alpha} = \mathbf{I}$;
(b) $\boldsymbol{\alpha}^n$ is obtained by adding $(n-1)$ rows of $[1\,0\,0 \cdots]$ to the top of $\boldsymbol{\alpha}$;
(c) $\boldsymbol{\beta}^n$ is obtained by adding $(n-1)$ columns of zeros to the front of $\boldsymbol{\beta}$;
(d) because of property (a), every product of $\boldsymbol{\alpha}$ and $\boldsymbol{\beta}$ can be written as $\boldsymbol{\alpha}^n\boldsymbol{\beta}^m$;
(e) in $\boldsymbol{\alpha}^n\boldsymbol{\beta}^m$, every element is either 0 or 1 and the first m columns are zeros.

Then (13) can be written in the form

(24)
$$\frac{d\mathbf{Q}(\tau)}{d\tau} = \mathbf{Q}(\tau)\left[\boldsymbol{\alpha} + \rho(\tau)\sum_{j=1}^{\infty} c_j\beta^j\right].$$

Let us assume that $\rho(\tau)$ can be expressed in the form

(25)
$$\rho(\tau) = \sum_{n=0}^{\infty} \rho_n\tau^n.$$

Then a formal solution of (24) satisfying (14) is

(26)
$$\mathbf{Q}(\tau) = \sum_{n=0}^{\infty} \mathbf{Q}_n\tau^n,$$

where

(27)
$$\mathbf{Q}_n = \sum_{q=0}^{\infty}\sum_{r=0}^{n} a_r(n, q)\boldsymbol{\alpha}^{n-r}\boldsymbol{\beta}^q, \qquad \mathbf{Q}_0 = \mathbf{I},$$

and $a_r(n, q)$ are given by

(28) $$a_r(n, q) = 0 \quad \text{if} \quad r < q \quad \text{or} \quad r > n,$$

(29) $$(n + 1)a_{r+1}(n + 1, 0) = a_r(n, 1) + a_{r+1}(n, 0),$$

(30) $$(n + 1)a_{r+1}(n + 1, q) = a_r(n, q + 1) + \sum_{j=0}^{r} \rho_j \sum_{m=1}^{q} c_m a_{r-j}(n - j, q - m)$$

$$(q = 1, 2, \ldots).$$

Moreover, the solution (26) is convergent if $\sum_{n=0}^{\infty} |\rho_n| \tau^n$ converges. Luchak has found moments of the queue length and waiting time distributions for the time-dependent case and observes that

(31) $$W_q(\tau) = L(\tau)/\mu,$$

where $W_q(\tau)$ and $L(\tau)$ are the time-dependent expressions for the mean time spent in the *queue* and the mean number in the *system*.

For the case when $\rho(\tau) \equiv \rho$, a constant, it is shown for the *steady-state* case that

(32) $$P_0 = 1 - i,$$

(33) $$L = \left(\frac{1}{2}\right)\left(\frac{i}{1 - i}\right)\left[1 + \left(\sum_{j=1}^{\infty} j^2 c_j \Big/ \sum_{j=1}^{\infty} j c_j\right)\right],$$

(34) $$\operatorname{Var}(n) = L^2 + L + \left(\frac{1}{3}\right)\left(\frac{i}{1 - i}\right)\left[\left(\sum_{j=1}^{\infty} j^3 c_j \Big/ \sum_{j=1}^{\infty} j c_j\right) - 1\right],$$

where

(35) $$i = \rho \sum_j j c_j$$

is the traffic intensity, and $\operatorname{Var}(n)$ is the variance of number in the system.

ρ Constant and k-Erlang Service Times

Setting

(36) $$c_j = \delta_{jk},$$

and noting that the first column of the matrix $\boldsymbol{\alpha}^m$ has $(m + 1)$ successive 1 entries, after which all entries are 0, it is evident that the $(a, 0)$ element of $\mathbf{Q}(\tau)$ is

(37) $$q_{a0}(\tau) = \sum_{n=a}^{\infty} \tau^n \sum_{r=0}^{[(n-a)/(k+1)]} \frac{n - (rk + r - 1)}{r!(n - r + 1)!} \rho^r,$$

where $[x]$ stands for the integral part of x. Also, we have

(38) $$P_0(\tau, a) = q_{a0}(\tau) \exp[-(1 + \rho)\tau],$$

where

$$P_0(\tau, a) = \text{Prob}\{N(\tau) = 0 \mid N(0) = a\}, \tag{39}$$

$N(\tau)$ being the number in the system at time τ. From (37), we have

$$\left(\frac{d}{d\tau}\right) q_{a0}(\tau) - (1 + \rho) q_{a0}(\tau) = a\rho^{-a/(k+1)} \left(\frac{I_a^k(z)}{z}\right) - \rho \sum_{n=a+1}^{a+k} n\rho^{-n/(k+1)} \left(\frac{I_n^k(z)}{z}\right), \tag{40}$$

where

$$z = 2\rho^{1/(k+1)}\tau \tag{41}$$

and

$$I_n^k(z) = \sum_{r=0}^{\infty} \frac{(z/2)^{n+(k+1)r}}{r!(rk+n)!} = (\tfrac{1}{2}z)^n \phi(k, n+1; (\tfrac{1}{2}z)^{k+1}), \tag{42}$$

$\phi(\alpha, \beta; z)$ being the function introduced by Wright (1933) in his study of the asymptotic theory of partitions [see also II.5(22)].

From (38) and (40), we have

$$\frac{dP_0(z,a)}{dz} = \frac{a\rho^{-a/(k+1)} I_a^k(z)}{z} - \rho \sum_{n=a+1}^{a+k} \frac{n\rho^{-n/(k+1)} I_n^k(z)}{z} \exp\left[-\frac{(1+\rho)z}{2\rho^{1/(k+1)}}\right]. \tag{43}$$

By regrouping of the expanded form of (37), we have

$$P_0(z,a) = 2 \sum_{n=1}^{\infty} (a+n)\rho^{-(a+n-1)/(k+1)} \left(\frac{I_{a+n}^k(z)}{z}\right) \exp\left[-\frac{(1+\rho)z}{2\rho^{1/(k+1)}}\right]. \tag{44}$$

Exponential Service Times and $\rho = \rho_1 \tau$

In this case Eq. (29) holds and (30) takes the form

$$(n+1)a_{r+1}(n+1, q) = a_r(n, q+1) + \rho_1 a_{r-1}(n-1, q-1). \tag{45}$$

Equations (29) and (45) can be used for computations as before. However, the results could not be expressed in a closed form, and Luchak has conjectured that the solution in this case may not be expressible in terms of known tabulated functions or the $I_n^k(z)$ functions.

Generalizations

The method described above is applicable to the more general case where μ is also a function of time. We assume that $\lambda(t)$ and $\mu(t)$ are both expandable in powers of t, viz.,

$$\lambda(t) = \sum_{n=0}^{\infty} \lambda_n t^n, \tag{46}$$

$$\mu(t) = \sum_{n=0}^{\infty} \mu_n t^n. \tag{47}$$

Then, if

$$P_n(t) = q_n(t) \exp\left[- \int_0^t \{\lambda(\tau) + \mu(\tau)\}\, d\tau \right] \tag{48}$$

and $\mathbf{Q}(t)$ is defined as before (as a function of t instead of τ), the problem reduces to the solution of

$$\frac{d\mathbf{Q}(t)}{dt} = \mathbf{Q}(t)\left[\mu(t)\boldsymbol{\alpha} + \lambda(t) \sum_{j=1}^{\infty} c_j \boldsymbol{\beta}^j \right], \tag{49}$$

with

$$\mathbf{Q}(0) = \mathbf{I}. \tag{50}$$

Equations (26) and (27) hold, with τ replaced by t, and the $a_r(n, q)$ are given by

$$(n+1)a_{r+1}(n+1, 0) = \sum_{j=0}^{r+1} \mu_j a_{r-j+1}(n-j, 0) + \sum_{j=0}^{r} \mu_j a_{r-j}(n-j, 1), \tag{51}$$

$$\begin{aligned} (n+1)a_{r+1}(n+1, q) = {} & \sum_{j=0}^{r} \mu_j a_{r-j}(n-j, q+1) \\ & + \sum_{j=0}^{r} \lambda_j \sum_{m=1}^{q} c_m a_{r-j}(n-j, q-m) \end{aligned} \tag{52}$$

$(q = 1, 2, \ldots)$.

Busy Period Distribution

Assuming $n = 0$ to be an absorbing state and μ to be constant, we have the following equations for the system:

$$\frac{d}{dt} P_0(t) = \mu P_1(t); \tag{53}$$

$$\frac{d}{dt} P_1(t) = -[\lambda(t) + \mu] P_1(t) + \mu P_2(t); \tag{54}$$

$$\begin{aligned} \frac{d}{dt} P_n(t) = {} & -[\lambda(t) + \mu] P_n(t) + \mu P_{n+1}(t) \\ & + \lambda(t) \sum_{j=1}^{n-1} c_j P_{n-j}(t) \qquad (n = 2, 3, \ldots). \end{aligned} \tag{55}$$

Using the notation of (3)–(5), we can write (53)–(55) as follows:

(56) $$\frac{d}{d\tau} q_0(\tau) = [1 + \rho(\tau)]q_0(\tau) + q_1(\tau);$$

(57) $$\frac{d}{d\tau} q_1(\tau) = q_2(\tau);$$

(58) $$\frac{d}{d\tau} q_n(\tau) = q_{n+1}(\tau) + \rho(\tau) \sum_{j=1}^{n-1} c_j q_{n-j}(\tau) \qquad (n = 2, 3, \ldots).$$

Equations (57) and (58) can now be written

(59) $$\frac{d\mathbf{q}(\tau)}{d\tau} = \mathbf{q}(\tau)\mathbf{A}(\tau),$$

where

(60) $$\mathbf{q}(\tau) = [q_1(\tau) \quad q_2(\tau) \quad \cdots],$$

(61) $$\mathbf{A}(\tau) = \begin{bmatrix} 0 & c_1\rho(\tau) & c_2\rho(\tau) & c_3\rho(\tau) & \cdots \\ 1 & 0 & c_1\rho(\tau) & c_2\rho(\tau) & \cdots \\ 0 & 1 & 0 & c_1\rho(\tau) & \cdots \\ 0 & 0 & 1 & 0 & \cdots \\ \vdots & \vdots & \vdots & \vdots & \end{bmatrix}.$$

To solve (59) with initial conditions

(62) $$P_n(0) = q_n(0) = \omega_n, \qquad 0 \leqq \omega_n \leqq 1 \qquad (n = 0, 1, 2, \ldots),$$

$$\omega_0 = 0, \qquad \sum_{n=1}^{\infty} \omega_n = 1,$$

we write

(63) $$q_n(\tau) = \sum_{i=1}^{\infty} \omega_i q_{in}(\tau) \qquad (n = 1, 2, \ldots)$$

and solve

(64) $$\frac{d\mathbf{Q}(\tau)}{d\tau} = \mathbf{Q}(\tau)\mathbf{A}(\tau), \qquad \mathbf{Q}(0) = \mathbf{I},$$

where $\mathbf{Q}(\tau)$ is the matrix $[q_{in}(\tau)]$, as before.

To solve (64), we use the matrix $\boldsymbol{\beta}$ given by (23), and another infinite matrix $\boldsymbol{\gamma}$ defined as

(65) $$\boldsymbol{\gamma} = \begin{bmatrix} 0 & 0 & 0 & 0 & \cdots \\ 1 & 0 & 0 & 0 & \cdots \\ 0 & 1 & 0 & 0 & \cdots \\ 0 & 0 & 1 & 0 & \cdots \\ \vdots & \vdots & \vdots & \vdots & \end{bmatrix}.$$

The following properties can be verified by direct multiplication:

(a) $\boldsymbol{\beta}\boldsymbol{\gamma} = \mathbf{I}$;
(b) $\boldsymbol{\gamma}^n$ is obtained by adding $n - 1$ rows of zeros to the top of $\boldsymbol{\gamma}$;
(c) every product of $\boldsymbol{\beta}$ and $\boldsymbol{\gamma}$ can be written in the form $\boldsymbol{\gamma}^n\boldsymbol{\beta}^m$;
(d) every element of $\boldsymbol{\gamma}^n\boldsymbol{\beta}^m$ is either 0 or 1, and the first m columns and n rows of the matrix are zeros.

Now, (64) can be written

$$\frac{d\mathbf{Q}(\tau)}{d\tau} = \mathbf{Q}(\tau)\left[\boldsymbol{\gamma} + \rho(\tau)\sum_{j=1}^{\infty} c_j\boldsymbol{\beta}^j\right]. \tag{66}$$

Hence, following the solution method outlined above, it is shown that in the special case of *constant traffic intensity* ρ *and k-Erlang service times*, we have

$$q_{a1}(\tau) = \sum_{n=0}^{\infty} \frac{a(\tau\rho^{1/(1+k)})^{n(k+1)+a-1}}{n!(nk+a)!}. \tag{67}$$

Thus, following the notation of (39) and (41), we have

$$P_1(z,a) = 2a\rho^{(1-a)/(1+k)}[I_a^k(z)/z]\exp[-(1+\rho)z/2\rho^{1/(1+k)}]. \tag{68}$$

Therefore, using (53), we see that the density function of the duration of a busy period starting with a phases is

$$\frac{d}{dt}P_0(t,a) = \mu P_1(t,a), \tag{69}$$

where $P_1(\cdot,\cdot)$ is given by (68).

4.3 Nonhomogeneous Poisson Input and General Service

Consider a single-server queuing system with general service and Poisson arrivals with a time-dependent parameter $\lambda(t)$ (Hasofer, 1964). Let the distribution function of the service time v be

$$\text{Prob}\{v \leqq x\} = B(x). \tag{1}$$

Denote by $\zeta(t)$ the time required to serve all customers in the system at time t, and let

$$W(t,x) = \text{Prob}\{\zeta(t) \leqq x\}, \tag{2}$$

$$P_0(t) = \text{Prob}\{\zeta(t) = 0\}, \tag{3}$$

i.e., $P_0(t)$ is the probability that the server is idle at time t. Also let

$$w(t, s) = \int_0^\infty e^{-sx} \, d_x W(t, x), \tag{4}$$

$$\hat{w}(s) = w(0, s), \tag{5}$$

$$\psi(s) = \int_0^\infty e^{-sx} \, dB(x). \tag{6}$$

We make the following assumptions:

(a) $\lambda(t)$ is a continuous function of t.
(b) $B(x)$ is absolutely continuous, i.e., there exists $b(x)$ such that

$$B(x) = \int_0^x b(\xi) \, d\xi. \tag{7}$$

(c) $b(x)$ is an L^2-function, i.e., $\int_0^\infty e^{-2pt}|f(t)|^2 \, dt$ converges for some real value p_0 of p.
(d) $W(0, x)$ is absolutely continuous.
(e) Both $\psi(s)$ and $\hat{w}(s)$ are analytic in a neighborhood of the origin.

Let us set

$$\int_0^t \lambda(\tau) \, d\tau = \Lambda(t) = t - za(t), \tag{8}$$

where z is a parameter. It follows that $a(t)$ is a differentiable function of t and that its derivative is continuous. Only values of z such that $za(t)$ is real and $za'(t) < 1$ for all t are relevant for the queuing problem. However, z is allowed to take complex values for obtaining an explicit solution for $P_0(t)$, which now becomes a function of z and hence is denoted by $P_0(t, z)$. Using the theory developed by Reich (1958, 1959), Hasofer (1964) shows that

$$P_0(t, z) = \sum_{n=0}^{\infty} F_n(t) z^n, \tag{9}$$

the series converging for all z uniformly in $0 \leqq t \leqq T$ for an arbitrary finite T. Moreover, each $F_n(t)$ is a continuous function of t.

Considering the special case of simple harmonic input

$$a(t) = \sin \omega t, \tag{10}$$

so that

$$\lambda(t) = 1 - z\omega \cos \omega t, \tag{11}$$

it is shown that the Laplace transforms of $F_0(t)$ and $F_1(t)$ are given by

$$F_0^*(s) = \hat{w}(\eta_s)/\eta_s, \tag{12}$$

$$F_1^*(s) = \frac{1 - \psi(\eta_s)}{2i}\left[\frac{\hat{w}(\eta_r)}{\eta_r} - \frac{\hat{w}(\eta_q)}{\eta_q}\right], \tag{13}$$

where $\eta_s = \eta(s)$ is the unique root, in $\text{Re}(s) > 0$, of the equation

$$\eta - 1 + \psi(\eta) = s, \tag{14}$$

and

$$q = s + i\omega, \tag{15}$$

$$r = s - i\omega. \tag{16}$$

Now for small z, from (11), we can regard the queue to have a Poisson input with parameter $\lambda = 1$ that is slightly perturbed periodically by the term $-z\omega\cos\omega t$, and $F_1(t)$ is then the leading term whose behavior determines the departure of $P_0(t)$ from $F_0(t)$, it being evident that $P_0(t) = F_0(t)$ for the unperturbed queue (Hasofer, 1965). From (12), it follows that

$$\frac{1}{2i}\left[\frac{\hat{w}(\eta_r)}{\eta_r} - \frac{\hat{w}(\eta_q)}{\eta_q}\right] = \mathscr{L}[F_0(t)\sin\omega t]. \tag{17}$$

Also, it is well known [see, for example, Takács (1962, p. 58)] that for the unperturbed queue the Laplace transform of the busy period density function $\Gamma(t)$ is given by

$$\mathscr{L}[\Gamma(t)] = \psi(\eta_s). \tag{18}$$

Therefore, inverting (13), we have

$$F_1(t) = F_0(t)\sin\omega t - \int_0^t \Gamma(t-\tau)F_0(\tau)\sin\omega\tau\, d\tau. \tag{19}$$

Considering the particular case of exponential service with parameter μ, and assuming that the server is idle initially, we have [see Saaty (1961a, pp. 93, 128)]

$$F_0(t) = e^{-(1+\mu)t}\sum_{n=0}^{\infty} a_n I_n(2t\sqrt{\mu}), \tag{20}$$

where

$$a_0 = 1, \qquad a_1 = \sqrt{\mu}, \qquad a_n = (1 - 1/\mu)\mu^{n/2} \qquad (n > 1); \tag{21}$$

$$\Gamma(t) = e^{-(1+\mu)t}\sqrt{\mu}\, t^{-1} I_1(2t\sqrt{\mu}). \tag{22}$$

Using the identity

$$\int_0^t \frac{I_1[2a(t-\tau)]I_n(2a\tau)}{t-\tau}\,d\tau = I_{n+1}(2at), \tag{23}$$

we have

$$F_1(t) = e^{-(1+\mu)t}\Bigg[\sin\omega t\{I_0(2at) - I_2(2at)\} \tag{24}$$

$$+ \sqrt{\mu}\sum_{n=1}^{\infty} a_{n-1}\int_0^t \frac{I_1[2\sqrt{\mu}(t-\tau)]I_{n-1}(2\sqrt{\mu}\,\tau)}{t-\tau}(\sin\omega t - \sin\omega\tau)\,d\tau\Bigg].$$

Hasofer considers the asymptotic behavior of $F_1(t)$ and shows that, for large t,

$$F_1(t) \sim (1-\rho)f_1(t), \tag{25}$$

where

$$\rho = -\psi'(0) < 1, \tag{26}$$

$$f_1(t) = x\sin\omega t + (y-\omega)\cos\omega t, \tag{27}$$

x and y being given by

$$\eta(i\omega) = x + iy, \tag{28}$$

so that we can write

$$P_0(t,z) \sim (1-\rho)[1 + zf_1(t) + o(z)]. \tag{29}$$

4.4 The Generalized Double-Ended Queue

We consider a generalization of the queuing system discussed in Section 1.3 by assuming a general interarrival time distribution, with probability density $A(x)$, for customers (passengers), while the taxis arrive in a Poisson stream with mean rate μ, as before. There is limited waiting space for M taxis and N passengers (Kashyap, 1965b,c). We use the probabilities

$$P_n(x,t)\,dx, \qquad -M \leqq n \leqq N, \tag{1}$$

whose meaning is similar to that given in 1.3(13) with the addition that the time since the last customer arrival lies in the interval $(x, x+dx)$. Let $\eta(x)\Delta$ be the first-order probability that a customer arrives in time $(x, x+\Delta)$, given that there was no customer arrival up to x, so that

$$A(x) = \eta(x)\exp\left[-\int_0^x \eta(u)\,du\right]. \tag{2}$$

Following usual probability arguments [see, for example, Keilson and Koohārian (1960)], we have the following differential difference equations for the system:

(3) $$\left(\frac{\partial}{\partial t}+\frac{\partial}{\partial x}\right)P_{-M}(x,t)+\eta(x)P_{-M}(x,t)=\mu P_{-M+1}(x,t);$$

(4) $$\left(\frac{\partial}{\partial t}+\frac{\partial}{\partial x}\right)P_n(x,t)+[\mu+\eta(x)]P_n(x,t)$$
$$=\mu P_{n+1}(x,t)\qquad(-M+1\leqq n\leqq N-1);$$

(5) $$\left(\frac{\partial}{\partial t}+\frac{\partial}{\partial x}\right)P_N(x,t)+[\mu+\eta(x)]P_N(x,t)=0.$$

These equations are to be solved subject to the boundary conditions

(6) $$P_{-M}(0,t)=0,$$

(7) $$P_n(0,t)=\int_0^\infty P_n(x,t)\eta(x)\,dx\qquad(-M+1\leqq n\leqq N-1),$$

(8) $$P_N(0,t)=\int_0^\infty [P_{N-1}(x,t)+P_N(x,t)]\eta(x)\,dx.$$

We define the generating functions

(9) $$H(x,\alpha,t)=\sum_{n=-M}^{N}\alpha^n P_n(x,t),$$

(10) $$\Pi(\alpha,t)=\sum_{n=-M}^{N}\alpha^n P_n(t),$$

where $P_n(t)$ is the probability that n customers are waiting at time t, irrespective of the value of x, so that

(11) $$P_n(t)=\int_0^\infty P_n(x,t)\,dx$$

and, therefore,

(12) $$\Pi(\alpha,t)=\int_0^\infty H(x,\alpha,t)\,dx.$$

From Eqs. (3)–(5), we have

(13) $$\left(\frac{\partial}{\partial t}+\frac{\partial}{\partial x}\right)H(x,\alpha,t)+\left[\mu+\eta(x)-\frac{\mu}{\alpha}\right]H(x,\alpha,t)$$
$$=\mu\alpha^{-M}\left(1-\frac{1}{\alpha}\right)P_{-M}(x,t).$$

The boundary conditions similarly yield

$$H(0, \alpha, t) = \alpha \int_0^\infty H(x, \alpha, t)\eta(x)\,dx + \alpha^N(1 - \alpha)\int_0^\infty P_N(x, t)\eta(x)\,dx. \tag{14}$$

From (13), taking Laplace transforms with respect to t, denoted by the corresponding lower case letters, we have

$$\frac{\partial h}{\partial x} + \left[\mu + s + \eta(x) - \frac{\mu}{\alpha}\right]h(x, \alpha, s) = \alpha^i\delta(x) + \mu\alpha^{-M}\left(1 - \frac{1}{\alpha}\right)p_{-M}(x, s), \tag{15}$$

where we assume that the system starts with the arrival of a customer which makes the number in the system equal to i $(-M < i < N)$.

Equations (3)–(5) yield

$$\frac{\partial}{\partial x}p_{-M}(x, s) + [s + \eta(x)]p_{-M}(x, s) = \mu p_{-M+1}(x, s), \tag{16}$$

$$\frac{\partial}{\partial x}p_n(x, s) + [\mu + s + \eta(x)]\,p_n(x, s) = \mu p_{n+1}(x, s) + \delta_{in}\delta(x) \qquad (-M + 1 \leqq n \leqq N - 1), \tag{17}$$

$$\frac{\partial}{\partial x}p_N(x, s) + [\mu + s + \eta(x)]\,p_N(x, s) = 0, \tag{18}$$

and (6)–(8) and (14) give

$$p_{-M}(0, s) = 0, \tag{19}$$

$$p_n(0, s) = \int_0^\infty p_{n-1}(x, s)\eta(x)\,dx \qquad (-M + 1 \leqq n \leqq N - 1), \tag{20}$$

$$p_N(0, s) = \int_0^\infty [p_{N-1}(x, s) + p_N(x, s)]\eta(x)\,dx, \tag{21}$$

$$h(0, \alpha, s) = \alpha\int_0^\infty h(x, \alpha, s)\eta(x)\,dx - \alpha^N(\alpha - 1)\int_0^\infty p_N(x, s)\eta(x)\,dx. \tag{22}$$

The solution of (18) is

$$p_N(x, s) = p_N(0, s)\exp\left[-(\mu + s)x - \int_0^x \eta(u)\,du\right]. \tag{23}$$

Substituting (23) in (17) for $n = N - 1$ and solving the resulting differential equation, we have

(24) $$p_{N-1}(x, s) = [\mu x p_N(0, s) + p_{N-1}(0, s)] \exp\left[-(\mu + s)x - \int_0^x \eta(u)\, du \right].$$

Proceeding similarly, and solving recursively for $p_n(x, s)$ $(n = N - 2, N - 3, \ldots, i, \ldots, -M)$, we finally have

(25) $$p_{-M}(x, s) = \mu \exp\left[-sx - \int_0^x \eta(u)\, du \right] \cdot \left[\sum_{r=1}^{M+N} \int_0^x p_{r-M}(0, s) e^{-\mu y} \left(\frac{(\mu y)^{r-1}}{(r-1)!} \right) dy + \int_0^x e^{-\mu y} \left(\frac{(\mu y)^{M+i-1}}{(M+i-1)!} \right) dy \right].$$

Substituting (25) in (15) and solving, we have

(26) $$h(x, \alpha, s) = \exp\left[-\left(\mu + s - \frac{\mu}{\alpha} \right) x - \int_0^x \eta(u)\, du \right] \cdot \left[\alpha^i + h(0, \alpha, s) + \mu^2 \alpha^{-M} \left(1 - \frac{1}{\alpha} \right) \cdot \left\{ \int_0^x \exp\left[\mu \left(1 - \frac{1}{\alpha} \right) z \right] \sum_{r=1}^{M+N} p_{r-M}(0, s) \cdot \int_0^z e^{-\mu y} \left(\frac{(\mu y)^{r-1}}{(r-1)!} \right) dy\, dz + \int_0^x \exp\left[\mu \left(1 - \frac{1}{\alpha} \right) z \right] \int_0^z e^{-\mu y} \left(\frac{(\mu y)^{M+i-1}}{(M+i-1)!} \right) dy\, dz \right\} \right].$$

Now, evaluating the double integral by changing the order of integration, we have

(27) $$\int_0^x \int_0^z \exp\left[\mu \left(1 - \frac{1}{\alpha} \right) z \right] \left(\frac{(\mu y)^{r-1}}{(r-1)!} \right) \exp[-\mu y]\, dy\, dz = \left\{ \exp\left[\mu \left(1 - \frac{1}{\alpha} \right) x \right] \gamma(r, \mu x) - \alpha^r \gamma\left(r, \frac{\mu x}{\alpha} \right) \right\} \Big/ \left[(r-1)! \left(1 - \frac{1}{\alpha} \right) \mu^2 \right],$$

where γ is the incomplete gamma function given by

(28) $$\gamma(\alpha, x) = \int_0^x e^{-z} z^{\alpha - 1}\, dz = \alpha^{-1} x^\alpha\, {}_1F_1(\alpha; \alpha + 1; -x).$$

Therefore, (26) gives

$$(29)\qquad h(x,\alpha,s) = \exp\left[-\left(\mu + s - \frac{\mu}{\alpha}\right)x - \int_0^x \eta(u)\,du\right]$$
$$\cdot\left[\alpha^i + h(0,\alpha,s) + \alpha^{-M}\left\{\sum_{r=1}^{M+N} \frac{1}{(r-1)!} p_{r-M}(0,s)\right.\right.$$
$$\cdot\left[\exp\left(\mu\left(1-\frac{1}{\alpha}\right)x\right)\gamma(r,\mu x) - \alpha^r\gamma\left(r,\frac{\mu x}{\alpha}\right)\right]$$
$$+\frac{1}{(M+i-1)!}\left[\exp\left(\mu\left(1-\frac{1}{\alpha}\right)x\right)\gamma(M+i,\mu x)\right.$$
$$\left.\left.\left.-\alpha^{M+i}\gamma\left(M+i,\frac{\mu x}{\alpha}\right)\right]\right\}\right].$$

Substituting for $h(x,\alpha,s)$ from (29) in (26), we have

$$(30)\qquad h(0,\alpha,s) = [\alpha^{i+1} + \alpha h(0,\alpha,s)]a\left(\mu + s - \frac{\mu}{\alpha}\right)$$
$$-\alpha^N(\alpha-1)p_N(0,s)a(\mu+s)$$
$$+\alpha^{-M+1}\left[\sum_{r=1}^{M+N}\left\{\frac{1}{(r-1)!}p_{r-M}(0,s)\int_0^\infty A(x)\right.\right.$$
$$\cdot\exp\left[-\left(\mu+s-\frac{\mu}{\alpha}\right)x\right]$$
$$\left.\cdot\left(\exp\left[\mu\left(1-\frac{1}{\alpha}\right)x\right]\gamma(r,\mu x) - \alpha^r\gamma\left(r,\frac{\mu x}{\alpha}\right)\right)dx\right\}$$
$$+\frac{1}{(M+i-1)!}\int_0^\infty A(x)\exp\left[-\left(\mu+s-\frac{\mu}{\alpha}\right)x\right]$$
$$\cdot\left\{\exp\left[\mu\left(1-\frac{1}{\alpha}\right)x\right]\gamma(M+i,\mu x)\right.$$
$$\left.\left.-\alpha^{M+1}\gamma\left(M+i,\frac{\mu x}{\alpha}\right)\right\}dx\right].$$

Now, substituting for $p_n(x,s)$ $(-M \leqq n \leqq N)$, from Eqs. (23)–(25) in (20) and (21), we have $M+N$ equations in the $M+N$ unknowns $p_n(0,s)$ $(-M+1 \leqq n \leqq N)$, which are sufficient to determine them. Then, eliminating $h(0,\alpha,s)$ between (29) and (30), we get $h(x,\alpha,s)$. Hence $\pi(\alpha,s)$, the L.T. of $\Pi(\alpha,t)$, is given by

$$(31)\qquad \pi(\alpha,s) = \int_0^\infty h(x,\alpha,s)\,dx.$$

We consider the following particular cases.

Case I—k-Erlang Arrivals of Customers Let

$$A(x) = \lambda^k x^{k-1} e^{-\lambda x}/(k-1)!, \tag{32}$$

so that

$$a(s) = [\lambda/(\lambda + s)]^k, \tag{33}$$

$$\eta(x) = \lambda^k x^{k-1} e^{-\lambda x}/\Gamma(k, \lambda x), \tag{34}$$

$$\exp\left[-\int_0^x \eta(u)\, du\right] = \frac{\Gamma(k, \lambda x)}{(k-1)!}, \tag{35}$$

where $\Gamma(k, \lambda x)$ is the incomplete gamma function defined by [cf. II.1(53)]

$$\Gamma(\alpha, x) = \int_x^\infty e^{-z} z^{\alpha-1}\, dz = \Gamma(\alpha) - \gamma(\alpha, x). \tag{36}$$

On using (32), (33), and Erdélyi *et al.* (1954, Vol. II, p. 308, Eq. (15)), viz.

$$\int_0^\infty x^{\mu-1} e^{-\beta x} \gamma(v, \alpha x)\, dx = \left(\frac{\alpha^v \Gamma(\mu + v)}{v(\alpha + \beta)^{\mu+v}}\right) {}_2F_1\left(1, \mu + v; v + 1; \frac{\alpha}{\alpha + \beta}\right) \tag{37}$$

$$[\mathrm{Re}(\alpha + \beta) > 0, \mathrm{Re}(\beta) > 0, \mathrm{Re}(\mu + v) > 0],$$

we have, from (30),

$$\begin{aligned} h(0, \alpha, s) = {} & \left[1 - \alpha\left\{\frac{\lambda}{\lambda + \mu + s - \lambda/\alpha}\right\}^k\right]^{-1} \left[\alpha^{i+1}\left\{\frac{\lambda}{\lambda + \mu + s - \lambda/\alpha}\right\}^k\right. \\ & - \alpha^N(\alpha - 1) p_N(0, s)\left\{\frac{\lambda}{\lambda + \mu + s}\right\}^k \\ & + \alpha^{-M+1} \sum_{r=1}^{M+N} p_{r-M}(0, s)\left[\frac{(r + k - 1)!\lambda^k \mu^r}{r!(k-1)!(\lambda + \mu + s)^{r+k}}\right] \\ & \cdot\left\{{}_2F_1\left[1, r + k; r + 1; \frac{\mu}{\lambda + \mu + s}\right]\right. \\ & \left. - {}_2F_1\left[1, r + k; r + 1; \frac{\mu/\alpha}{\lambda + \mu + s}\right]\right\} \\ & + \left[\frac{\alpha^{-M+1}(M + i + k - 1)!\lambda^k \mu^{M+i}}{(M + i)!(k-1)!(\lambda + \mu + s)^{M+i+k}}\right] \\ & \cdot\left\{{}_2F_1\left[1, M + i + k; M + i + 1; \frac{\mu}{\lambda + \mu + s}\right]\right. \\ & \left.\left. - {}_2F_1\left[1, M + i + k; M + i + 1; \frac{\mu/\alpha}{\lambda + \mu + s}\right]\right\}\right]. \end{aligned} \tag{38}$$

Substituting for $h(0, \alpha, s)$ from (38) in (29), we get $h(x, \alpha, s)$. Then (31) yields

$$(39) \quad \pi(\alpha, s) = \{\alpha^i + h(0, \alpha, s)\}[\{1 - [\lambda/(\lambda + \mu + s - \mu/\alpha)]^k\}/\{\mu + s - \mu/\alpha\}] + \phi(\alpha, s),$$

where ϕ is given by

$$(40) \quad \phi(\alpha, s) = \alpha^{-M} \sum_{r=1}^{M+N} \frac{1}{(r-1)!} p_{r-M}(0, s)\left[\left[\int_0^\infty \exp\left\{-sx - \int_0^x \eta(u)\,du\right\}\right.\right.$$
$$\cdot \gamma(r, \mu x)\,dx$$
$$\left. - \alpha^r \int_0^\infty \exp\left\{-\left(\mu + s - \frac{\mu}{\alpha}\right)x - \int_0^x \eta(u)\,du\right\}\gamma\left(r, \frac{\mu x}{\alpha}\right)dx\right]$$
$$+ \left\{\frac{\alpha^{-M}}{(M + i - 1)!}\right\}\left[\left[\int_0^\infty \exp\left\{-sx - \int_0^x \eta(u)\,du\right\}\right.\right.$$
$$\cdot \gamma(M + i, \mu x)\,dx$$
$$- \alpha^{M+i} \int_0^\infty \exp\left\{-\left(\mu + s - \frac{\mu}{\alpha}\right)x - \int_0^x \eta(u)\,du\right\}$$
$$\left.\cdot \gamma\left(M + i, \frac{\mu x}{\alpha}\right)dx\right].$$

Now, using (35) and Erdélyi *et al.* (1954, Vol. I, p. 179, Eq. (34)), we have

$$(41) \quad \frac{1}{(r-1)!}\int_0^\infty \exp\left[-sx - \int_0^x \eta(u)\,du\right]\gamma(r, \mu x)\,dx$$
$$= \left(\frac{1}{s}\right)\left(\frac{\mu}{\mu + s}\right)^r - \left\{\frac{\lambda^k \mu^r}{k!r!}\right\}$$
$$\cdot \int_0^\infty e^{-sx} x^{k+r} {}_1F_1(k; k+1; -\lambda x)\, {}_1F_1(r; r+1; -\mu x)\,dx.$$

Using a special case of Erdélyi *et al.* (1954, Vol. I, p. 216, Eq. (14)), we have

$$(42) \quad \int_0^\infty e^{-sx} x^{k+r} {}_1F_1(k; k+1; -\lambda x)\, {}_1F_1(r; r+1; -\mu x)\,dx$$
$$= \frac{\Gamma(k + r + 1)}{(s + \lambda + \mu)^{k+r+1}}$$
$$\cdot F_2\left[k + r + 1, 1, 1; k + 1, r + 1; \frac{\lambda}{s + \lambda + \mu}, \frac{\mu}{s + \lambda + \mu}\right],$$

where $\operatorname{Re}(k + r + 1) > 0$, $\operatorname{Re}(s) > 0$, $\operatorname{Re}(s + \lambda + \mu) > 0$, and F_2 is Appell's function defined by [see also II.6(5)]

$$F_2(\alpha, \beta, \beta'; \gamma, \gamma'; x, y) = \sum_{m=0}^{\infty} \sum_{n=0}^{\infty} \frac{(\alpha)_{m+n}(\beta)_m(\beta')_n}{(\gamma)_m(\gamma')_n} \frac{x^m}{m!} \frac{y^n}{n!}. \tag{43}$$

Thus, (40) simplifies to

$$\begin{aligned}
\phi(\alpha, s) = \alpha^{-M} \sum_{r=1}^{M+N} p_{r-M}(0, s) \Bigg[&\left(\frac{1}{s} - \frac{1}{\mu + s - \mu/\alpha}\right)\left(\frac{\mu}{\mu + s}\right)^r \\
&- \frac{(k + r)!\lambda^k \mu^r}{k!r!(s + \lambda + \mu)^{k+r+1}} \\
&\cdot \left\{F_2\left(k+r+1, 1, 1; k+1, r+1; \frac{\lambda}{s+\lambda+\mu}, \frac{\mu}{s+\lambda+\mu}\right)\right. \\
&\left.- F_2\left(k+r+1, 1, 1; k+1, r+1; \frac{\lambda}{s+\lambda+\mu}, \frac{\mu/\alpha}{s+\lambda+\mu}\right)\right\}\Bigg] \\
&+ \alpha^{-M}\Bigg[\left(\frac{1}{s} - \frac{1}{\mu + s - \mu/\alpha}\right)\left(\frac{\mu}{\mu + s}\right)^{M+i} \\
&- \frac{(k + M + i)!\lambda^k \mu^{M+i}}{k!(M + i)!(s + \lambda + \mu)^{k+M+i+1}} \\
&\cdot \left\{F_2\left(k+M+i+1, 1, 1; k+1, M+i+1; \frac{\lambda}{s+\lambda+\mu}, \frac{\mu}{s+\lambda+\mu}\right)\right. \\
&\left.- F_2\left(k+M+i+1, 1, 1; k+1, M+i+1; \frac{\lambda}{s+\lambda+\mu}, \frac{\lambda/\alpha}{s+\lambda+\mu}\right)\right\}\Bigg].
\end{aligned} \tag{44}$$

Case II—Poisson Arrivals of Customers Setting $k = 1$ in (38), and simplifying, we have

$$\begin{aligned}
h(0, \alpha, s) = \left(\frac{1}{\lambda + \mu + s - \mu/\alpha - \lambda\alpha}\right)\Bigg\{ &\mu\alpha^{i+1} - \alpha^N(\alpha - 1)p_N(0, s) \\
&\cdot \frac{\lambda(\lambda + \mu + s - \mu/\alpha)}{\lambda + \mu + s} \\
&+ \frac{\lambda\mu\alpha^{-M}(\alpha - 1)}{\lambda + s}\Bigg[\sum_{r=1}^{M+N}\left(\frac{\mu}{\lambda + \mu + s}\right)^r p_{r-M}(0, s) \\
&+ \left(\frac{\mu}{\lambda + \mu + s}\right)^{M+i}\Bigg]\Bigg\}.
\end{aligned} \tag{45}$$

Similarly, from (39) and (44),

$$(46)\quad \pi(\alpha, s) = \frac{\alpha^i + h(0, \alpha, s)}{\lambda + \mu + s - \mu/\alpha} + \alpha^{-M}\left\{\frac{1}{s+\lambda} - \frac{1}{s + \lambda + \mu - \mu/\alpha}\right\}$$
$$\cdot\left[\sum_{r=1}^{M+N}\left(\frac{\mu}{\lambda+\mu+s}\right)^r p_{r-M}(0, s) + \left(\frac{\mu}{\lambda+\mu+s}\right)^{M+i}\right].$$

Equation (46) contains two unknowns $p_N(0, s)$ and

$$\sum_{r=1}^{M+N} \{\mu/(\lambda + \mu + s)\}^r p_{r-M}(0, s),$$

which can be determined by requiring that in (45) the numerator vanish at the two zeros of the denominator, i.e., the roots α_1, α_2 of the equation

$$(47)\qquad \lambda\alpha^2 - (s + \lambda + \mu)\alpha + \mu = 0.$$

Substituting these values in $\pi(\alpha, s)$, we obtain the generating function $\pi(\alpha, s)$, which agrees with 1.3(14).

Further Results

Let us now define two generating functions (Kashyap, 1967) as follows:

$$(48)\qquad F(x, \alpha, t) = \sum_{n=-M}^{-1} \alpha^n P_n(x, t);$$

$$(49)\qquad G(x, \alpha, t) = \sum_{n=0}^{N} \alpha^n P_n(x, t).$$

Then, the probabilities that at time t there is a queue of taxis, and of customers, are given, respectively, by

$$(50)\qquad P_T(t) = \Pi_1(1, t),$$

$$(51)\qquad P_C(t) = \Pi_2(1, t) - P_0(t),$$

where

$$(52)\qquad \Pi_1(\alpha, t) = \sum_{n=-M}^{-1} \alpha^n P_n(t) = \int_0^\infty F(x, \alpha, t)\, dx,$$

$$(53)\qquad \Pi_2(\alpha, t) = \sum_{n=0}^{N} \alpha^n P_n(t) = \int_0^\infty G(x, \alpha, t)\, dx,$$

and $P_n(t)\,(-M \leq n \leq N)$ is the marginal probability that the system is in state n at time t. Also, the mean numbers of taxis and of customers in the queue at time t are given by

$$L_T(t) = -\frac{\partial}{\partial\alpha}\Pi_1(\alpha, t)\bigg|_{\alpha=1}, \tag{54}$$

$$L_C(t) = \frac{\partial}{\partial\alpha}\Pi_2(\alpha, t)\bigg|_{\alpha=1}. \tag{55}$$

Considering the particular case of *k-Erlang arrivals of customers* and using the same procedure as described above, we show that the L.T. of $P_T(t)$ is

$$\begin{aligned}
(56)\quad p_T(s) &= \pi_1(1, s) \\
&= f(0,1,s)\left(\frac{1}{s}\right)\left[1 - \left(\frac{\lambda}{\lambda+s}\right)^k\right] \\
&\quad + \sum_{r=0}^{N} p_r(0,s)\left[\left(\frac{1}{s}\right)\left(\frac{\mu}{\mu+s}\right)^{r+1} - \left(\frac{(k+r+1)!\lambda^k\mu^{r+1}}{k!(r+1)!(\lambda+\mu+s)^{k+r+2}}\right)\right. \\
&\quad \left.\cdot F_2\left(k+r+2, 1, 1; k+1, r+2; \frac{\lambda}{\lambda+\mu+s}, \frac{\mu}{\lambda+\mu+s}\right)\right] \\
&\quad + \frac{\{\mu/(\mu+s)\}^{i+1}}{s} - \left\{\frac{(k+i+1)!\lambda^k\mu^{i+1}}{k!(i+1)!(\lambda+\mu+s)^{k+i+2}}\right\} \\
&\quad \cdot F_2\left(k+i+2, 1, 1; k+1, i+2; \frac{\lambda}{\lambda+\mu+s}, \frac{\mu}{\lambda+\mu+s}\right),
\end{aligned}$$

where

$$\begin{aligned}
(57)\quad f(0,1,s) &= \left[1 - \left(\frac{\lambda}{\lambda+s}\right)^k\right]^{-1}\left[-p_0(0,s)\right. \\
&\quad + \sum_{r=0}^{N} p_r(0,s)\left(\frac{(k+r)!\lambda^k\mu^{r+1}}{(k-1)!(r+1)!(\lambda+\mu+s)^{k+r+1}}\right) \\
&\quad \cdot {}_2F_1\left(1, k+r+1; r+2; \frac{\mu}{\lambda+\mu+s}\right) \\
&\quad + \left(\frac{(k+i)!\lambda^k\mu^{i+1}}{(k-1)!(i+1)!(\lambda+\mu+s)^{k+i+1}}\right) \\
&\quad \left.\cdot {}_2F_1\left(1, k+i+1; i+2; \frac{\mu}{\lambda+\mu+s}\right)\right],
\end{aligned}$$

and the L.T. of $P_C(t)$ is

$$p_C(s) = \pi_2(1, s) - p_0(s), \tag{58}$$

where

$$(59)\quad \pi_2(1,s) = \{1 + g(0,1,s)\}\frac{1 - \{\lambda/(\lambda+s)\}^k}{s}$$

$$- \sum_{r=0}^{N} p_r(0,s)\left[\left(\frac{1}{s}\right)\left(\frac{\mu}{\mu+s}\right)^{r+1} - \frac{(k+r+1)!\lambda^k\mu^{r+1}}{k!(r+1)!(\lambda+\mu+s)^{k+r+2}}\right.$$

$$\left.\cdot F_2\left(k+r+2,1,1;k+1,r+2;\frac{\lambda}{\lambda+\mu+s},\frac{\mu}{\lambda+\mu+s}\right)\right]$$

$$-\left[\left(\frac{1}{s}\right)\left(\frac{\mu}{\mu+s}\right)^{i+1} - \frac{(k+i+1)!\lambda^k\mu^{i+1}}{k!(i+1)!(\lambda+\mu+s)^{k+i+2}}\right.$$

$$\left.\cdot F_2\left(k+i+2,1,1;k+1,i+2;\frac{\lambda}{\lambda+\mu+s},\frac{\mu}{\lambda+\mu+s}\right)\right],$$

$$(60)\quad g(0,1,s) = \left[1 - \left(\frac{\lambda}{\lambda+s}\right)^k\right]^{-1}\left[\left(\frac{\lambda}{\lambda+s}\right)^k\right.$$

$$- \sum_{r=0}^{N} p_r(0,s)\left[\frac{(k+r)!\lambda^k\mu^{r+1}}{(k-1)!(r+1)!(\lambda+\mu+s)^{k+r+1}}\right]$$

$$\cdot {}_2F_1\left(1,k+r+1;r+2;\frac{\mu}{\lambda+\mu+s}\right)$$

$$- \frac{(k+i)!\lambda^k\mu^{i+1}}{(k-1)!(i+1)!(\lambda+\mu+s)^{k+i+1}}$$

$$\left.\cdot {}_2F_1\left(1,k+i+1;i+2;\frac{\mu}{\lambda+\mu+s}\right)\right],$$

$$(61)\qquad p_0(s) = \sum_{r=0}^{N}\left(\frac{1}{\mu}\right)p_r(0,s)\left[\left(\frac{\mu}{\mu+s}\right)^{r+1} - \frac{\lambda^k\mu^{r+1}(k+r)!}{k!r!(\lambda+\mu+s)^{k+r+1}}\right.$$

$$\left.\cdot {}_2F_1\left(1,k+r+1;k+1;\frac{\lambda}{\lambda+\mu+s}\right)\right]$$

$$+\left(\frac{1}{\mu}\right)\left[\left(\frac{\mu}{\mu+s}\right)^{i+1} - \frac{\lambda^k\mu^{i+1}(k+i)!}{k!i!(\lambda+\mu+s)^{k+i+1}}\right.$$

$$\left.\cdot {}_2F_1\left(1,k+i+1;k+1;\frac{\lambda}{\lambda+\mu+s}\right)\right].$$

Using (54) and (55), the Laplace transforms $l_T(s)$ and $l_C(s)$ have been found in terms of the functions ${}_2F_1$ and F_2.

In the special case of *Poisson arrivals of customers*, for $k = 1$, all the functions ${}_2F_1$ and F_2 occurring in the above analysis simplify in terms of convergent binomial series. In particular, we have

$$(62)\quad f(0, \alpha, s) = \frac{\lambda A + \lambda\mu\alpha^{-M}(\alpha - 1)B/(\lambda + s) - (\lambda + \mu + s - \mu/\alpha)C}{\lambda + \mu + s - \lambda\alpha - \mu/\alpha},$$

$$(63)\quad \begin{aligned} g(0, \alpha, s) = {} & [\lambda\alpha^{i+1} - \lambda A + (\lambda + \mu + s - \mu/\alpha)C \\ & + \alpha^N(1 - \alpha)\{\lambda/(\lambda + \mu + s)\}(\lambda + \mu + s - \mu/\alpha)D] \\ & \cdot (\lambda + \mu + s - \lambda\alpha - \mu/\alpha)^{-1}, \end{aligned}$$

where

$$(64)\quad A = \sum_{r=0}^{N} p_r(0, s)\left(\frac{\mu}{\lambda + \mu + s}\right)^{r+1} + \left(\frac{\mu}{\lambda + \mu + s}\right)^{i+1},$$

$$(65)\quad B = \sum_{r=1}^{M+N} p_{r-M}(0, s)\left(\frac{\mu}{\lambda + \mu + s}\right)^{r} + \left(\frac{\mu}{\lambda + \mu + s}\right)^{M+i},$$

$$(66)\quad C = p_0(0, s),$$

$$(67)\quad D = p_N(0, s).$$

Now, the numerators of (62) and (63) must vanish at the zeros α_1 and α_2 of the denominator, α_2 having the + sign before the radical. Thus, we have four equations, solving which we obtain

$$(68)\quad B = \left(\frac{\lambda(\lambda + s)}{\mu s}\right)(\alpha_1\alpha_2)^{M+i+1}\frac{\alpha_2^{N-i}(\alpha_2 - 1) + \alpha_1^{N-i}(1 - \alpha_1)}{\alpha_2^{M+N+1} - \alpha_1^{M+N+1}},$$

$$(69)\quad D = \left(\frac{\lambda + \mu + s}{s}\right)\frac{\alpha_2^{M+i+1}(1 - \alpha_1) + \alpha_1^{M+i+1}(\alpha_2 - 1)}{\alpha_2^{M+N+1} - \alpha_1^{M+N+1}},$$

$$(70)\quad A = \left(\frac{\mu}{\lambda + s}\right)\left(\frac{\alpha_1^{M+1}(\alpha_2 - 1) + \alpha_2^{M+1}(1 - \alpha_1)}{(\alpha_1\alpha_2)^M(\alpha_2 - \alpha_1)}\right)B,$$

$$(71)\quad C = \left(\frac{\mu}{\lambda + s}\right)\left(\frac{\alpha_1^{M}(\alpha_2 - 1) + \alpha_2^{M}(1 - \alpha_1)}{(\alpha_1\alpha_2)^M(\alpha_2 - \alpha_1)}\right)B.$$

Setting $k = 1$ in the results obtained above, we have

$$(72)\quad p_T(s) = (A - C)/s,$$

$$(73)\quad p_C(s) = (1 + C - A)/s - A/\mu,$$

$$(74)\quad p_0(s) = A/\mu,$$

$$(75)\quad l_T(s) = (\mu + s - \lambda)A/s^2 - (\mu - \lambda)C/s^2 - B\mu/\{s(\lambda + s)\},$$

$$(76)\quad \begin{aligned} l_C(s) = {} & i/s + (\lambda - \mu)/s^2 + (\mu - \lambda + s)A/s^2 \\ & + (\lambda - \mu)C/s^2 - \lambda D/\{s(\lambda + \mu + s)\}. \end{aligned}$$

4.5 The Double-Ended Queue with Bulk Service

Consider the generalization of the queuing system described in Section 4.4 to the case of S-seated taxis, where a taxi takes a fixed number S of customers from the queue, or the whole queue, whichever is less, the arrivals of customers and taxis being, respectively, general and Poisson as before. With the notation of Section 4.4, the system is represented by the equations

$$(1)\qquad \left(\frac{\partial}{\partial t}+\frac{\partial}{\partial x}+\eta(x)\right)P_{-M}(x,t)=\mu P_{-M+1}(x,t),$$

$$(2)\qquad \left(\frac{\partial}{\partial t}+\frac{\partial}{\partial x}+\{\mu+\eta(x)\}\right)P_n(x,t)=\mu P_{n+1}(x,t)$$

$$(-M+1\leqq n\leqq -1),$$

$$(3)\qquad \left(\frac{\partial}{\partial t}+\frac{\partial}{\partial x}+\{\mu+\eta(x)\}\right)P_0(x,t)=\mu\sum_{j=1}^{S}P_j(x,t),$$

$$(4)\qquad \left(\frac{\partial}{\partial t}+\frac{\partial}{\partial x}+\{\mu+\eta(x)\}\right)P_n(x,t)=\mu P_{n+S}(x,t)\qquad (1\leqq n\leqq N-S),$$

$$(5)\qquad \left(\frac{\partial}{\partial t}+\frac{\partial}{\partial x}+\{\mu+\eta(x)\}\right)P_n(x,t)=0\qquad (N-S+1\leqq n\leqq N),$$

with the same boundary and initial conditions as before. We assume that the initial state i of the system is such that $0<i<S$. We define the following generating functions:

$$(6)\qquad F(x,\alpha,t)=\sum_{n=-M}^{-1}\alpha^nP_n(x,t);$$

$$(7)\qquad G(x,\alpha,t)=\sum_{n=1}^{N}\alpha^nP_n(x,t);$$

$$(8)\qquad \Pi_1(\alpha,t)=\sum_{n=-M}^{-1}\alpha^nP_n(t)=\int_0^\infty F(x,\alpha,t)\,dx;$$

$$(9)\qquad \Pi_2(\alpha,t)=\sum_{n=1}^{N}\alpha^nP_n(t)=\int_0^\infty G(x,\alpha,t)\,dx.$$

Using the generating functions and taking Laplace transforms, we have

$$(10)\qquad \frac{\partial f(x,\alpha,s)}{\partial x}+\left\{\mu+s+\eta(x)-\frac{\mu}{\alpha}\right\}f(x,\alpha,s)$$

$$=\mu\alpha^{-1}p_0(x,s)+\mu\alpha^{-M}\left(1-\frac{1}{\alpha}\right)p_{-M}(x,s),$$

$$\text{(11)}\qquad \frac{\partial g(x,\alpha,s)}{\partial x} + \{\mu + s + \eta(x) - \mu\alpha^{-S}\}g(x,\alpha,s) = -\mu\alpha^{-S}\sum_{j=1}^{S}\alpha^j p_j(x,s) + \alpha^i\delta(x),$$

$$\text{(12)}\qquad \frac{\partial p_0(x,s)}{\partial x} + \{\mu + s + \eta(x)\}\,p_0(x,s) = \mu\sum_{j=1}^{S} p_j(x,s).$$

Solving recursively Eqs. (1)–(5), after taking Laplace transforms, starting from the last equation and working backward, we obtain $p_n(x,s)$ $(n = N, N-1, \ldots, -M)$. Thus, assuming

$$\text{(13)}\qquad cS \leqq N < (c+1)S \qquad (c \text{ an integer}),$$

we have

$$\text{(14)}\qquad \sum_{j=1}^{S} p_j(x,s) = \exp\left[-(\mu+s)x - \int_0^x \eta(u)\,du\right] \cdot\left[1 + \sum_{r=1}^{(c+1)^*}\sum_{j=(r-1)S+1}^{rS}\frac{p_j(0,s)(\mu x)^{r-1}}{(r-1)!}\right],$$

$$\text{(15)}\qquad \sum_{j=1}^{S}\alpha^j p_j(x,s) = \exp\left[-(\mu+s)x - \int_0^x \eta(u)\,du\right] \cdot\left[\alpha^i + \sum_{r=1}^{(c+1)^*}\sum_{j=(r-1)S+1}^{rS}\frac{\alpha^j p_j(x,s)(\mu x\alpha^{-S})^{r-1}}{(r-1)!}\right],$$

where the asterisk signifies that, for the case $r = c+1$, the summation over j extends up to $j = N$ only. In what follows the double summation involved in (14) and (15) will be abbreviated as $\sum\sum$. Also,

$$\text{(16)}\qquad p_0(x,s) = \exp\left[-(\mu+s)x - \int_0^x \eta(u)\,du\right] \cdot\left\{p_0(0,s) + \mu x + \sum\sum\frac{p_j(0,s)(\mu x)^r}{r!}\right\},$$

$$\text{(17)}\qquad p_{-M}(x,s) = \mu\exp\left[-sx - \int_0^x \eta(u)\,du\right]\left[\int_0^x e^{-\lambda y}\left\{\frac{(\lambda y)^M}{M!}\right\}dy + \sum_{r=1}^{M} p_{r-M}(0,s)\int_0^x e^{-\lambda y}\left\{\frac{(\lambda y)^{r-1}}{(r-1)!}\right\}dy + \sum\sum p_j(0,s)\int_0^x e^{-\lambda y}\left\{\frac{(\lambda y)^{r+M-1}}{(r+M-1)!}\right\}dy\right].$$

We solve (10) and (11) after substitution from (14)–(17) as in Section 4.4. Then, $\pi_1(\alpha, s)$ and $\pi_2(\alpha, s)$ are obtained by using (8) and (9). Considering the case of k-Erlang arrivals of customers and proceeding as before, we have

$$
\begin{aligned}
(18)\quad \pi_1(1,s) &= \mathscr{L}[\text{Prob}\{\text{taxis waiting at time } t\}] \\
&= \{f(0,1,s) + p_0(0,s) + 1\}\frac{1-\{\lambda/(\lambda+s)\}^k}{s} \\
&\quad - \{p_0(0,s)+1\}\frac{1-\{\lambda/(\lambda+\mu+s)\}^k}{\mu+s} - \frac{\mu}{(\mu+s)^2} \\
&\quad + \frac{(k+1)\mu\lambda^k}{(\lambda+\mu+s)^{k+2}}\, {}_2F_1\left(1, k+2; k+1; \frac{\lambda}{\lambda+\mu+s}\right) \\
&\quad + \sum\sum p_j(0,s)\left[\left(\frac{1}{s}\right)\left\{\frac{\mu}{\mu+s}\right\}^k\right. \\
&\quad - \left\{\frac{\lambda^k\mu^{r+1}(r+k+1)!}{(r+1)!k!(\lambda+\mu+s)^{r+k+2}}\right\} \\
&\quad \left.\cdot F_2\left(r+k+2, 1, 1; k+1, r+2; \frac{\lambda}{\lambda+\mu+s}, \frac{\mu}{\lambda+\mu+s}\right)\right],
\end{aligned}
$$

where

$$
\begin{aligned}
(19)\quad f(0,1,s) &= \left[1-\left\{\frac{\lambda}{\lambda+s}\right\}^k\right]^{-1}\left[\{p_0(0,s)+1\}\left\{\left(\frac{\lambda}{\lambda+s}\right)^k - \left(\frac{\lambda}{\lambda+\mu+s}\right)^k\right\}\right. \\
&\quad - \frac{k\mu\lambda^k}{(\lambda+\mu+s)^{k+1}} - p_0(0,s) \\
&\quad + \sum\sum p_j(0,s)\left\{\frac{\lambda^k\mu^{r+1}(r+k)!}{(\lambda+\mu+s)^{r+k+1}(r+1)!(k-1)!}\right\} \\
&\quad \left.\cdot\, {}_2F_1\left(1, r+k+1; r+2; \frac{\mu}{\lambda+\mu+s}\right)\right],
\end{aligned}
$$

and

$$
\begin{aligned}
(20)\quad \pi_2(1,s) &= \mathscr{L}[\text{Prob}\{\text{customers waiting at time } t\}] \\
&= g(0,1,s)\frac{1-\{\lambda/(\lambda+s)\}^k}{s} + \frac{1-\{\lambda/(\lambda+\mu+s)\}^k}{\mu+s} \\
&\quad - \sum\sum p_j(0,s)\left[\frac{\{\mu/(\mu+s)\}^k}{s} - \left\{\frac{\lambda^k\mu^r(r+k)!}{(\lambda+\mu+s)^{r+k+1}r!k!}\right\}\right. \\
&\quad \left.\cdot F_2\left(r+k+1, 1, 1; k+1, r+1; \frac{\lambda}{\lambda+\mu+s}, \frac{\mu}{\lambda+\mu+s}\right)\right],
\end{aligned}
$$

where

(21) $$g(0,1,s) = \left[1 - \left\{\frac{\lambda}{\lambda + s}\right\}^k\right]^{-1} \left[\left\{\frac{\lambda}{\lambda + \mu + s}\right\}^k + p_1(0,s) - \sum\sum p_j(0,s)\left\{\frac{(r+k-1)!\lambda^k\mu^r}{r!(k-1)!(\lambda+\mu+s)^{r+k}}\right\} \cdot {}_2F_1\left(1, r+k; r+1; \frac{\mu}{\lambda+\mu+s}\right)\right].$$

4.6 Occupation Times and Busy Periods

Consider an $M/G/1$ queue and let $\eta(t)$ be the occupation time of the server at time t, i.e., the time needed to complete the service of all customers who are present in the system at time t (Takács, 1964). Let $\chi(t)$ be the total service time of all those customers who arrive in the time interval $[0, t]$. If $\alpha(t)$ is the total idle time of the server in the time interval $(0, t)$, it is easy to see that

(1) $$\alpha(t) = \sup\{0 \text{ and } u - \chi(u) - \eta(0) \text{ for } 0 \leqq u \leqq t\}.$$

Also,

(2) $$\eta(t) = \sup\{\chi(t) - \chi(u) - (t - u) \text{ for } 0 \leqq u \leqq t, \text{ and } \eta(0) + \chi(t) - t\}.$$

Denote by $\beta(t)$ the total occupation time of the server in $(0, t)$; then obviously

(3) $$\alpha(t) + \beta(t) = t,$$

and, from (1) and (2),

(4) $$\beta(t) = \eta(0) + \chi(t) - \eta(t).$$

Consider the case of additive input where $\{\chi(t), 0 \leqq t < \infty\}$ is a separable stochastic process with nonnegative, stationary, independent increments such that $\chi(t)$ increases only in jumps and is continuous to the right, and let $\chi(0) = 0$. For such a process, we have (Takács, 1962)

(5) $$\text{Prob}\{\chi(u) \leqq u \text{ for } 0 \leqq u \leqq t \mid \chi(t) = y\} = \begin{cases} 1 - y/t & \text{if } 0 \leqq y \leqq t \\ 0 & \text{otherwise,} \end{cases}$$

where the conditional probability is defined up to an equivalence. Using (5) and probability arguments, Takács (1964) proves that if $c < x \leqq t$, then

(6) $$\text{Prob}\{\alpha(t) \geqq x - c \mid \eta(0) = c\} = \int_x^t \left(\frac{x}{y}\right) d_y \text{Prob}\{\chi(y) \leqq y - x\}.$$

Also, using the strong law of large numbers and the central limit theorem, he shows that if

(7) $$E\{\chi(t)\} = \rho t,$$

(8) $$\operatorname{Var}\{\chi(t)\} = \sigma^2 t,$$

where $\rho < 1$ and σ^2 is a finite positive number, then

(9) $$\lim_{t\to\infty} \operatorname{Prob}\left\{\frac{\beta(t) - \rho t}{\sqrt{\sigma^2 t}} \leqq x\right\} = \Phi(x),$$

independently of the distribution of $\eta(0)$, $\Phi(x)$ being the normal distribution function with mean 0 and variance 1.

In the particular case when

(10) $$\operatorname{Prob}\{\chi(t) \leqq x\} = \frac{1}{\Gamma(t)}\int_0^{\mu x} e^{-y} y^{t-1}\, dy,$$

where μ is a positive number, we have, from (6),

(11) $$\operatorname{Prob}\{\alpha(t) \geqq x - c \mid \eta(0) = c\}$$
$$= \int_0^{t-x} \frac{\mu x}{\Gamma(x + y + 1)} e^{-\mu y}(\mu y)^{x+y-1}\, dy, \qquad c < x \leqq t,$$

and, from (9),

(12) $$\lim_{t\to\infty} \operatorname{Prob}\left\{\frac{\beta(t) - (t/\mu)}{\sqrt{t/\mu^2}} \leqq x\right\} = \Phi(x).$$

Let θ_0 be the length of the initial busy period and let $\xi(t)$ denote the number in the system at time t. If $\xi(0) = 0$ or $\eta(0) = 0$, then $\theta_0 = 0$; otherwise, θ_0 is a positive random variable that may be infinite with a positive probability (Takács, 1976). Let us write

(13) $$P(x, k|i) = \operatorname{Prob}\{\theta_0 \leqq x \text{ and } \xi(t) \leqq k \text{ for } 0 \leqq t \leqq \theta_0 \mid \xi(0) = i\}$$
$$(x \geqq 0; 0 \leqq i \leqq k),$$

i.e., $P(x, k|i)$ is the joint probability distribution for the length of the initial busy period and the maximal queue (system) size during that period, conditioned on the initial queue size i. Let us define the Laplace–Stieltjes transform (L.S.T.)

(14) $$\Pi(s, k|i) = \int_0^{\infty} e^{-sx}\, d_x P(x, k|i), \qquad \operatorname{Re}(s) \geqq 0.$$

Then, it is proved (Takács, 1976, Theorem 1) that

$$\Pi(s, k|i) = Q_{k-i}(s)/Q_k(s), \tag{15}$$

where

$$\sum_{k=0}^{\infty} Q_k(s)z^k = \frac{\psi(s + \lambda - \lambda z)}{\psi(s + \lambda - \lambda z) - z}, \qquad |z| < |\gamma(s)|, \tag{16}$$

$$\psi(s) = \int_0^{\infty} e^{-sx}\, dB(x) \tag{17}$$

is the L.S.T. of the service time distribution $B(x)$, and we define $z = \gamma(s)$ as the root of the equation

$$z = \psi(s + \lambda - \lambda z) \tag{18}$$

that has the smallest absolute value. The proof of (15) is outlined below.

We observe that

$$P(x, k|k - i) = \int_0^x P(x - u, k|k - j)\, d_u P(u, j|j - i), \tag{19}$$

$$(0 \leqq i \leqq j \leqq k; x \geqq 0),$$

which is obviously true for $j = i$, because

$$P(u, j|0) = 1 \qquad \text{for} \quad u \geqq 0. \tag{20}$$

That (19) is true for $j > i$ is seen by considering the circumstance that the queue size decreases from $k - j + 1$ to $k - j$ for the first time at time u, where $0 \leqq u \leqq x$. Taking L.S.T.s in (19), we have

$$\Pi(s, k|k - i) = \Pi(s, k|k - j)\Pi(s, j|j - i). \tag{21}$$

Setting $i = 0$ in (21), we have

$$\Pi(s, k|k - j) = Q_j(s)/Q_k(s), \tag{22}$$

where

$$Q_k(s) = 1/\Pi(s, k|k), \tag{23}$$

$$Q_0(s) \equiv 1. \tag{24}$$

Taking into consideration that during the first service time of the initial busy period the number of arrivals may be $j = 0, 1, 2, \ldots$, we can write

$$\Pi(s, k + i|i) = \sum_{j=0}^{k} \pi_j(s)\Pi(s, k + i|i + j - 1) \qquad (i \geqq 1; k \geqq 0), \tag{25}$$

where

$$\pi_j(s) = \frac{\lambda^j}{j!} \int_0^\infty \exp[-(s+\lambda)x]\, x^j \, dB(x) \qquad (j = 0, 1, 2, \ldots). \tag{26}$$

From (22) and (25) we have the following recurrence formula for $Q_k(s)$:

$$Q_k(s) = \sum_{j=0}^{k} \pi_j(s) Q_{k+1-j}(s) \qquad (k = 0, 1, 2, \ldots). \tag{27}$$

Also, since

$$\sum_{j=0}^{\infty} \pi_j(s) z^j = \int_0^\infty \exp[-(s+\lambda-\lambda z)x]\, dB(x)$$
$$= \psi(s+\lambda-\lambda z) \qquad [\mathrm{Re}(s) \geqq 0, |z| \leqq 1], \tag{28}$$

from (28) we obtain (16). The function $Q_k(s)$ can be expressed explicitly as a polynomial of $1/\pi_0(s)$ and $\pi_j(s)$, $j = 1, 2, \ldots, k-1$.

Consider the particular case of the $M/M/1$ queue, where

$$\psi(s) = \mu/(\mu + s), \tag{29}$$

and, therefore, from (16),

$$\sum_{k=0}^{\infty} Q_k(s) z^k = \frac{\mu}{\lambda z^2 - (\lambda + \mu + s)z + \mu} \tag{30}$$

for $\mathrm{Re}(s) \geqq 0$ and sufficiently small $|z|$. If we write

$$Q_k(s) = (\lambda/\mu)^{k/2} U_k[(\lambda + \mu + s)/2\sqrt{\lambda\mu}] \qquad (k \geqq 0), \tag{31}$$

then (30) gives

$$\sum_{k=0}^{\infty} U_k(x) z^k = \frac{1}{z^2 - 2xz + 1} \qquad (|z| \text{ sufficiently small}), \tag{32}$$

and, therefore, $U_k(x)$, $k = 0, 1, 2, \ldots$, are Chebyshev polynomials of the second kind (Erdélyi *et al.*, 1953, Vol. II, p. 186, Eq. (31)). Hence, we have

$$U_k(\cos\phi) = \sin(k+1)\phi/\sin\phi \qquad (k = 0, 1, 2, \ldots). \tag{33}$$

Since

$$x = \cos[r\pi/(k+1)], \qquad r = 1, 2, \ldots, k, \tag{34}$$

are the k roots of the equation

$$U_k(x) = 0, \tag{35}$$

by Lagrange's formula we obtain

$$(36)\qquad U_{k-i}(x)/U_k(x)$$
$$= \sum_{r=1}^{k} \frac{U_{k-i}[\cos(r\pi/(k+1))]}{U_k'[\cos(r\pi/(k+1))]\{x - \cos(r\pi/(k+1))\}}$$
$$= \frac{1}{k+1} \sum_{r=1}^{k} \frac{\sin(r\pi/(k+1))\sin(ri\pi/(k+1))}{x - \cos(r\pi/(k+1))}$$
$$(1 \leqq i \leqq k; |x| \geqq 1).$$

From (15) and (31), we have

$$(37)\quad \Pi(s, k|i) = (\mu/\lambda)^{i/2} U_{k-i}[(\lambda + \mu + s)/2\sqrt{\lambda\mu}]/U_k[(\lambda + \mu + s)/2\sqrt{\lambda\mu}].$$

Now, (36) and (37) give

$$(38)\qquad P(x, k|i) = \left(\frac{2}{k+1}\right)(\lambda\mu)^{1/2}\left(\frac{\mu}{\lambda}\right)^{i/2}$$
$$\cdot \sum_{r=1}^{k} \sin\left(\frac{r\pi}{k+1}\right)\sin\left(\frac{ri\pi}{k+1}\right)(1 - e^{g_r})/g_r,$$

where

$$(39)\qquad g_r = \lambda + \mu - 2\sqrt{\lambda\mu}\cos(r\pi/(k+1)) \qquad (1 \leqq i \leqq k;\ x \geqq 0).$$

Also,

$$(40)\qquad P(\infty, k|i) = \lim_{s\to 0}\,[Q_{k-i}(s)/Q_k(s)]$$
$$= \begin{cases} \mu^i(\lambda^{k-i+1} - \mu^{k-i+1})/(\lambda^{k+1} - \mu^{k+1}) & (\lambda \neq \mu) \\ (k - i + 1)/(k + 1) & (\lambda = \mu). \end{cases}$$

Maximal Waiting Time

Let us define (Takács, 1976)

$$(41)\qquad F(x, y|c) = \text{Prob}\{\theta_0 \leqq x \text{ and } \eta(t) \leqq y \text{ for } 0 \leqq t \leqq \theta_0 \,|\, \eta(0) = c\}$$
$$(x \geqq 0; 0 \leqq c \leqq y),$$

so that $F(x, y|c)$ is the joint probability distribution of the length of the initial busy period and the maximal virtual waiting time, given that the initial occupation time of the server is c. $\eta(t)$ is the virtual waiting time at time t, i.e., the time a customer would have to wait if he arrived at time t. We define the L.S.T.

$$(42)\qquad \phi(s, y|c) = \int_0^\infty e^{-sx}\,d_xF(x, y|c) \qquad [\text{Re}(s) \geqq 0; 0 \leqq c \leqq y].$$

Then, we have (Takács, 1976, Theorem 2)

$$\phi(s, y|c) = W(s, y-c)/W(s, y), \tag{43}$$

where

$$\int_0^\infty e^{-\omega y}\, d_y W(s, y) = \frac{\omega}{\omega - s - \lambda\{1 - \psi(\omega)\}} \tag{44}$$

$$(\mathrm{Re}(\omega) > \mathrm{Re}[s + \lambda - \lambda\gamma(s)]).$$

The proof of (43) is outlined below.

The process $\{\eta(t),\ 0 \leqq t < \infty\}$ is a homogeneous Markov process; therefore,

$$\phi(s, y|y-c) = \phi(s, u|u-c)\phi(s, y|y-u) \qquad (0 \leqq c \leqq u;\ \mathrm{Re}(s) \geqq 0). \tag{45}$$

Setting $c = 0$ in (45), we have

$$\phi(s, y|y-u) = W(s, u)/W(s, y) \qquad (0 \leqq u \leqq y), \tag{46}$$

where

$$W(s, y) = 1/\phi(s, y|y), \tag{47}$$

$$\phi(s, y|0) \equiv 1. \tag{48}$$

For $u \geqq t$ and $y \geqq 0$, we have

$$\phi(s, u+y|u) = \exp[-(s+\lambda)t]\Big\{\phi(s, u+y|u-t) + \lambda t \int_0^y \phi(s, u+y|u+v-t)\, dB(v) + o(t)\Big\}, \tag{49}$$

which has been written by the considerations that during $(0, t)$, either 0 or 1 or more than 1 customer may arrive. Multiplying (49) by $W(s, u+y)$ and using (46), we have

$$W(s, y) = \exp[-(s+\lambda)t]\Big\{W(s, y+t) + \lambda t \int_0^y W(s, y-v+t)\, dB(v) + o(t)\Big\}, \tag{50}$$

whence, on dividing by t and taking limits as $t \to 0$,

$$\frac{\partial W(s, y)}{\partial y} = (\lambda + s)W(s, y) - \lambda \int_0^y W(s, y-v)\, dB(v). \tag{51}$$

Let $\Omega(s, \omega)$ be the L.S.T. of $W(s, y)$, which is defined by the left-hand side of

(44); then (51) gives

$$\omega[\Omega(s,\omega) - W(s,0)] = (\lambda + s)\Omega(s,\omega) - \lambda\Omega(s,\omega)\psi(\omega) \tag{52}$$

$$[\mathrm{Re}(\omega) > \mathrm{Re}(s + \lambda - \lambda\gamma(s));\ \mathrm{Re}(s) \geqq 0],$$

which shows that (44) holds. This completes the proof.

For the $M/M/1$ queue, it is shown that

$$F(x,y|c) = \sum_{j=0}^{(x-c)/y} \left(\frac{\lambda}{\mu}\right)^j \exp[(\lambda-\mu)jy]\, H_{2j}(x+jy, 2jy+c) \tag{53}$$

$$- \sum_{j=1}^{x/y} \left(\frac{\lambda}{\mu}\right)^j \exp[(\lambda-\mu)(jy-c)]$$

$$\cdot\, H_{2j}(x+jy-c, 2jy-c) \qquad (0 \leqq c \leqq y; x \geqq 0),$$

where the $H_r(x,a)$ are given by

$$\int_0^\infty e^{-sx}\, d_x H_r(x,a) = \exp[-a\{s+\lambda-\lambda\gamma(s)\}]\,[\gamma(s)]^r \tag{54}$$

$$(r = 0, 1, 2, \ldots; a \geqq 0).$$

If we write

$$Q_r(x,a) = \int_0^\infty u^{-1} \exp[-(\lambda+\mu)u] \left(\frac{\mu u}{\lambda(u+a)}\right)^{r/2} \tag{55}$$

$$\cdot\, I_r[(4\lambda\mu u(u+a))^{1/2}]\, du,$$

where $I_r(z)$ is the modified Bessel function, then it is shown that, for $0 \leqq x < a$,

$$H_r(x,a) = 0, \tag{56}$$

and, for $x \geqq a$,

$$H_0(x,a) = e^{-\lambda a}[1 + \lambda a Q_1(x-a, a)], \tag{57}$$

$$H_r(x,a) = e^{-\lambda a}[rQ_r(x-a,a) + \lambda a Q_{r+1}(x-a,a)] \qquad (r \geqq 1). \tag{58}$$

Also,

$$F(\infty, y|c) = \lim_{s\to 0} W(s, y-c)/W(y) \tag{59}$$

$$= \begin{cases} \exp(\mu c)\dfrac{[\lambda\exp\{\lambda(y-c)\} - \mu\exp\{\mu(y-c)\}]}{\lambda\exp(\lambda y) - \mu\exp(\mu y)} & (\lambda \neq \mu) \\[2ex] \dfrac{1+\lambda y - \lambda c}{1+\lambda y} & (\lambda = \mu). \end{cases}$$

4.7 Busy Periods for the Double-Ended Queue

We study busy periods for the double queues, studied in Section 4.4, under two headings: (1) customer periods and (2) taxi periods, as defined in Section 1.3, under the assumption of general and Poisson arrivals for customers and taxis, respectively (Kashyap, 1965c, 1970a).

Customer Periods

With the notation of Section 4.4, we shall assume the system to start with the arrival of a customer, which makes $n = 1$, and stops when n drops to the value zero for the first time. We have the following equations,

(1) $$\left(\frac{\partial}{\partial t} + \frac{\partial}{\partial x}\right) P_0(x, t) = \mu P_1(x, t),$$

(2) $$\left(\frac{\partial}{\partial t} + \frac{\partial}{\partial x} + \{\mu + \eta(x)\}\right) P_n(x, t) = \mu P_{n+1}(x, t) \qquad (1 \leqq n \leqq N - 1),$$

(3) $$\left(\frac{\partial}{\partial t} + \frac{\partial}{\partial x} + \{\mu + \eta(x)\}\right) P_N(x, t) = 0,$$

with the boundary conditions

(4) $$P_0(0, t) = 0,$$

(5) $$P_1(0, t) = \delta(t),$$

(6) $$P_n(0, t) = \int_0^\infty P_{n-1}(x, t)\eta(x)\,dx \qquad (2 \leqq n \leqq N - 1),$$

(7) $$P_N(0, t) = \int_0^\infty [P_{N-1}(x, t) + P_N(x, t)]\eta(x)\,dx.$$

Also, the initial condition is

(8) $$P_n(x, 0) = \delta_{1n}\delta(x).$$

The marginal probabilities $P_n(t)$ are given by

(9) $$P_n(t) = \int_0^\infty P_n(x, t)\,dx \qquad (0 \leqq n \leqq N).$$

If $\Gamma_C(t)$ is the probability density function of the duration of a customer period,

we have

$$\Gamma_C(t) = \frac{d}{dt} P_0(t). \tag{10}$$

Denoting Laplace transforms by lowercase letters, we have, from (10),

$$\gamma_C(s) = sp_0(s) = s \int_0^\infty p_0(x, s)\, dx. \tag{11}$$

From (1), on taking L.T.s and solving, we have

$$p_0(x, s) = e^{-sx} \int_0^x \mu e^{sy} p_1(y, s)\, dy. \tag{12}$$

Also, taking L.T.s in (2) and (3) and solving the various equations for $p_n(x, s)$ ($n = N, N - 1, \ldots, 1$), as in Section 4.4, we get

$$p_1(x, s) = \exp\left[-(\mu + s)x - \int_0^x \eta(u)\, du\right]\left[1 + \sum_{r=1}^{N} \frac{p_r(0, s)(\mu x)^{r-1}}{(r-1)!}\right]. \tag{13}$$

From (11)–(13), we have

$$\gamma_C(s) = \mu \int_0^\infty \exp\left[-(\mu + s)x - \int_0^x \eta(u)\, du\right] \cdot \left\{1 + \sum_{r=1}^{N} \frac{p_r(0, s)(\mu x)^{r-1}}{(r-1)!}\right\} dx. \tag{14}$$

For *k-Erlang arrivals of customers*, we have

$$\exp\left[-\int_0^x \eta(u)\, du\right] = \frac{\Gamma(k, \lambda x)}{(k-1)!} = 1 - \frac{\gamma(k, \lambda x)}{(k-1)!}, \tag{15}$$

and, evaluating the integrals as in Section 4.4, we have

$$\gamma_C(s) = \sum_{r=1}^{N} p_r(0, s)\left\{\left(\frac{\mu}{\mu + s}\right)^r - \frac{(k + r - 1)!\lambda^k \mu^r}{k!(r-1)!(\lambda + \mu + s)^{k+r}} \cdot {}_2F_1\left(1, k + r; k + 1; \frac{\lambda}{\lambda + \mu + s}\right)\right\} + \frac{\mu}{\mu + s}\left\{1 - \left(\frac{\lambda}{\lambda + \mu + s}\right)^k\right\}, \tag{16}$$

where $p_r(0, s)$ ($r = 1, \ldots, N$) can be determined as in Section 4.4.

For *Poisson arrivals* of customers, setting $k = 1$, (16) simplifies to

$$\gamma_C(s) = \sum_{r=1}^{N} \left(\frac{\lambda}{\lambda + \mu + s}\right)^r p_r(0, s) + \frac{\mu}{\lambda + \mu + s}, \tag{17}$$

where the $p_r(0, s)$ are given as before.

Alternatively, in the Poisson-arrival case, we may proceed as follows. Using the generating function

$$G(x, \alpha, t) = \sum_{n=1}^{N} \alpha^n P_n(x, t), \tag{18}$$

and taking Laplace transforms, we have, from (2), (3), and (8),

$$\frac{\partial g(x, \alpha, s)}{\partial x} + \left\{\mu + s + \eta(x) - \frac{\mu}{\alpha}\right\} g(x, \alpha, s) = \alpha\delta(x) - \mu p_1(x, s). \tag{19}$$

Solving (19) after substituting for $p_1(x, s)$ from (13), we have

$$\begin{aligned} g(x, \alpha, s) = \exp&\left[-\left(\mu + s - \frac{\mu}{\alpha}\right)x - \int_0^x \eta(u)\,du\right] \\ &\cdot \left\{g(0, \alpha, s) + \alpha e^{-\mu x/\alpha} - \sum_{r=1}^{N} p_r(0, s)\left\{\frac{\alpha^r}{(r-1)!}\right\}\gamma\left(r, \frac{\mu x}{\alpha}\right)\right\}, \end{aligned} \tag{20}$$

and, from the boundary conditions (5)–(7),

$$\begin{aligned} g(0, \alpha, s) = \alpha &\int_0^\infty g(x, \alpha, s)\eta(x)\,dx \\ &+ \alpha^N(1 - \alpha)\int_0^\infty p_N(x, s)\eta(x)\,dx. \end{aligned} \tag{21}$$

From (3), on taking L.T.s and solving, we get

$$p_N(x, s) = p_N(0, s)\exp\left[-(\mu + s)x - \int_0^x \eta(u)\,du\right]. \tag{22}$$

Substituting for $g(x, \alpha, s)$ and $p_N(x, s)$ from (20) and (22) in (21), and noting that for the Poisson case $\eta(x) \equiv \lambda$, we get, after simplification and solving for $g(0, \alpha, s)$,

$$\begin{aligned} g(0, \alpha, s) = &\left(\frac{1}{\lambda + \mu + s - \mu/\alpha - \lambda\alpha}\right) \\ &\cdot \left[\lambda\alpha^2 - \lambda\alpha\left\{\sum_{r=1}^{N}\left(\frac{\mu}{\lambda + \mu + s}\right)^r p_r(0, s) + \frac{\mu}{\lambda + \mu + s}\right\}\right. \\ &\left. + \frac{\lambda}{\lambda + \mu + s}\alpha^N(1 - \alpha)p_N(0, s)\left(\lambda + \mu + s - \frac{\mu}{\alpha}\right)\right]. \end{aligned} \tag{23}$$

By requiring that the numerator of (23) vanish at the two zeros α_1 and α_2 of the denominator, which are the same as those given by 1.1(5), we get two equations, which are sufficient to determine

$$\sum_{r=1}^{N} \left\{\frac{\mu}{\lambda + \mu + s}\right\}^r p_r(0, s) + \frac{\mu}{\lambda + \mu + s} \qquad \text{and} \qquad p_N(0, s).$$

These equations give

$$\sum_{r=1}^{N} \left\{\frac{\mu}{\lambda + \mu + s}\right\}^r p_r(0, s) + \frac{\mu}{\lambda + \mu + s} = \frac{(\alpha_1\alpha_2)[\alpha_2^{N-1}(\alpha_2 - 1) + \alpha_1^{N-1}(1 - \alpha_1)]}{\alpha_2^N(\alpha_2 - 1) + \alpha_1^N(1 - \alpha_1)}, \tag{24}$$

which, from (17), is the required expression for $\gamma_C(s)$ and tallies with 1.3(27).

Taxi Periods

Assuming the system to start with the arrival of a taxi, which makes $n = -1$, and to stop when n becomes zero for the first time, we have the following equations for the system,

$$\left(\frac{\partial}{\partial t} + \frac{\partial}{\partial x}\right) P_0(x, t) = 0, \tag{25}$$

$$\left(\frac{\partial}{\partial t} + \frac{\partial}{\partial x} + \{\mu + \eta(x)\}\right) P_{-1}(x, t) = 0, \tag{26}$$

$$\left(\frac{\partial}{\partial t} + \frac{\partial}{\partial x} + \{\mu + \eta(x)\}\right) P_n(x, t) = \mu P_{n+1}(x, t) \quad (-2 \geqq n \geqq -M + 1), \tag{27}$$

$$\left(\frac{\partial}{\partial t} + \frac{\partial}{\partial x} + \eta(x)\right) P_{-M}(x, t) = \mu P_{-M+1}(x, t), \tag{28}$$

with the boundary conditions

$$P_n(0, t) = \int_0^\infty P_{n-1}(x, t)\eta(x)\, dx \qquad (0 \geqq n \geqq -M + 1) \tag{29}$$

$$P_{-M}(0, t) = 0. \tag{30}$$

Also, the initial conditions are

$$P_n(x, 0) = \delta_{-1,n}\delta(x - x_0), \tag{31}$$

where x_0 is the initial elapsed time since the last customer arrival.

From (25), on taking L.T.s and solving, we have

$$p_0(x, s) = p_0(0, s)e^{-sx}, \tag{32}$$

whence

$$p_0(s) = \int_0^\infty p_0(x, s)\,dx = \frac{p_0(0, s)}{s}. \tag{33}$$

Therefore,

$$\gamma_T(s) = sp_0(s) = p_0(0, s), \tag{34}$$

where $\gamma_T(s)$ is the L.T. of $\Gamma_T(t)$, the probability density function of the duration of a taxi period. Now, $p_0(0, s)$ can be found as explained in Section 4.4. The analysis can be simplified in the case of k-Erlang arrivals of customers by using (15).

In the case of Poisson arrivals of taxis, we may use a generating function

$$F(x, \alpha, t) = \sum_{n=-M}^{-1} \alpha^n P_n(x, t), \tag{35}$$

and, proceeding as above, we obtain

$$p_0(0, s) = \frac{\alpha_1^M(\alpha_2 - 1)\exp(\mu\alpha_1 x_0) + \alpha_2^M(1 - \alpha_1)\exp(\mu\alpha_2 x_0)}{\alpha_1^{M+1}(\alpha_2 - 1) + \alpha_2^{M+1}(1 - \alpha_1)}, \tag{36}$$

which, from (34), is the expression for $\gamma_T(s)$, conditional on an initial elapsed time x_0. Since, for the Poisson distribution, x_0 is irrelevant, we set $x_0 = 0$ to obtain the analog of (24).

4.8 Busy Periods in a Priority Queue

Consider a queuing system with preemptive priority discipline with two priority classes, which we shall call priority and nonpriority classes (Gaur and Kashyap, 1975). The distribution of interarrival times is taken to be general with density function $A(x)$, given by

$$A(x) = \eta(x)\exp\left[-\int_0^x \eta(u)\,du\right], \tag{1}$$

where $\eta(x)$ has the same meaning as in Section 4.4 [cf. 4.4(2)]. Further, let an arrival be a priority customer with probability p and a nonpriority customer with probability q, where

$$p + q = 1, \tag{2}$$

and let

$$\eta_1(x) = p\eta(x), \qquad \eta_2(x) = q\eta(x), \tag{3}$$

so that

$$\eta_1(x) + \eta_2(x) = \eta(x). \tag{4}$$

Thus, the two types of customers are assumed to emanate from the same source, in which the proportion of priority and nonpriority members are p and q, respectively. Let the service times be exponentially distributed with mean rates μ_1 and μ_2 for the priority and nonpriority customers. We denote by $P_{nm}(x, t)\,dx$ the probability that there are n priority and m nonpriority customers in the system at time t, the time since the last arrival lying in the interval $(x, x + dx)$. We assume that there is limited waiting space for both classes, so that there cannot be more than N priority and M nonpriority customers in the system. To find the distribution of the busy periods for the system initiated by the arrival of a priority unit, we let the system start with $(n, m) = (1, 0)$, $x = 0$, and take the state $(n, m) = (0, 0)$ to be an absorbing state. We have the following differential difference equations for the system:

$$\left(\frac{\partial}{\partial t} + \frac{\partial}{\partial x}\right) P_{00}(x, t) = \mu_1 P_{10}(x, t) + \mu_2 P_{01}(x, t), \tag{5}$$

$$\left(\frac{\partial}{\partial t} + \frac{\partial}{\partial x} + \{\mu_2 + \eta(x)\}\right) P_{0m}(x, t) = \mu_1 P_{1m}(x, t) + \mu_2 P_{0,m+1}(x, t) \qquad (1 \leqq m \leqq M - 1), \tag{6}$$

$$\left(\frac{\partial}{\partial t} + \frac{\partial}{\partial x} + \{\mu_2 + \eta(x)\}\right) P_{0M}(x, t) = \mu_1 P_{1M}(x, t), \tag{7}$$

$$\left(\frac{\partial}{\partial t} + \frac{\partial}{\partial x} + \{\mu_1 + \eta(x)\}\right) P_{nm}(x, t) = \mu_1 P_{n+1,m}(x, t) \qquad (1 \leqq n \leqq N - 1; 0 \leqq m \leqq M), \tag{8}$$

$$\left(\frac{\partial}{\partial t} + \frac{\partial}{\partial x} + \{\mu_1 + \eta(x)\}\right) P_{Nm}(x, 0) = 0 \qquad (0 \leqq m \leqq M), \tag{9}$$

with the boundary conditions

(10) $$P_{0m}(0, t) = 0 \qquad (m = 0, 1),$$

(11) $$P_{10}(0, t) = \delta(t),$$

(12) $$P_{0m}(0, t) = \int_0^\infty P_{0,m-1}(x, t)\eta_2(x)\,dx \qquad (2 \leqq m \leqq M - 1),$$

(13) $$P_{0M}(0, t) = \int_0^\infty [P_{0,M-1}(x, t) + P_{0M}(x, t)]\eta_2(x)\,dx,$$

(14) $$P_{n0}(0, t) = \int_0^\infty P_{n-1,0}(x, t)\eta_1(x)\,dx \qquad (2 \leqq n \leqq N - 1),$$

(15) $$P_{nm}(0, t) = \int_0^\infty [P_{n-1,m}(x, t)\eta_1(x)\,dx + P_{n,m-1}(x, t)\eta_2(x)]\,dx$$
$$(1 \leqq n \leqq N - 1; 1 \leqq m \leqq M - 1),$$

(16) $$P_{nM}(0, t) = \int_0^\infty [P_{n-1,M}(x, t)\eta_1(x) + \{P_{n,M-1}(x, t) + P_{nM}(x, t)\}\eta_2(x)]\,dx$$
$$(1 \leqq n \leqq N - 1),$$

(17) $$P_{N0}(0, t) = \int_0^\infty [P_{N-1,0}(x, t) + P_{N0}(x, t)]\eta_1(x)\,dx,$$

(18) $$P_{Nm}(0, t) = \int_0^\infty [\{P_{N-1,m}(x, t) + P_{Nm}(x, t)\}\eta_1(x) + P_{N,m-1}(x, t)\eta_2(x)]\,dx \qquad (1 \leqq m \leqq M - 1),$$

(19) $$P_{NM}(0, t) = \int_0^\infty [P_{NM}(x, t)\eta(x) + P_{N-1,M}(x, t)\eta_1(x) + P_{N,M-1}(x, t)\eta_2(x)]\,dx$$

and the initial conditions

(20) $$P_{nm}(x, 0) = \delta_{1n}\delta_{0m}\delta(x).$$

As in Section 4.7, the L.T. of the p.d.f. of the busy period duration is given by

(21) $$\gamma(s) = sp_{00}(s) = s\int_0^\infty p_{00}(x, s)\,dx,$$

where we use the corresponding lowercase letters to denote Laplace transforms. To determine $p_{00}(x, s)$ from the L.T. of (1), we need $p_{10}(x, s)$ and

$p_{01}(x, s)$. We introduce the following notation:

$$H_j(x) = (s + \mu_j)x + \int_0^x \eta(u)\,du \qquad (j = 1, 2), \tag{22}$$

$$h = \mu_1 - \mu_2. \tag{23}$$

Solving the L.T. of (9) for $m = 0$, we have

$$p_{N0}(x, s) = \exp[-H_1(x)]\, p_{N0}(0, s). \tag{24}$$

Now, taking $n = N - 1, m = 0$ in the L.T. of (8), substituting for $p_{N0}(x, s)$ from (24) and solving, we have

$$p_{N-1,0}(x, s) = \exp[-H_1(x)]\,\{\mu_1 x p_{N0}(0, s) + p_{N-1,0}(0, s)\}.$$

Proceeding similarly, we finally have

$$p_{10}(x, s) = \exp[-H_1(x)]\left\{1 + \sum_{n=1}^{N} \frac{p_{n0}(0, s)(\mu_1 x)^{n-1}}{(n-1)!}\right\}. \tag{25}$$

For finding $p_{01}(s, x)$, we first solve the L.T. of (9) for $m = M$ to get

$$p_{NM}(x, s) = \exp[-H_1(x)]\, p_{NM}(0, s). \tag{26}$$

Solving the L.T. of (8) for $n = N - 1$, $m = M$ after substituting from (26), we have

$$p_{N-1,M}(x, s) = \exp[-H_1(x)]\,\{\mu_1 x p_{NM}(0, s) + p_{N-1,M}(0, s)\}. \tag{27}$$

Proceeding similarly, we have

$$p_{1M}(x, s) = \exp[-H_1(x)] \sum_{n=1}^{N} \frac{p_{nM}(0, s)(\mu_1 x)^{n-1}}{(n-1)!}. \tag{28}$$

Now, substituting for $p_{1M}(x, s)$ from (28) in the L.T. of (7) and solving, we have

$$p_{0M}(x, s) = \exp[-H_2(x)]\left\{p_{0M}(0, s) + \sum_{n=1}^{N} \frac{(\mu_1 x)^n}{n!h^n} p_{nM}(0, s)\, {}_1F_1(n; n+1; -hx)\right\}. \tag{29}$$

Thus solving recursively the equations for $p_{0m}(x, s)$ for $m = M - 1$, $M - 2$, $\ldots, 1$, we have

$$p_{01}(x, s) = \exp[-H_2(x)]\left[\sum_{m=1}^{M} \frac{(\mu_2 x)^{m-1}}{(m-1)!} p_{0m}(0, s) + \sum_{m=1}^{M}\sum_{n=1}^{N} \frac{(\mu_1 x)^n(\mu_2 x)^{m-1}}{(n+m-1)!} p_{nm}(0, s)\, {}_1F_1(n; n+m; -hx)\right]. \tag{30}$$

Solving the L.T. of (5) after substitution from (25) and (26), we obtain

$$(31)\quad p_{00}(x, s) = e^{-sx}\Bigg[\sum_{n=1}^{N} \frac{\mu_1^n}{(n-1)!} p_{n0}(0, s) \int_0^x y^{n-1} e^{-H_1(y)}\, dy + \mu_1 \int_0^x e^{-H_1(y)}\, dy + \sum_{m=1}^{M} \frac{\mu_2^m}{(m-1)!} p_{0m}(0, s) \int_0^x y^{m-1} e^{-H_2(y)}\, dy + \sum_{m=1}^{M}\sum_{n=1}^{N} \frac{\mu_1^n \mu_2^m}{(n+m-1)!h^n} p_{nm}(0, s) \int_0^x y^{n+m-1} e^{-H_2(y)} \cdot {}_1F_1(n; n+m; -hy)\, dy\Bigg].$$

Using (31) and (21), we have an expression for $\gamma(s)$, which is in the form of a Laplace transform; therefore, taking the inverse transform, we readily have

$$(32)\quad \Gamma(t) = \sum_{n=1}^{N} \frac{\mu_1^n}{(n-1)!} P_{n0}(0, t) * \left[t^{n-1} \exp\left\{-\mu_1 t - \int_0^t \eta(u)\, du\right\}\right] + \mu \exp\left\{-\mu_1 t - \int_0^t \eta(u)\, du\right\} + \sum_{m=1}^{M} \frac{\mu_2^m}{(m-1)!} P_{0m}(0, t) * \left[t^{m-1} \exp\left\{-\mu_2 t - \int_0^t \eta(u)\, du\right\}\right] + \sum_{m=1}^{M}\sum_{n=1}^{N} \frac{\mu_1^n \mu_2^m}{(n+m-1)!} P_{nm}(0, t) * \left[t^{n+m-1} \exp\left\{-\mu_2 t - \int_0^t \eta(u)\, du\right\} \cdot {}_1F_1(n; n+m; -ht)\right],$$

where the asterisk stands for convolution. The $P_{nm}(0, t)$ occurring in (32) can be obtained from the boundary conditions as explained in Section 4.4.

For an ordinary (nonpriority) queue, we set $M = 0$, $\mu_2 = 0$, $\mu_1 = \mu$ in (32) to obtain

$$(33)\quad \Gamma(t) = \sum_{n=1}^{N} \frac{\mu^n}{(n-1)!} P_{n0}(0, t) * \left[t^{n-1} \exp\left\{-\mu t - \int_0^t \eta(u)\, du\right\}\right] + \mu \exp\left[-\mu t - \int_0^t \eta(u)\, du\right],$$

which is in agreement with 4.7(14).

4.9 A Diffusion Approximation for the Busy Period

Heyman (1974) uses a diffusion model for obtaining an approximation to the distribution of busy periods in an $M/G/1$ system. Let W_t be the virtual delay (or virtual waiting time) at time t, and let a busy period start at time t_0 with the arrival of a customer whose service time is x. Gaver (1968) suggested that the virtual delay process $\{W_t : t \geqq 0\}$ be approximated by a diffusion process, say $\{D_t : t \geqq 0\}$, with infinitesimal mean and variance given, respectively, by

(1) $$c = \rho - 1,$$

(2) $$V = \lambda E(x^2) = \lambda b_2,$$

where

(3) $$\rho = \lambda E(x) = \lambda b_1,$$

and x stands for the service time random variable, λ being the mean arrival rate as usual. Let $f(t; x)$ be the probability density function (p.d.f.) of the first passage time from x to zero of $\{D_t : t \geqq t_0\}$. Then, the busy period p.d.f. for the approximate process is

(4) $$f(t) = \int_0^\infty f(t; x)\, dB(x),$$

where $B(x)$ is the service time distribution function. We shall denote the corresponding density function by $b(x)$. Now, $f(t; x)$ may be regarded as the first passage time from zero to x of the Wiener process with infinitesimal mean and variance equal to $-c$ and V, respectively; therefore, its L.T. is given [see Cox and Miller (1965)] by

(5) $$f^*(s; x) = \exp[-cV^{-1}\{1 - (1 + 2Vs/c^2)^{1/2}\}x].$$

Heyman (1974) uses (5) to show that the first two moments of the approximate distribution of the busy period are equal to those of the actual distribution $\Gamma(t)$, whose L.T. is given by the well-known result of Kendall, viz.

(6) $$\Gamma^*(s) = B^*[\lambda + s - \lambda\Gamma^*(s)],$$

where $B^*(s)$ stands for the L.S.T. of the service time distribution $B(t)$.

From (4) and (5), the L.T. of the busy period p.d.f. is given by

(7) $$f^*(s) = \int_0^\infty f^*(s; x)\, dB(x)$$
$$= B^*\left[\left\{\frac{\rho - 1}{\lambda b_2}\right\}\{1 - [1 + 2s\lambda b_2(1 - \rho)^2]^{1/2}\}\right].$$

Note that (7) gives $f^*(s)$ directly in terms of $B^*(s)$ and not as an implicit equation as in (6). Heyman points out that (7) is valid for $\rho \geqq 1$, but this is the case also with (6) [see Ledermann and Reuter (1954)].

It has been shown [see, e.g., Cox and Miller (1965)] that

$$f(t; x) = x(2\pi V t^3)^{-1/2} \exp[-(x + ct)^2/(2Vt)], \tag{8}$$

whence, on using (4), we have

$$f(t) = (2\pi V t^3)^{-1/2} \int_0^\infty x \exp[-(x + ct)^2/(2Vt)]\, dB(x). \tag{9}$$

We consider the following particular cases.

Case I—Constant Service Times In the case of a constant service time equal to S, we have

$$b(x) = \delta(x - S), \tag{10}$$

$$c = \lambda S - 1, \qquad V = \lambda S^2, \tag{11}$$

and, therefore, (9) gives

$$f(t) = (2\pi\lambda t^3)^{-1/2} \exp[-\{S + (\lambda S - 1)t\}^2/(2\lambda S^2 t)]. \tag{12}$$

Case II—Negative-Exponential Service Times In this case, we have

$$b(x) = \mu e^{-\mu x}, \tag{13}$$

$$\rho = \lambda/\mu, \qquad c = \rho - 1, \qquad V = 2\lambda/\mu^2. \tag{14}$$

Using (13) in (9), we have

$$f(t) = (2\pi V t^3)^{-1/2} \mu \int_0^\infty x \exp\left(-\frac{(x + ct)^2}{2Vt}\right) e^{-\mu x}\, dx. \tag{15}$$

Denote the integral in (15) by I and let

$$\beta = ct, \tag{16}$$

$$\gamma = 2Vt, \tag{17}$$

$$\alpha = (\mu\gamma - 2\beta)/\gamma; \tag{18}$$

then

$$I = e^{-\beta^2/\gamma} \int_0^\infty x \exp\left(-\alpha x - \frac{x^2}{\gamma}\right) dx, \tag{19}$$

and, using Erdélyi *et al.* (1954, Vol. I, p. 146, Eq. (22)), we have

$$I = e^{-\beta^2/\gamma}[\tfrac{1}{2}\gamma - \tfrac{1}{4}\pi^{1/2}\alpha\gamma^{3/2} \exp(\tfrac{1}{4}\gamma\alpha^2)\, \mathrm{Erfc}(\tfrac{1}{2}\alpha\gamma^{1/2})], \tag{20}$$

where Erfc(x) is the complementary error function, given by [cf. II.1(67)]

$$\text{Erfc}(x) = 2\pi^{-1/2} \int_x^\infty e^{-u^2}\, du. \tag{21}$$

From (15), using (20) and (14), we have

$$\begin{aligned} f(t) = {} & \mu\rho(\pi\rho\mu t)^{-1/2} \exp[-(1-\rho)^2 \mu t/(4\rho)] \\ & - \{(3\rho - 1)/2\}\mu \exp[(2\rho - 1)\mu t]\, \text{Erfc}[\{(3\rho - 1)/2\}(\mu t/\rho)^{1/2}]. \end{aligned} \tag{22}$$

Case III – Gamma-Distributed Service Times Here we have

$$b(x) = \mu^\nu x^{\nu - 1} e^{-\mu x}/\Gamma(\nu), \tag{23}$$

where the shape parameter ν is not necessarily an integer, and

$$\rho = \lambda\nu\mu^{-1}, \qquad c = \lambda\nu\mu^{-1} - 1, \qquad V = \lambda\nu(\nu + 1)\mu^{-2}. \tag{24}$$

Hence

$$f(t) = (2\pi V t^3)^{-1/2} \left\{\frac{\mu^\nu}{\Gamma(\nu)}\right\} \int_0^\infty x^\nu \exp\left(-\frac{(x + ct)^2}{2Vt}\right) e^{-\mu x}\, dx. \tag{25}$$

Denoting the integral in (25) by I_ν and using Erdélyi *et al.* (1954, Vol. I, p. 146, Eq. (24)), we have, with the notation given in (16)–(18),

$$\begin{aligned} I_\nu = {} & \exp(-\beta^2/\gamma)\,\Gamma(\nu + 1) 2^{\nu + 1} (\gamma/8)^{(\nu + 1)/2} \exp(\gamma\alpha^2/8) \\ & \cdot D_{-\nu - 1}[2\alpha(\gamma/8)^{1/2}], \end{aligned} \tag{26}$$

where $D_n(z)$ is the parabolic cylinder function and, for negative integer n, is given (Erdélyi *et al.*, 1953, Vol. II, p. 119, Eq. (19)) by

$$D_{-m-1}(z) = \sqrt{2}\left(\frac{(-1)^m}{m!}\right) e^{-z^2/4} \frac{d^m}{dz^m} \left[e^{z^2/2}\, \hat{\text{E}}\text{rfc}\left(\frac{z}{\sqrt{2}}\right)\right], \tag{27}$$

where [cf. II.1(70)]

$$\hat{\text{E}}\text{rfc}(x) = \int_x^\infty e^{-u^2}\, du = \tfrac{1}{2}\sqrt{\pi}\, \text{Erfc}(x) \tag{28}$$

is Kramp's form of the complementary error function.

From (25) and (26), we have

$$f(t) = (2\pi V t^3)^{-1/2} \nu\mu^\nu e^{ct/(2V)} (Vt)^{(\nu + 1)/2} D_{-\nu - 1}[(\mu + c\alpha^{-1})\sqrt{Vt}]. \tag{29}$$

Therefore, in the case when ν is a positive integer, m say, we have, from (27) and (29),

$$f(t) = \frac{(Vt)^{m/2}}{t\sqrt{2\pi}} \frac{m(-\mu)^m}{(m - 1)!} e^{-\beta^2/(2Vt)} \frac{d^m}{dz^m}\left[e^{z^2/2}\, \hat{\text{E}}\text{rfc}\left(\frac{z}{\sqrt{2}}\right)\right], \tag{30}$$

where the derivative is evaluated at $z = \alpha(Vt/2)^{1/2}$.

The parabolic cylinder functions are tabulated in Abramowitz and Stegun (1964), and they can be evaluated by expressing them in terms of the confluent hypergeometric and gamma functions, which are readily computable [see, e.g., Magnus *et al.* (1966)]. For computation of $f(t)$ from (30), we may proceed as follows.

Let us write

$$D^{(m)}(z) = \frac{d^m}{dz^m}\left[e^{z^2/2}\,\hat{\mathrm{E}}\mathrm{rfc}\left(\frac{z}{\sqrt{2}}\right)\right]. \tag{31}$$

From Erdélyi *et al.* (1953, Vol. II, p. 147, Eq. (2)), or II.3(61), we have

$$\hat{\mathrm{E}}\mathrm{rfc}(x) = \tfrac{1}{2}e^{-x^2}\Psi(\tfrac{1}{2}, \tfrac{1}{2}; x^2), \tag{32}$$

where [cf. II.3(14)]

$$\Psi(a, c; x) = \frac{1}{\Gamma(a)}\int_0^\infty e^{-xt}t^{a-1}(1+t)^{c-a-1}\,dt, \qquad \mathrm{Re}(a) > 0. \tag{33}$$

From (31) and (32), we have

$$D^{(m)}(z) = \frac{1}{2}\frac{d^m}{dz^m}\Psi(\tfrac{1}{2}, \tfrac{1}{2}; \tfrac{1}{2}z^2). \tag{34}$$

Now, from Erdélyi *et al.* (1953, Vol. I, p. 257, Eq. (7)), or II.3(18), we get

$$\Psi(a, c; x) = \frac{\Gamma(1-c)}{\Gamma(a-c+1)}\,{}_1F_1(a; c; x) + \frac{\Gamma(c-1)}{\Gamma(a)}x^{1-c}\,{}_1F_1(a-c+1; 2-c; x), \tag{35}$$

where ${}_1F_1(a; c; x)$ is the confluent hypergeometric function, and, therefore,

$$\Psi(\tfrac{1}{2}, \tfrac{1}{2}; x) = \sqrt{\pi}\,e^x - 2\sqrt{x}\,{}_1F_1(1; \tfrac{3}{2}; x). \tag{36}$$

Now, on using Erdélyi *et al.* (1953, Vol. I, p. 255, Eq. (12)), viz.,

$$\frac{d^m}{dx^m}[x^{c-1}\,{}_1F_1(a; c; x)] = (-1)^m(1-c)_m x^{c-1-m}\,{}_1F_1(a; c-m; x), \tag{37}$$

we have

$$\frac{d^m}{dx^m}\Psi(\tfrac{1}{2}, \tfrac{1}{2}; x) = \sqrt{\pi}\,e^x + (-1)^{m+1}2(\tfrac{1}{2})_m x^{(1/2)-n}\,{}_1F_1(1; \tfrac{3}{2}; x). \tag{38}$$

It can be seen that (38) and the chain rule for differentiation suffice for evaluation of $D^{(m)}(z)$ and hence $f(t)$. Heyman tabulates the values of the actual and approximate distribution functions for the busy period, $\Gamma(t)$ and $F(t)$, and the corresponding relative error for constant and exponential service times.

4.10 The $M/G/1$ Queue with Paretian Service Times

Harris (1968b) considers a queuing system in which the service time distribution follows the exponential distribution

$$(1) \qquad B(t|\mu) = 1 - e^{-\mu t},$$

subject to the condition that μ is distributed according to the three-parameter gamma density function

$$(2) \qquad f(\mu) = (\mu - \gamma)^{\alpha} \exp\{-(\mu - \gamma)/\beta\}/\beta^{\alpha+1}\Gamma(\alpha + 1)$$
$$(\alpha > -1, \beta > 0, \mu > \gamma),$$

so that the unconditional distribution of service times is given by

$$(3) \qquad B_{\mathrm{T}}(t) = \int_0^{\infty} B(t|\mu) f(\mu)\, d\mu = 1 - \frac{e^{-\gamma t}}{(\beta t + 1)^{\alpha+1}} \qquad (t > 0),$$

where T stands for the random variable representing the service time. If we take $\gamma = 0$, we get

$$(4) \qquad B_{\mathrm{T}}(t) = 1 - 1/(\beta t + 1)^{\alpha+1} \qquad (t > 0).$$

Equation (4) is the Pareto distribution (Fama, 1965), and can be put in the standard form (Lévy, 1925), given by the density function

$$(5) \qquad g(x) = \begin{cases} 0 & (x \leqq b), \\ (a/b)(b/x)^{a+1} & (x > b), \end{cases}$$

using the transformation

$$(6) \qquad x = \beta t + 1, \qquad a = \alpha + 1, \qquad b = 1.$$

From (4), we have the expectation

$$(7) \qquad E(T) = 1/(\alpha\beta)$$

and the variance

$$(8) \qquad \mathrm{Var}(T) = (\alpha + 1)/[\alpha^2\beta^2(\alpha - 1)] \qquad (\alpha > 1, \beta > 0).$$

Harris (1968b) uses (7) and (8) to find expressions for the mean queue length and mean waiting time, using the Pollaczek–Khintchine formula, and also discusses the problem of estimation of μ, α, and β in a practical situation, using maximum likelihood and the Bayesian approach.

4.11 Simultaneous Service with Dependent Service Times

Consider a queuing system in which each customer is served by a number of servers simultaneously, the service rendered by each server being regarded as a phase. Fujisawa (1962) studies this process with n independent phases, while Alagaraja (1970) considers the process with dependent phases, dealing in particular with the two-phase case. The interarrival times and service times in each phase are taken to be negative exponentially distributed with means $1/\lambda$ and $1/\mu$, respectively. Considering the case of n phases with constant correlation coefficient r ($r < 1$), the joint probability density function (p.d.f.) of the service times in the n phases is $g(x_1, \ldots, x_n)$, say, whose characteristic function is given (Gurland, 1955) by

$$(1) \qquad \Phi(t_1, \ldots, t_n) = \begin{vmatrix} 1 - it_1/\mu & -it_1r/\mu & \cdots & -it_1r/\mu \\ -it_2r/\mu & 1 - it_2/\mu & \cdots & -it_2r/\mu \\ \vdots & \vdots & & \vdots \\ -it_nr/\mu & -it_nr/\mu & \cdots & 1 - it_n/\mu \end{vmatrix}^{-1}.$$

Alagaraja (1970) inverts the characteristic function in the case $n = 2$, and shows that the joint p.d.f.

$$(2) \qquad g(x_1, x_2) = [R\mu^2/(1 - r^2)] \exp[-\mu(x_1 + x_2)/(1 - r^2)],$$

where

$$(3) \qquad R = \sum_{n=0}^{\infty} \left[\frac{x_1 x_2 \mu^2 r^2}{(1 - r^2)^2}\right]^n \frac{1}{(n!)^2}.$$

The service time distribution for the queuing process described above is the distribution of $\max(x_1, x_2)$ and is therefore given by

$$(4) \qquad B(x) = \int_0^x \int_0^x g(x_1, x_2)\, dx_1\, dx_2$$

$$= (1 - r^2) \sum_{n=0}^{\infty} \frac{r^{2n}}{(n!)^2} \left\{\gamma\left(n + 1, \frac{\mu x}{1 - r^2}\right)\right\}^2,$$

on using (2) and (3) and integrating, where, as usual, $\gamma(n, x)$ stands for the incomplete gamma function given by [cf. II.1(52)]

$$(5) \qquad \gamma(n, x) = \int_0^x e^{-z} z^{n-1}\, dz.$$

Thus, $B(x)$ can be regarded as a mixture, where the mixing distribution is given by

$$(6) \qquad p(n) = r^{2n}(1 - r^2), \qquad n = 0, 1, \ldots,$$

and the kernel is

$$F(x, n) = \frac{1}{(n!)^2} \left\{\gamma\left(n + 1, \frac{\mu x}{1 - r^2}\right)\right\}^2.$$

For the case of three phases, the service time distribution function is shown by Alagaraja to be

$$(7) \qquad B(x) = \frac{\mu^3}{b}\sum_{n=0}^{\infty}\left(\frac{a\mu r}{b}\right)^{2n}\frac{1}{(n!)^2}\sum_{j=0}^{2n}(-1)^j\binom{2n}{j}\left(\frac{a^2-b}{a}\right)^j$$
$$\cdot\sum_{m=0}^{\infty}(-1)^m\left(\frac{\mu}{b}\right)^{j+2m}\frac{(a^2-b)^n}{(j+m)!m!}\left\{\frac{b}{a\mu}\right\}^{j+m+1}$$
$$\cdot\gamma\left(j+m+1,\frac{x}{c}\right)\frac{d^m}{dx^m}\left[c^{2n+2}\left\{\gamma\left(n+1,\frac{x}{c}\right)\right\}^2\right],$$

where

$$(8) \qquad a = 1 - r^2, \qquad b = 1 - 3r^2 + 2r^3, \qquad c = b/a\mu.$$

From (4), the Laplace–Stieltjes transform of the service time distribution is

$$(9) \qquad b(s) = \int_0^{\infty} e^{-sx}\,dB(x)$$
$$= 2(1-r^2)\sum_{n=0}^{\infty}\frac{r^{2n}}{(n!)^2}\cdot\left(\frac{\mu}{1-r^2}\right)^{n+1}$$
$$\cdot\int_0^{\infty} x^n \exp\left[-\left\{s+\frac{\mu}{1-r^2}\right\}x\right]\gamma\left(n+1,\frac{\mu x}{1-r^2}\right)dx,$$

whence, using Erdélyi *et al.* (1954, Vol. II, p. 308, Eq. (15)), [see 4.4(37)], we have

$$(10) \qquad b(s) = 2(1-r^2)\sum_{n=0}^{\infty}\frac{r^{2n}}{(n!)^2}\frac{(2n+1)!}{(n+1)\{2+s(1-r^2)/\mu\}^{2n+2}}$$
$$\cdot\,{}_2F_1[1, 2n+2; n+2; 1/\{2+(1-r^2)s/\mu\}].$$

By writing the series ${}_2F_1$ in the expanded form, and differentiating at $s = 0$, Alagaraja obtains the first three moments of the service time distribution about the origin. However, his results appear to be in error. Writing

$$(11) \qquad b^{(k)}(0) = \left.\frac{d^k}{ds^k}b(s)\right|_{s=0},$$

it can be shown that the kth moment of $B(x)$ can be expressed as

$$(12) \quad (-1)^k b^{(k)}(0) = (k+1)!\{\tfrac{1}{2}(1-r^2)\}^{k+1}(1/\mu^k)H_3(k+2, 1; 2; \tfrac{1}{4}r^2, \tfrac{1}{2}),$$

where [see also II.6(22)–II.6(26)]

$$(13) \qquad H_3(\alpha, \beta; \gamma; x, y) = \sum_{m,n=0}^{\infty}\frac{(\alpha)_{2m+n}(\beta)_n}{(\gamma)_{m+n}}\frac{x^m}{m!}\frac{y^n}{n!}$$

is Horn's function (Erdélyi *et al.*, 1953, Vol. I, p. 225, Eq. (15)).

4.12 An $M/G/1$ Queue with Balking

Consider an $M/G/1$ system with a finite waiting room of size m such that when the waiting room is full, the mean arrival rate changes from λ to λp ($p < 1$), implying that $100(1 - p)\%$ of the arrivals balk (Dick, 1970). The special cases $m = 0$ and $p = 0$ have been studied by Takács (1961) and Keilson (1966), respectively. Service does not begin until there are at least r_1 customers in the queue, and then r_2 customers or the whole queue, whichever is less, are taken into service.

Let $B(x)$ be the service time density function. Consider first the case $r_1 = r_2 = 1, m = 0$. Using the imbedded Markov chain method, we define the transition probabilities

(1) $$P_{ij} = \text{Prob}\{\xi_{n+1} = j \mid \xi_n = i\}, \qquad n = 1, 2, \ldots,$$

where

(2) $$\xi_n = \xi(\tau_n + 0),$$

$\xi(t)$ stands for the number in the system at time t, and τ_n is the epoch of the nth departure. Then, we have

(3) $$P_{ij} = \begin{cases} k_{j-i+1} & (j \geqq i - 1;\ i = 1, 2, \ldots) \\ k_j & (j \geqq 0;\ i = 0) \\ 0 & (j < i - 1;\ i = 2, 3, \ldots), \end{cases}$$

where

(4) $$k_j = \int_0^\infty e^{-\lambda p x}\left(\frac{(\lambda p x)^j}{j!}\right) B(x)\,dx \qquad (j = 0, 1, 2, \ldots)$$

is the probability that j customers join the queue during a service time.

For the case $m \geqq 1$, we divide the transition probabilities into four classes; thus,

(5) $$P_{ij}^A(m) = \text{Prob}\{\xi_{n+1} = j \mid \xi_n = i\} \qquad (i = 0, \ldots, m;\ j = 0, \ldots, m - 1),$$

(6) $$P_{ij}^B(m) = \text{Prob}\{\xi_{n+1} = j \mid \xi_n = i\} \qquad (i = 0, \ldots, m;\ j = m, m + 1, \ldots),$$

(7) $$P_{ij}^C(m) = \text{Prob}\{\xi_{n+1} = j \mid \xi_n = i\}$$
$$(i = m + 1, m + 2, \ldots;\ j = m, m + 1, \ldots),$$

(8) $$P_{ij}^0(m) = 0 \qquad (i = m + 1, m + 2, \ldots;\ j = 0, \ldots, m - 1).$$

We have

(9) $$P_{ij}^A(m) = \begin{cases} k_{j-i+1}^A(m) & (j \geqq i - 1; i = 1, \ldots, m) \\ k_j^A(m) & (j \geqq 0; i = 0) \\ 0 & (j < i - 1; i = 2, 3, \ldots), \end{cases}$$

where

$$(10)\quad k_j^A(m) = \int_0^\infty e^{-\lambda x}\left(\frac{(\lambda x)^j}{j!}\right) B(x)\,dx \qquad (j = 0, 1, \ldots, m-1)$$

is the probability defined as in (4), but is different since, during the transitions considered here, the arrival rate remains λ throughout. Also,

$$(11)\qquad P_{ij}^C(m) = \begin{cases} k_{j-i+1}^C(m) & (j \geqq i-1;\ i = m+1, m+2, \ldots) \\ 0 & (j < i-1;\ i = m+1, m+2, \ldots), \end{cases}$$

where $k_j^C(m)$ is the probability that j customers join the queue during a service time when $i > m$, and is given by

$$(12)\quad k_j^C(m) = \int_0^\infty \sum_{n=j}^\infty e^{-\lambda x}\left(\frac{(\lambda x)^n}{n!}\right)\binom{n}{j} p^j q^{n-j} B(x)\,dx \qquad (q = 1 - p),$$

which reduces to the right-hand side of (4). During the transitions over class B, there is an effective change in input rate during a service time; so we have

$$(13)\qquad P_{ij}^B(m) = \begin{cases} k^B(m, j-m) & (j \geqq m;\ i = 0), \\ k^B(m+1-i, j-m) & (j \geqq m;\ i = 1, \ldots, m) \end{cases}$$

and $k^B(a, b)$ $(a \geqq 1)$ is the probability that, during a service time, $(a + b)$ customers join the queue in such a way that the first a customers all join, and out of the next $n \geqq b$ arrivals, b join and $(n - b)$ balk. Thus

$$(14)\qquad k^B(a, b) = \int_0^\infty P(a, b, x) B(x)\,dx,$$

where

$$(15)\qquad P(a, b, x) = \int_0^x e^{-\lambda y}\left(\frac{\lambda^a y^{a-1}}{(a-1)!}\right) \cdot \left\{\sum_{n=b}^\infty e^{-\lambda(x-y)}[\lambda(x-y)]^n \left(\frac{1}{n!}\right)\binom{n}{b} p^b q^{n-b}\right\} dy.$$

Let $z = y/x$; then

$$(16)\qquad P(a, b, x) = \left[\frac{e^{-\lambda p x}(\lambda x)^{a+b} p^b}{(a-1)!\,b!}\right] \int_0^1 e^{-\lambda q x z} z^{a-1}(1-z)^b\,dz,$$

which, on using Erdélyi *et al.* (1954, Vol. I, p. 139, Eq. (23)), simplifies to

$$(17)\quad P(a, b, x) = [e^{-\lambda p x}(\lambda x)^{a+b} p^b/(a+b)!]\ {}_1F_1(a; a+b+1; -\lambda q x).$$

Equation (14), after substitution from (17), gives $k^B(a, b)$. It may be noted that for $a = 0$,

(18) $$k^B(0, b) = k_b,$$

where k_b is given by (4).

The case of general r_1 and r_2 can be dealt with similarly, the only difference being that the transition probability matrix is more involved in terms of dimensionality. Dick uses the above results to obtain the generating function of the steady-state queue length probabilities for the queuing model in question in terms of the Laplace transform $b(s)$ of the service time distribution. As shown in Section 4.4., the various results, e.g. (14), can be stated in terms of ${}_2F_1$, the hypergeometric function.

4.13 PERT-Type Approach in Queuing Theory

Greenberg (1973) considered the problem of handling $M/G/1$ and $G/M/1$ queuing systems for which the form of the general, service or input, distribution is unknown and suggested their replacement by the empirical distribution given by the observed service or interarrival times. The authors suggested an approximate analysis by using a beta distribution instead of the empirical one (Kashyap and Srivastava, 1974). This approach may be followed with advantage in situations similar to those handled by the program evaluation and review technique (PERT), viz., where we have little or no empirical data to fall back upon. Let us first take the case of an $M/G/1$ queue. Using the PERT approach [see, for example, MacCrimmon and Ryavec (1964)], we require three estimates for the duration of a service time. Let a be the least (or optimistic) estimate of a service time, b the greatest (or pessimistic) estimate, and m the most likely estimate, which, by definition, should be the modal value in the distribution. However, if it is desired to fit a beta distribution to observed data, a and b may be chosen to equal the least and greatest observed values. m is to be fixed on the basis of judgment. With the chosen a and b, we take the service time density function to be

(1) $$b(t) = \begin{cases} C(t-a)^{\alpha-1}(b-t)^{\beta-1} & (\alpha, \beta \geqq 1; a \leqq t \leqq b), \\ 0 & \text{otherwise}, \end{cases}$$

where C is the normalizing factor and has the value

(2) $$C = [(b-a)^{\alpha+\beta-1}B(\alpha, \beta)]^{-1},$$

$B(\alpha, \beta)$ being the beta function, defined by [see II.1(43)–II.1(48)]

(3) $$B(\xi, \eta) = \int_0^1 x^{\xi-1}(1-x)^{\eta-1}\,dx \qquad (\mathrm{Re}(\xi), \mathrm{Re}(\eta) > 0).$$

A complete specification of (1) requires knowledge of α and β, which we determine from the requirements that the distribution (1) has

(i) a modal value m,
(ii) a mean value $m_e = (a + 4m + b)/6$,

the requirement (ii) being one of the PERT assumptions. Equating the mode and the mean of the distribution (1) to m and m_e, respectively, we have

(4) $$m = \{(\beta - 1)a + (\alpha - 1)b\}/(\alpha + \beta - 2),$$

(5) $$m_e = (\beta a + \alpha b)/(\alpha + \beta).$$

Solving (4) and (5) for α and β, we obtain

(6) $$\alpha = (m_e - a)(a + b - 2m)/\{(m_e - m)(b - a)\},$$

(7) $$\beta = (b - m_e)(a + b - 2m)/\{(m_e - m)(b - a)\}.$$

Using Erdélyi *et al.* (1954, Vol. I, p. 139, Eq. (23)), we find that $b(t)$ has the Laplace transform

(8) $$b^*(s) = e^{-as}\,{}_1F_1\{\alpha; \alpha + \beta; -(b - a)s].$$

Now, the generating function of the steady-state probabilities P_j is [see, e.g., Prabhu (1965a, pp. 40–42)]

(9) $$P(z) = \sum_{j=0}^{\infty} P_j z^j = \frac{(1 - \rho)(1 - z)K(z)}{K(z) - z},$$

where ρ is the traffic intensity, and

(10) $$K(z) = b^*(\lambda - \lambda z),$$

λ being the mean arrival rate. Equations (8)–(10) yield

(11) $$P(z) = \frac{(1 - \rho)(1 - z)\,{}_1F_1[\alpha; \alpha + \beta; -\lambda(b - a)(1 - z)]}{{}_1F_1[\alpha; \alpha + \beta; -\lambda(b - a)(1 - z)] - z\exp[\lambda a(1 - z)]}.$$

Further, we use Skinner's (1968) method of inversion of the Pollaczek–Khintchine formula to find lower and upper bounds of $F(w)$, the waiting time distribution function. We have [cf. Greenberg (1973)]

(12) $$F(w) = (1 - \rho)\sum_{j=0}^{\infty} \rho^j [G(w)]^{j*},$$

where $[G(w)]^{j*}$ is the j-fold convolution of

(13) $$G(w) = \mu \int_0^w [1 - B(t)]\,dt,$$

μ being the mean service rate and $B(t)$ the cumulative distribution function of service time, given by

(14) $$B(t) = \int_a^t b(u)\,du = I_\tau(\alpha, \beta),$$

where

$$\tau = (t - a)/(b - a) \tag{15}$$

and $I_\tau(\alpha, \beta)$ is the incomplete beta function ratio, defined by [see, e.g., II.1(59), II.1(61)–II.1(65); see also Abramowitz and Stegun (1964, p. 263)]

$$I_\tau(\alpha, \beta) = \frac{B_\tau(\alpha, \beta)}{B(\alpha, \beta)} = \frac{1}{B(\alpha, \beta)} \int_0^\tau x^{\alpha - 1}(1 - x)^{\beta - 1}\, dx. \tag{16}$$

Therefore, we have (Abramowitz and Stegun, 1964, p. 263)

$$1 - B(t) = 1 - I_\tau(\alpha, \beta) = I_{1-\tau}(\beta, \alpha), \tag{17}$$

whence on using the relation (Abramowitz and Stegun, 1964, p. 263, Eq. (6.6.8)), viz.,

$$B_x(a, b) = a^{-1} x^a\, {}_2F_1(a, 1 - b; a + 1; x), \tag{18}$$

we have

$$\begin{aligned} \int_{r_1}^{r_2} [1 - B(t)]\, dt = {} & \left[\frac{\Gamma(\alpha + \beta)}{\Gamma(\alpha)\Gamma(\beta + 2)(b - a)^\beta}\right] \\ & \cdot \left[(b - r_1)^{\beta + 1}\, {}_2F_1\left(1 - \alpha, \beta; \beta + 2; \frac{b - r_1}{b - a}\right)\right. \\ & \left. - (b - r_2)^{\beta + 1}\, {}_2F_1\left(1 - \alpha, \beta; \beta + 2; \frac{b - r_2}{b - a}\right)\right]. \end{aligned} \tag{19}$$

Alternatively, the integral in (19) may be evaluated thus:

$$\begin{aligned} \int_{r_1}^{r_2} [1 - B(t)]\, dt &= \int_{r_1}^{r_2} I_{1-\tau}(\beta, \alpha)\, dt \\ &= [B(\beta, \alpha)]^{-1} \int_{r_1}^{r_2} \int_0^{1-\tau} x^{\beta - 1}(1 - x)^{\alpha - 1}\, dx\, dt, \end{aligned} \tag{20}$$

whence, on changing the order of integration, we have

$$\begin{aligned} \int_{r_1}^{r_2} [1 - B(t)]\, dt = {} & [B(\beta, \alpha)]^{-1} \int_0^{R_2} \int_{r_1}^{r_2} x^{\beta - 1}(1 - x)^{\alpha - 1}\, dt\, dx \\ & + \int_{R_2}^{R_1} \int_{r_1}^{b - (b - a)x} x^{\beta - 1}(1 - x)^{\alpha - 1}\, dt\, dx \\ = {} & (b - r_1) I_{R_1}(\beta, \alpha) - (b - r_2) I_{R_2}(\beta, \alpha) \\ & - \left\{\frac{(b - a)\beta}{\alpha + \beta}\right\} [I_{R_1}(\beta + 1, \alpha) - I_{R_2}(\beta + 1, \alpha)], \end{aligned} \tag{21}$$

where

$$R_j = (b - r_j)/(b - a) \qquad (j = 1, 2). \tag{22}$$

In case the form (21) is used for calculations, the values of the various incomplete beta functions can be evaluated from Pearson (1968), but as one has to use trivariate interpolation, the calculations are more involved than the evaluation of the hypergeometric functions in (19). The equivalence of (19) and (21) follows on using Rainville (1960, p. 15, Eq. (11)), viz.,

$$(a - c + 1)\,{}_2F_1(a, b; c; x) = a\,{}_2F_1(a + 1, b; c; x) - (c - 1)\,{}_2F_1(a, b; c - 1; x). \tag{23}$$

Besides the points cited in favor of the assumption of a beta distribution in the PERT literature [see, e.g., MacCrimmon and Ryavec (1964)], it may be noted that this distribution contains the two extreme cases, an impulse function and the uniform distribution, which can be obtained from (1) by setting $a = b = m$ and $\alpha = \beta = 1$, respectively, and therefore can be expected to give a good fit to other distributions, which must all be intermediate cases between these extremes. Thus, we have the following special cases.

For $M/D/1$ Setting $a = b = m$ in (11),

$$P(z) = (1 - \rho)(1 - z)/[1 - z\exp\{\lambda m(1 - z)\}] \tag{24}$$

[cf. Prabhu (1965a, p. 33, Eq. (1.152))].

For $M/U/1$ Setting $\alpha = \beta = 1$ in (11),

$$P(z) = \frac{(1 - \rho)(1 - z)[\exp\{-\lambda a(1 - z)\} - \exp\{-\lambda b(1 - z)\}]}{\exp\{-\lambda a(1 - z)\} - \exp\{-\lambda b(1 - z)\} - \lambda(b - a)z(1 - z)}. \tag{25}$$

The results obtained here are convenient for computation even on a small calculator, and numerical results on queue length probabilities and bounds of waiting time were obtained in Kashyap and Srivastava (1974), taking a numerical example. The case of $G/M/1$ queues can be treated similarly, using the imbedded Markov chain theory for such systems [see, e.g., Prabhu (1965a), pp. 42–44)] and taking a beta distribution for the interarrival times.

4.14 The $M/G/1$ Queue in Heavy Traffic

Consider a sequence of stable $M/G/1$ queuing systems for which the nth system has an arrival rate λ_n and service time distribution B_n (Ott, 1977). Let the kth moment of B_n be denoted by b_{kn}, and let us write b_n for b_{1n}. Since all systems are assumed to be stable, we have

(1) $$\lambda_n b_n < 1 \qquad \text{for all} \quad n.$$

Let Γ_n be the probability distribution function of the length of busy periods and $X_n(t)$ the virtual waiting time process in the nth system. Further, let $W(t)$ be a Brownian motion process on $[0, \infty)$ with an impenetrable barrier at 0, drift $-a$ $(a > 0)$ and dispersion σ^2, and $W^*(t)$ the Brownian motion process on $(-\infty, \infty)$ with the same drift and dispersion and with $W^*(0) = 0$. Define R_n and R to be the covariance functions of the stationary processes $X_n(t)$ and $W(t)$, respectively, so that

(2) $$R_n(t) = \text{Cov}[X_n(\tau), X_n(\tau + t)],$$

(3) $$R(t) = \text{Cov}[W(\tau), W(\tau + t)],$$

where $X_n(\tau)$ and $W(\tau)$ have the stationary distributions of these processes. Let

(4) $$\tilde{X}_n(t) = X_n(\eta_n t) \qquad (\eta_n > 0)$$

be a time-rescaled version of $X_n(t)$. Ott (1977) studies the limiting behavior, as $n \to \infty$, of $\tilde{X}_n(t)$ under the heavy traffic situation, so that

(5) $$\lambda_n b_n \uparrow 1 \qquad \text{as} \quad n \to \infty.$$

Considering the special case of $M/M/1$ systems, where the nth system has mean arrival and service rates λ_n and μ_n, so that

(6) $$b_n = 1/\mu_n, \qquad b_{2,n} = 2/\mu_n^2,$$

Ott shows that

(7) $$\int_0^\infty e^{-st}\, dR(t) = \int_0^\infty e^{-st} R'(t)\, dt = \frac{-\sigma^4}{(\sqrt{2\sigma^2 s + a^2} + a)^2},$$

where a and σ^2 are given by

(8) $$\eta_n\{1 - (\lambda_n/\mu_n)\} = a \qquad \text{for all } n \quad (0 < a < \infty),$$

(9) $$\eta_n \lambda_n/\mu_n^2 = \sigma^2/2 \qquad \text{for all } n \quad (0 < \sigma^2 < \infty),$$

and, therefore,

(10) $$\eta_n = a + \sigma^2/(2\mu_n).$$

Now, using Erdélyi *et al.* (1954, Vol. I, p. 132, Eq. (34)), we have

(11) $$\begin{aligned} &\mathscr{L}^{-1}\{(\sqrt{2\sigma^2 s + a^2} + a)^{-n}\} \\ &= \frac{n}{\sigma^2}\left(\frac{2t}{\sigma^2}\right)^{(1/2)n-1} \frac{2}{\sqrt{\pi}} \int_{\sqrt{a^2 t/(2\sigma^2)}}^\infty \{y - \sqrt{a^2 t/(2\sigma^2)}\}^n \left(\frac{1}{n!}\right) e^{-y^2}\, dy \\ &= \frac{n}{\sigma^2}\left(\frac{2t}{\sigma^2}\right)^{(1/2)n-1} i^n \operatorname{erfc}\left(\sqrt{\frac{a^2 t}{2\sigma^2}}\right) \qquad (n = 1, 2, \ldots), \end{aligned}$$

where [see Abramowitz and Stegun (1964, p. 299)]

(12) $$i^n \operatorname{erfc}(z) = \int_z^\infty i^{n-1} \operatorname{erfc}(t)\,dt \qquad (n = 0, 1, \ldots),$$

$$i^{-1} \operatorname{erfc}(z) = (2/\sqrt{\pi})e^{-z^2},$$

$$i^0 \operatorname{erfc}(z) = \operatorname{erfc}(z) = \frac{2}{\sqrt{\pi}} \int_z^\infty e^{-t^2}\,dt.$$

Therefore, (7) gives

(13) $$R'(t) = -2\sigma^2 i^2 \operatorname{erfc}(\sqrt{a^2 t/2\sigma^2}).$$

Using (11) and Abramowitz and Stegun (1964, p. 1026, Eq. (29.3.82)), viz.,

(14) $$\mathscr{L}^{-1}(e^{-k\sqrt{s}}) = \{k/(2\sqrt{\pi t^3})\} \exp[-k^2/(4t)] \qquad (k > 0),$$

it is shown by Ott that

(15) $$R'(t) = -\frac{1}{\pi}\sqrt{\frac{a^2\sigma^2}{2}} \int_{a^2/(2\sigma^2)}^\infty e^{-st} \left\{\frac{s - \sqrt{a^2/2\sigma^2}}{s^2}\right\} ds,$$

which expresses $R'(t)$ as a Laplace transform and hence proves that $-R'(t)$ is monotone, a result that also follows from the fact that $W(t)$ is time reversible.

Further, it is proved that

(16) $$\lim_{n\to\infty} \int_0^\infty e^{-st} \eta_n E_n(\eta_n t)\,dt = \frac{a + \sqrt{2\sigma^2 s + a^2}}{2s},$$

whence, on using Erdélyi *et al.* (1954, Vol. I, p. 235, Eq. (22)), it follows that

(17) $$\lim_{n\to\infty} \eta_n E_n(\eta_n t)$$
$$= a\left[1 + \frac{1}{\sqrt{\pi}} \int_{\sqrt{a^2 t/\sigma^2}}^\infty x^{-2} \exp\left(-\frac{x^2}{2}\right) dx\right] \qquad \text{for all} \quad t \geqq 0,$$

where

(18) $$E_n(t) = \operatorname{Prob}\{X_n(t) = 0 \mid X_n(0) = 0\}.$$

Also,

(19) $$\lim_{n\to\infty} \int_0^\infty e^{-st} \lambda_n [1 - \Gamma_n(\eta_n t)]\,dt = \frac{2}{a + \sqrt{2\sigma^2 s + a^2}},$$

whence[2]

(20) $$\lim_{n\to\infty} \lambda_n [1 - \Gamma_n(\eta_n t)] = (2/\sqrt{2\sigma^2 t})\, i \operatorname{erfc}[\sqrt{a^2 t/(2\sigma^2)}]).$$

[2] An obvious error in Ott (1977, p. 184, Eq. (6.15)) has been corrected here.

CHAPTER 5

Queues with General Arrival and General Service Distributions

In this chapter we discuss queuing systems for which arrival and service mechanisms are both characterized by general distributions. We describe the work on double queues with general arrivals by Kashyap (1965c, 1970b) and the work of Daley (1969) and Daley and Jacobs (1969) on the total waiting time in a busy period.

5.1 The Double-Ended Queue with General Arrivals

Consider a further generalization of the queuing system discussed in Section 4.4, where the arrival distribution of taxis is general and is simulated by introducing an arrival timing channel consisting of j phases, the arrival distribution of customers being general with density function $A(x)$, as before (Kashyap, 1965c). The time spent in each phase is exponentially distributed with mean $1/\mu$. An interarrival time consists of r phases with probability C_r ($r = 1, 2, \ldots, j$). The phases are numbered in the reverse order, so that after completion of the rth phase, the $(r-1)$th phase starts, and so on until the first phase. The "arrival" takes place at the end of the first phase. The remaining features of this model are the same as in Section 4.4. Let $P_{nr}(x,t)\,dx$ ($-M \leqq n \leqq N$; $1 \leqq r \leqq j$) be the probabilities that at time t there are n customers (or $-n$ taxis, $n < 0$) waiting, the last customer having arrived a time between $(x, x+dx)$ earlier and the "arriving" taxi being in the rth phase. The system is represented by the following equations:

$$(1)\qquad \left(\frac{\partial}{\partial t} + \frac{\partial}{\partial x} + \eta(x)\right) P_{-M,1}(x,t) = \mu P_{-M,2}(x,t) + \mu C_1 P_{-M+1,1}(x,t),$$

$$(2)\qquad \left(\frac{\partial}{\partial t} + \frac{\partial}{\partial x} + \{\mu + \eta(x)\}\right) P_{nr}(x, t)$$

$$= \mu P_{n,r+1}(x, t) + \mu C_r P_{n+1,1}(x, t),$$

$$(n = -M, 1 < r < j; -M < n < N, 1 \leqq r < j),$$

$$(3)\qquad \left(\frac{\partial}{\partial t} + \frac{\partial}{\partial x} + \{\mu + \eta(x)\}\right) P_{nj}(x, t)$$

$$= \mu C_j P_{n+1,1}(x, t) \qquad (-M \leqq n < N),$$

$$(4)\qquad \left(\frac{\partial}{\partial t} + \frac{\partial}{\partial x} + \{\mu + \eta(x)\}\right) P_{Nr}(x, t)$$

$$= \mu P_{N,r+1}(x, t) \quad (1 \leqq r < j),$$

$$(5)\qquad \left(\frac{\partial}{\partial t} + \frac{\partial}{\partial x} + \{\mu + \eta(x)\}\right) P_{Nj}(x, t)$$

$$= 0,$$

where $\eta(x)$ is defined as in Section 4.4. The boundary conditions are, for $1 \leqq r \leqq j$,

$$(6)\qquad P_{-M,r}(0, t) = 0,$$

$$(7)\qquad P_{nr}(0, t) = \int_0^\infty P_{n-1,r}(x, t)\eta(x)\, dx \qquad (-M < n < N),$$

$$(8)\qquad P_{Nr}(0, t) = \int_0^\infty [P_{N-1,r}(x, t) + P_{Nr}(x, t)]\eta(x)\, dx,$$

and let the initial conditions be

$$(9)\qquad P_{nr}(x, 0) = \delta_{in}\delta_{mr}\delta(x) \qquad (-M < i \leqq N, 1 \leqq m \leqq j).$$

We define the generating functions

$$(10)\qquad F_r(x, \alpha, t) = \sum_{n=-M}^{N} \alpha^n P_{nr}(x, t),$$

$$(11)\qquad F(x, \alpha, \beta, t) = \sum_{r=1}^{j} \beta^r F_r(x, \alpha, t).$$

Using the generating functions (10) and (11) and taking Laplace transforms, denoted by corresponding lowercase letters, with the initial conditions (9), we have, from Eqs. (1)–(5),

$$\left(\frac{\partial}{\partial x} + \left\{\mu + s + \eta(x) - \frac{\mu}{\beta}\right\}\right) f(x, \alpha, \beta, s) \tag{12}$$
$$= \mu[\alpha^{-1}C(\beta) - 1] f_1(x, \alpha, s)$$
$$+ \mu\alpha^{-M}[\beta - \alpha^{-1}C(\beta)] p_{-M,1}(x, s) + \alpha^i \beta^m \delta(x),$$

where

$$C(\beta) = \sum_{r=1}^{j} C_r \beta^r. \tag{13}$$

Similarly, from the boundary conditions,

$$f(0, \alpha, \beta, s) = \alpha \int_0^\infty f(x, \alpha, \beta, s)\eta(x)\, dx \tag{14}$$
$$+ \alpha^N(1 - \alpha) \sum_{r=1}^{j} \beta^r \int_0^\infty P_{Nr}(x, s)\eta(x)\, dx.$$

The method of solution further is the same as in Section 4.4, but the analysis is more involved due to the generalization introduced. For the solution of (12), we require $f_1(x, \alpha, s)$ and $p_{-M,1}(x, s)$, which we evaluate by solving the Laplace transforms of Eqs. (1)–(5) recursively, as before (Section 4.4). Let us use the following notation:

$$\Pi_l = \prod_{k=1}^{l} C_{r_k}; \qquad \Pi_0 = 1, \tag{15}$$

$$R_l = \sum_{k=1}^{l} r_k; \qquad R_0 = 0, \tag{16}$$

and let

$$\sum^{*} = \sum_{1 \leqq r, r_k \leqq j} \qquad (k = 1, 2, \ldots), \tag{17}$$

i.e., the summation over all the relevant r from 1 to j. Then, we have

$$p_{-M,1}(x, s) = \mu \exp\left[-sx - \int_0^x \eta(u)\, du\right] \tag{18}$$
$$\cdot \left[\sum_{l=1}^{M+N} \sum^{*} \Pi_l p_{l-M,r}(0, s) \int_0^x \frac{(\mu y)^{(r+R_l-2)}}{(r + R_l - 2)!} e^{-\mu y}\, dy\right.$$
$$\left. + \sum^{*} \Pi_{M+i} \int_0^x \frac{(\mu y)^{(m+R_{M+i}-2)}}{(m + R_{M+i} - 2)!} e^{-\mu y}\, dy\right],$$

$$(19)\qquad f_1(x,\alpha,s)=\alpha^{-M}p_{-M,1}(x,s)+\exp\left[-(\mu+s)x-\int_0^x \eta(u)\,du\right]$$

$$\cdot\left[\sum_{L=1}^{M+N}\sum_{l=1}^{L}\sum^{*}\alpha^{L-M-l+1}p_{L-M,r}(0,s)\right.$$

$$\cdot\,\Pi_{l-1}\frac{(\mu x)^{(r+R_{l-1}-1)}}{(r+R_{l-1}-1)!}$$

$$\left.+\sum_{l=1}^{M+i}\sum^{*}\alpha^{i-l+1}\Pi_{l-1}\frac{(\mu x)^{(m+R_{l-1}-1)}}{(m+R_{l-1}-1)!}\right].$$

Solving (12) after substitution from (18) and (19), we have the solution containing terms of the type

$$(20)\qquad \int_0^\infty \exp[-(s+a)x]\,A(x)\gamma(b,cx)\,dx,$$

where $\gamma(\alpha, z)$ is the incomplete gamma function. Proceeding as in Section 4.4, and considering the special case where the arrivals of taxis and customers are according to j-Erlang and k-Erlang distributions, i.e., where

$$(21)\qquad C_r=\delta_{rj},$$

$$(22)\qquad \eta(x)=\lambda^k x^{k-1}e^{-\lambda x}/(k-1)!,$$

we can express the generating function

$$(23)\qquad \pi(\alpha,1,s)=\int_0^\infty f(x,\alpha,1,s)\,dx=\sum_{n=-M}^{N}\alpha^n P_n(t)$$

of the marginal probabilities $P_n(t)$ completely in terms of the functions ${}_2F_1$ and F_2, which are convergent for the relevant values of the arguments.

Busy Periods

For the system described above, we study busy periods under two categories, viz., customer periods and taxi periods (cf. Sections 1.3 and 4.4). What follows was reported in Kashyap (1970b), of which we present a slightly simplified form below.

For finding the distribution of *customer periods*, we postulate an absorbing barrier at $n = 0$. We shall use the probabilities $P_{nr}(x, t)\,dx$ ($n = 1, \ldots, N$; $r = 1, \ldots, j$) as defined above, while for $n = 0$, we use simply the marginal probability $P_0(t)$. We have the following equations for the system:

$$\frac{d}{dt}P_0(t) = \mu \int_0^\infty P_{11}(x,t)\,dx; \tag{24}$$

$$\left(\frac{\partial}{\partial t} + \frac{\partial}{\partial x} + \{\mu + \eta(x)\}\right) P_{nr}(x,t) = \mu P_{n,r+1}(x,t) + \mu C_r P_{n+1,1}(x,t) \quad (1 \leqq n \leqq N;\ 1 \leqq r \leqq j-1), \tag{25}$$

$$\left(\frac{\partial}{\partial t} + \frac{\partial}{\partial x} + \{\mu + \eta(x)\}\right) P_{nj}(x,t) = \mu C_j P_{n+1,1}(x,t) \quad (1 \leqq n \leqq N-1); \tag{26}$$

$$\left(\frac{\partial}{\partial t} + \frac{\partial}{\partial x} + \{\mu + \eta(x)\}\right) P_{Nr}(x,t) = \mu P_{N,r+1}(x,t) \quad (1 \leqq r \leqq j-1); \tag{27}$$

$$\left(\frac{\partial}{\partial t} + \frac{\partial}{\partial x} + \{\mu + \eta(x)\}\right) P_{Nj}(x,t) = 0. \tag{28}$$

At time $t = 0$, let a customer arrive to initiate the busy period and let the arrival channel then be in the phase m, so that the initial conditions are

$$P_{nr}(x,0) = \delta_{1n}\delta_{mr}\delta(x), \tag{29}$$

and the boundary conditions are, for $1 \leqq r \leqq j$,

$$P_0(0,t) = 0, \tag{30}$$

$$P_{1r}(0,t) = \delta_{mr}\delta(t), \tag{31}$$

$$P_{nr}(0,t) = \int_0^\infty P_{n-1,r}(x,t)\eta(x)\,dx \qquad (2 \leqq n \leqq N-1), \tag{32}$$

$$P_{Nr}(0,t) = \int_0^\infty [P_{N-1,r}(x,t) + P_{Nr}(x,t)]\eta(x)\,dx. \tag{33}$$

Now, solving the L.T.s of Eqs. (25)–(28) recursively in the reverse order, as before, we have

$$p_{11}(x,s) = \exp\left[-(\mu+s)x - \int_0^x \eta(u)\,du\right] \cdot \left[\sum_{l=1}^{N} \sum^{*} \Pi_{l-1} \frac{(\mu x)^{(r+R_{l-1}-1)}}{(r+R_{l-1}-1)!} p_{lr}(0,s) + \frac{(\mu x)^{m-1}}{(m-1)!}\right]. \tag{34}$$

The probability density function of the duration of a customer period is

$$\Gamma_C(t) = \frac{d}{dt} P_0(t), \tag{35}$$

and, therefore, its Laplace transform, on using (24), is given by

$$\gamma_C(s) = sp_0(s) = \mu \int_0^\infty p_{11}(x, s)\, dx. \tag{36}$$

Substituting from (34) into (36), and inverting the transform, we have

$$\begin{aligned} \Gamma_C(t) &= \frac{\mu^m t^{m-1}}{(m-1)!} \exp\left\{-\mu t - \int_0^t \eta(u)\, du\right\} \\ &\quad + \mu \sum_{l=1}^{N} \sum^{*} \Pi_{l-1} P_{lr}(0, t) \\ &\quad * \left[\frac{(\mu t)^{(r+R_{l-1}-1)}}{(r+R_{l-1}-1)!} \exp\left\{-\mu t - \int_0^t \eta(u)\, du\right\}\right]. \end{aligned} \tag{37}$$

Taxi Periods

Letting the system start with the arrival of a taxi which makes $n = -1$, the arrival timing channel then being in the phase m, and taking $n = 0$ as an absorbing state, we have the following equations for the system:

$$\frac{d}{dt} P_0(t) = \int_0^\infty \sum_{r=1}^{j} P_{-1,r}(x, t)\eta(x)\, dx, \tag{38}$$

$$\begin{aligned} &\left(\frac{\partial}{\partial t} + \frac{\partial}{\partial x} + \{\mu + \eta(x)\}\right) P_{-1,r}(x, t) \\ &\qquad = \mu P_{-1,r+1}(x, t) \quad (1 \leqq r \leqq j-1), \end{aligned} \tag{39}$$

$$\begin{aligned} &\left(\frac{\partial}{\partial t} + \frac{\partial}{\partial x} + \{\mu + \eta(x)\}\right) P_{-1,j}(x, t) \\ &\qquad = 0, \end{aligned} \tag{40}$$

$$\begin{aligned} &\left(\frac{\partial}{\partial t} + \frac{\partial}{\partial x} + \{\mu + \eta(x)\}\right) P_{nr}(x, t) \\ &\qquad = \mu P_{n,r+1}(x, t) + \mu C_r P_{n+1,1}(x, t) \\ &\qquad (-2 \geqq n \geqq -M+1, 1 \leqq r \leqq j; n = -M, 2 \leqq r \leqq j), \end{aligned} \tag{41}$$

$$\begin{aligned} &\left(\frac{\partial}{\partial t} + \frac{\partial}{\partial x} + \{\mu + \eta(x)\}\right) P_{nj}(x, t) \\ &\qquad = \mu C_j P_{n+1,1}(x, t) \qquad (-2 \geqq n \geqq -M), \end{aligned} \tag{42}$$

$$\begin{aligned} &\left(\frac{\partial}{\partial t} + \frac{\partial}{\partial x} + \eta(x)\right) P_{-M,1}(x, t) \\ &\qquad = \mu P_{-M,2}(x, t) + \mu C_1 P_{-M+1,1}(x, t), \end{aligned} \tag{43}$$

with the boundary conditions, for $1 \leqq r \leqq j$,

(44) $$P_{nr}(0, t) = \int_0^\infty P_{n-1,r}(x, t)\eta(x)\,dx \qquad (-1 \geqq n \geqq -M+1),$$

(45) $P_{-M,r}(0, t) = 0,$

and the initial conditions

(46) $$P_{nr}(x, 0) = \delta_{-1,n}\delta_{mr}\delta(x - x_0),$$

where x_0 is the elapsed time, at $t = 0$, since the last customer arrival. Solving the L.T.s of (40) and (39) for $r = j-1, j-2, \ldots, 1$, it can be shown that

(47) $$\sum_{r=1}^{j} p_{-1,r}(x, s) = \exp\left(-(\mu+s)x - \int_0^x \eta(u)\,du\right)\left\{\delta(x - x_0) + \sum_{r=1}^{j}\sum_{r_1=r}^{j} \frac{(\mu x)^{r_1-1}}{(r_1-1)!} p_{-1,r_1}(0, s)\right\}.$$

Using (47) in the L.T. of (38), we have the L.T. of the density function of taxi periods, which can be readily inverted in terms of convolutions.

5.2 The Total Waiting Time in a Busy Period

Daley (1969) and Daley and Jacobs (1969) study the total waiting time in a busy period in a $GI/G/1$ queuing system, which is a random variable with behavior similar to the total waiting time of all vehicles in a given hold-up (Gaver, 1969).

Let the interarrival times $\{T_n\}$ and service times $\{S_n\}$ have distribution functions (D.F.) $A(x)$ and $B(x)$, respectively, and let

(1) $$U_n = S_n - T_n.$$

Then, the D.F. $U(x)$ of U_n is given by

(2) $$U(x) = \int_0^\infty B(x+t)\,dA(t),$$

where $\{T_n\}$ and $\{S_n\}$ are assumed to be mutually independent. Let the first customer C_0 have a waiting time W_0 and the customer C_n a waiting time W_n; then

(3) $$W_{n+1} = (W_n + U_n)^+ = \max(0, W_n + U_n).$$

Define

(4) $$N_w = \inf\left\{ j \,\middle|\, w + \sum_{i=1}^{j} U_{i-1} < 0 \right\},$$

so that N_w is the number of customers in a busy period, in which the first customer waits a time $W_0 = w$. It may be noted that the term *busy period* here means the *first* or *generalized* busy period in which the first customer may have to wait. Since we have

$$(5) \qquad \begin{aligned} W_n &= W_{n-1} + U_{n-1} \\ &= w + U_0 + U_1 + \cdots + U_{n-1} \qquad \text{for} \quad 1 \leqq n < N_w, \end{aligned}$$

it follows that the total waiting time of all the customers in this busy period is

$$(6) \qquad X_w = \sum_{j=0}^{N_w - 1} \left(w + \sum_{i=1}^{j} U_{i-1} \right) = N_w w + \sum_{j=1}^{N_w - 1} (N_w - j) U_{j-1}.$$

In a nontrivial stable queue,

$$(7) \qquad \begin{aligned} E(U) &< 0, \\ \text{Prob}\{N_w < \infty\} &= 1, \end{aligned}$$

so that

$$(8) \qquad \text{Prob}\{X_w < \infty\} = 1.$$

Let us further assume that $E(t) < \infty$, so that

$$(9) \qquad 0 < \rho < 1,$$

where

$$(10) \qquad \rho = E(S)/E(T).$$

In the above and in what follows, the subscript in S_n, T_n, etc., is dropped for convenience.

Let χ_A denote the indicator random variable of the set A. Then, from the definitions of X_w and W_1, we have

$$(11) \qquad X_w = w + X_{w + U\chi_{(-w,\infty)}(U)},$$

equality being in the sense that both sides have the same distribution. Let $F(x|w)$ denote the distribution of X_w; then, from (11),

$$(12) \qquad F(x|w) = \begin{cases} 0 & (x < w), \\ U(-w) + \displaystyle\int_{-w}^{\infty} F(x - w|w + u)\, dU(u) & (x \geqq w). \end{cases}$$

Define

$$(13) \qquad \phi(\theta|w) = E(e^{-\theta X_w}) = \int_0^{\infty} e^{-\theta x}\, dF(x|w),$$

$$(14) \qquad \phi(\theta, \psi) = \int_0^{\infty} e^{-\psi w} \phi(\theta|w)\, dw, \qquad \text{Re}(\theta) \geqq 0, \quad \text{Re}(\psi) > -\text{Re}(\theta);$$

then, from (12), we have

$$\phi(\theta, \psi - \theta) = \int_0^\infty e^{-\psi y} U(-y)\, dy + \int_0^\infty \phi(\theta|u)\, du \int_0^\infty e^{-\psi y} U(u - y)\, dy. \tag{15}$$

Considering the special case of the $GI/M/1$ queue, we have

$$B(x) = 1 - e^{-\mu x}. \tag{16}$$

Also, let

$$a(\theta) = E(e^{-\theta T}) = \int_0^\infty e^{-\theta t}\, dA(t). \tag{17}$$

Then, using (2), (16), and (17), we have

$$\int_0^\infty e^{-\psi y} U(u - y)\, dy = \frac{1 - e^{-\psi u} a(\psi)}{\psi} + \frac{e^{-\psi u} a(\psi) - e^{-\mu u} a(\mu)}{\psi - \mu} \qquad [\operatorname{Re}(\psi) > 0, u \geqq 0], \tag{18}$$

so that, after some algebra, (15) gives

$$\begin{aligned} &(\psi - \mu)\phi(\theta, \psi - \theta) - (\psi - \mu)[1 - a(\psi)]\psi^{-1} + a(\mu) - a(\psi) \\ &\quad = \mu a(\mu)\phi(\theta, \mu) - a(\psi)\phi(\theta, \psi) \qquad [\operatorname{Re}(\theta) \geqq 0, \operatorname{Re}(\psi) > 0]. \end{aligned} \tag{19}$$

Now, by the Tauberian theorem (Widder, 1941, Chapter V),

$$\lim_{\psi \to \infty} \psi\phi(\theta, \psi) = \phi(\theta|0) = \phi_0(\theta), \tag{20}$$

say. Letting $\psi \to \infty$ in (19) and using (20), we have

$$\mu a(\mu)\phi(\theta, \mu) = \phi_0(\theta) - [1 - a(\mu)]. \tag{21}$$

Thus, (19) can be rewritten in the form

$$(\psi - \mu)\phi(\theta, \psi - \theta) + \mu a(\psi)\phi(\theta, \psi) = \phi_0(\theta) - [1 - a(\psi)]\psi^{-1}. \tag{22}$$

After further algebraic manipulations, Daley is able to show that

$$\begin{aligned} \frac{(\psi - \mu)\phi(\theta, \psi - \theta)}{\Gamma(\omega + 1)} = \left\{\frac{\phi_0(\theta)}{\mu}\right\} \Bigg[&\sum_{n=0}^{\infty} \left\{\prod_{j=0}^{n-1} a(\psi + j\theta)\right\} \frac{\zeta^n}{\Gamma(\omega + n + 1)} \\ &- \sum_{n=0}^{\infty} \left\{\prod_{j=0}^{n-1} a(\psi + j\theta)\right\} \\ &\cdot \left\{\frac{\{1 - a(\psi + n\theta)\}\zeta^n}{\Gamma(\omega + n + 1)(\psi + n\theta)}\right\}\Bigg], \end{aligned} \tag{23}$$

where

(24) $$\zeta = -\mu/\theta, \qquad \omega = (\psi - \mu)/\theta.$$

In the case of an $M/M/1$ queue, where

(25) $$A(x) = 1 - e^{-\lambda x},$$

(23) gives

(26) $$(\psi - \mu)\phi(\theta, \psi - \theta) = \left\{\frac{\phi_0(\theta)}{\mu}\right\} {}_1F_2\left(1; \frac{\psi + \lambda}{\theta}, \frac{\psi - \mu}{\theta} + 1; -\frac{\lambda\mu}{\theta^2}\right)$$
$$- \frac{1}{\psi + \lambda} {}_1F_2\left(1; \frac{\psi + \lambda}{\theta} + 1, \frac{\psi - \mu}{\theta} + 1; -\frac{\lambda\mu}{\theta^2}\right).$$

Setting $\psi = \mu$ in (26), the functions ${}_1F_2$ simplify to ${}_0F_1$, which are Bessel functions, and hence we have

(27) $$\phi_0(\theta) = (\mu/\lambda)^{1/2} J_\eta(\tau)/J_{\eta-1}(\tau),$$

where

(28) $$\eta = (\mu + \lambda)/\theta, \qquad \tau = 2\sqrt{\lambda\mu}/\theta.$$

It is difficult to invert $\phi_0(\theta)$ given by (27), and as Daley points out, even to find moments of X_0 (and X_w); it is simpler to use (22), from which, using differentiation, he shows that for $GI/M/1$,

(29) $$E(X_0) = \delta/\{\mu(1 - \delta)^2\},$$

(30) $$\operatorname{Var}(X_0) = \frac{\delta(1 + \delta)\alpha_2}{(1 + \mu\alpha_1)^2(1 - \delta)^3} + \frac{2\delta(2 + \delta)}{\mu^2(1 - \delta)^4(1 + \mu\alpha_1)} - \frac{\delta(2 - \delta)}{\mu^2(1 - \delta)^4},$$

where δ is the unique root in $0 < \delta < 1$ of the equation

(31) $$\delta = a[\mu(1 - \delta)]$$

and

(32) $$\alpha_1 = a'[\mu(1 - \delta)], \qquad \alpha_2 = a''[\mu(1 - \delta)].$$

It is also shown that for $GI/G/1$, as $w \to \infty$,

(33) $$E(X_w) \sim w^2/\{2E(-U)\},$$

(34) $$\operatorname{Var}(X_w) \sim w^3 \operatorname{Var}(U)/\{3(E(-U))^3\},$$

and explicit expressions for $E(X_w)$ and $\operatorname{Var}(X_w)$ are obtained in the special case of the $M/M/1$ queue.

It is pointed out that if the initial delay $w = W_0$ is in fact a random variable having a negative exponential distribution with parameter v, then the transform

$$E[\exp(-\theta X_{W_0})] = \int_0^\infty E[\exp(-\theta X_w)]ve^{-vw}\,dw = v\phi(\theta, v), \tag{35}$$

which provides a probabilistic interpretation of $\phi(\theta, \psi)$.

Daley and Jacobs (1969) consider the total waiting time in a busy period for an $M/G/1$ queue. Let λ be the mean arrival rate and $B(x)$ the distribution function of service times. Let X_i' denote the total waiting time in a busy period which starts at a time t_0, say, with $i\ (\geqq 1)$ customers in the system, and let Z_i' be the length of this busy period. Daley and Jacobs begin by obtaining the mean and variance of the distributions of X_i', Z_i', and of $X_1' = X$, $Z_1' = Z$, and the corresponding covariances.

For the special case of $M/M/1$, using the Markovian property of the queue-size process, we have

$$X_0' = 0 \quad \text{a.s.}^1 \tag{36}$$

and for $i = 1, 2, \ldots,$

$$X_i' = \begin{cases} (i-1)\min(S, T) + X_{i+1}' & \text{with probability } \lambda/(\lambda+\mu) \\ (i-1)\min(S, T) + X_{i-1}' & \text{with probability } \mu/(\lambda+\mu), \end{cases} \tag{37}$$

where S, T denote generic service and interarrival time random variables. Writing

$$\phi_i(\theta) = E[\exp(-\theta X_{i+1}')] \qquad [\operatorname{Re}(\theta) \geqq 0], \tag{38}$$

we have, from (37) and (36),

$$(\lambda + \mu + i\theta)\phi_i(\theta) = \lambda\phi_{i+1}(\theta) + \mu\phi_{i-1}(\theta) \qquad (i = 0, 1, \ldots), \tag{39}$$
$$\phi_{-1}(\theta) = 1.$$

For solving Eqs. (39), use the generating function

$$\sum_{k=i+1}^{\infty} \phi_k(\theta) z^k \qquad (|z| < 1); \tag{40}$$

then solving the resulting equation and equating coefficients of powers of z, we have

$$\phi_i(\theta) = (\mu/\lambda)^{(i+1)/2} J_{\eta+i}(\tau)/J_{\eta-1}(\tau), \tag{41}$$

where, as also in (27) and (28),

$$\eta = (\lambda + \mu)/\theta, \qquad \tau = 2\sqrt{\lambda\mu}/\theta. \tag{42}$$

Equation (41) is not tractable, as before [cf. (27)], but as Daley and Jacobs

[1] Almost surely.

point out, the result is interesting from the point of view of the appearance of Bessel functions "in a guise in which we have not encountered them before in probability theory." For moments of X_i, (39) and (40) may be used directly. They also give a simpler method of establishing (41) as follows.

Define

$$\chi_i(\theta) = (\lambda/\mu)^{1/2}\phi_i(\theta)/\phi_{i-1}(\theta) \qquad (i = 0, 1, \dots; \operatorname{Re}(\theta) \geqq 0); \tag{43}$$

then, from (39),

$$\lambda + \mu + i\theta = (\lambda\mu)^{1/2}\chi_{i+1}(\theta) + (\lambda\mu)^{1/2}/\chi_i(\theta). \tag{44}$$

Therefore

$$\chi_i(\theta) = (\tau/2)[\eta + i - (\tau/2)\chi_{i+1}(\theta)]^{-1}, \tag{45}$$

which leads to the continued fraction representation

$$\chi_i(\theta) = \frac{\tau/2}{\eta + i -}\,\frac{(\tau/2)^2}{\eta + i + 1 -}\,\frac{(\tau/2)^3}{\eta + i + 2 -}\cdots \tag{46}$$

$$= \frac{a_1}{1 +}\,\frac{a_2}{1 +}\,\frac{a_3}{1 +}\cdots,$$

where

$$a_1 = \tfrac{1}{2}\tau/(\eta + i), \tag{47}$$

$$a_n = (\tfrac{1}{2}\tau)^2/\{(\eta + n - 2)(\eta + n - 1)\} \qquad (n \geqq 2).$$

Convergence of (46) follows from Worpitzky's theorem (Wall, 1948, p. 42). Now, (46) is the known representation for the ratio of Bessel functions $J_{\eta+i}(\tau)/J_{\eta+i-1}(\tau)$ (Wall, 1948, p. 349; Whittaker and Watson, 1927, p. 380); hence, using (43), we have (41).

CHAPTER 6

Miscellaneous Results

We devote this chapter to the discussion of the application of special functions in other areas mostly allied to queues, like management science, stochastic processes, and generalized probability distributions. We include the work of Kimura (1964) on population genetics, Bhattacharya (1966) on life testing, Schweitzer (1967) on spares planning, Sengupta (1972) on chance-constrained linear programming, Jones (1973) on consumer behavior, Nishimura (1974) on collision processes, and Das (1977) on inventories. Further, we describe the work on probability distributions by various authors including Pearson (1924a,b), Hansmann (1934), Ayyangar (1934), Laha (1954), Kabe (1958, 1962), Ruben (1960, 1962), Stacy (1962), Kemp and Kemp (1965), Mathai and Saxena (1966, 1968), Manocha (1968), McNolty (1968), Srivastava and Singhal (1972), and Gupta and Jain (1974). We conclude this chapter by describing the recent work of Getz (1975, 1976), Kapur (1978a,b; 1979), and Kapur and Kumar (1978) on generalized birth and death processes.

6.1 Population Genetics

Consider the problem of finding the change in gene frequency, i.e., the proportion of genes in an isolated population of N breeding diploid individuals under the assumption of random mating and random sampling of gametes from the population to form the next generation. The following differential equation holds for the model (Kimura, 1964) with two alleles A_1 and A_2:

$$(1) \qquad \frac{\partial \phi}{\partial t} = \frac{1}{4N} \frac{\partial^2}{\partial x^2} \{x(1-x)\phi\} \qquad (0 < x < 1),$$

where $\phi = \phi(p, x; t)$ is the probability density that the gene frequency becomes

x in the tth generation, given that it is p at $t = 0$, so that

(2) $$\phi(p, x; 0) = \delta(x - p),$$

$\delta(\cdot)$ being the Dirac delta function. Trying a solution of the form

(3) $$\phi = TX,$$

where T is a function of t only, and X is a function of x only, we have

(4) $$\frac{dT}{dt} = -\lambda T,$$

(5) $$x(1 - x)\frac{d^2X}{dx^2} + 2(1 - 2x)\frac{dX}{dx} - (2 - 4N\lambda)X = 0,$$

where λ is a constant. From (4), we have

(6) $$T \propto e^{-\lambda t},$$

and (5) is the hypergeometric equation (see also Section II.2)

(7) $$x(1 - x)X'' + [\gamma - (\alpha + \beta + 1)x]X' - \alpha\beta X = 0,$$

where

(8) $$\gamma = 2, \qquad \alpha + \beta = 3, \qquad \alpha\beta = 2 - 4N\lambda.$$

We are interested in the solution of (7), which is finite at the singular points $x = 0, 1$. Kimura shows that such a solution is

(9) $$X = {}_2F_1(2 + i, 1 - i; 2; x) \qquad (i = 1, 2, 3, \ldots),$$

provided that

(10) $$\lambda = i(i + 1)/4N,$$

which can be written

(11) $$X = T_{i-1}^1(z),$$

where $z = 1 - 2x$ and the right-hand member is the Gegenbauer polynomial, given by (Morse and Feshbach, 1953, p. 783)

(12) $$T_n^\beta(z) = \frac{\Gamma(n + 2\beta + 1)}{2^\beta n!\Gamma(\beta + 1)}\,{}_2F_1\left(-n, n + 2\beta + 1; 1 + \beta; \frac{1 - z}{2}\right).$$

It may be noted that with Gegenbauer's notation, $C_n^\lambda(z)$, using Erdélyi *et al.* (1953, Vol. II, p. 176, Eq. (20)), we can write

(13) $$X = T_{i-1}^1(z) = C_{i-1}^{3/2}(z).$$

From (6), (10), and (11), the complete solution of (1) is

$$\phi(p, x; t) = \sum_{i=1}^{\infty} C_i T_{i-1}^1(z) \exp\left[-\frac{i(i+1)t}{4N}\right], \qquad z = 1 - 2x. \tag{14}$$

The constants C_i are evaluated by using the initial condition (2) and the orthogonality property (Morse and Feshbach, 1953), viz.,

$$\int_{-1}^{1} (1 - z^2) T_m^1(z) T_{i-1}^1(z)\, dz = \delta_{m,i-1} \frac{2i(i+1)}{2i+1}. \tag{15}$$

Thus, finally we have

$$\begin{aligned}\phi(p, x; t) = \sum_{i=1}^{\infty} &\frac{(2i+1)(1-r^2)}{i(i+1)} T_{i-1}^1(r) T_{i-1}^1(z) \\ &\cdot \exp[-i(i+1)t/4N],\end{aligned} \tag{16}$$

where

$$r = 1 - 2p, \qquad z = 1 - 2x, \tag{17}$$

$$\begin{aligned} &T_0^1(r) = 1, && T_1^1(r) = 3r, \\ &T_2^1(r) = \tfrac{3}{2}(5r^2 - 1), && T_3^1(r) = \tfrac{5}{2}(7r^3 - 3r), \ldots . \end{aligned} \tag{18}$$

In terms of the hypergeometric function, we can write

$$\begin{aligned}\phi(p, x; t) &= \sum_{i=1}^{\infty} p(1-p) i(i+1)(2i+1)\, {}_2F_1(1-i, i+2; 2; p) \\ &\qquad \cdot {}_2F_1(1-i, i+2; 2; x) \exp[-i(i+1)t/4N] \\ &= 6p(1-p) \exp[-t/(2N)] \\ &\qquad + 30p(1-p)(1-2p)(1-2x) \exp[-3t/(2N)] + \cdots .\end{aligned} \tag{19}$$

The series in (19) is uniformly convergent in x and p, for $t > 0$. The probability that the two alleles A_1 and A_2 coexist in the population in the tth generation is represented by the area under the probability density curve and is, therefore, given by

$$\begin{aligned}\Omega_t &= \int_0^1 \phi(p, x; t)\, dx \\ &= \sum_{j=0}^{\infty} [P_{2j}(r) - P_{2j+2}(r)] \exp\left[-\frac{(2j+1)(2j+2)t}{4N}\right],\end{aligned} \tag{20}$$

where r is given by (17) and $P_n(r)$ stands for the Legendre polynomials [see Erdélyi *et al.* (1953, Vol. I, p. 150, Eq. (15))], given by

$$P_{2n}(z) = \frac{(-1)^n (2n)!}{2^{2n}(n!)^2}\, {}_2F_1(-n, n + \tfrac{1}{2}; \tfrac{1}{2}; z^2). \tag{21}$$

The boundaries $x = 0$ and $x = 1$ are absorbing barriers, and Kimura shows that the probability of absorption, i.e., of genes going into fixation (loss) by a given generation t, starting from an intermediate gene frequency p, is

$$(22) \qquad u(p, t) = p + \sum_{i=1}^{\infty} (2i + 1)p(1 - p)(-1)^i \cdot {}_2F_1(1 - i, i + 2; 2; p) \exp[-i(i + 1)t/4N],$$

or, in terms of Legendre polynomials,

$$(23) \qquad u(p, t) = p + \sum_{i=1}^{\infty} \frac{(-1)^i}{2} \{P_{i-1}(r) - P_{i+1}(r)\} \exp\left[-\frac{i(i + 1)t}{4N}\right],$$

where r is given as above.

Kimura (1964) gives the following generalization of (1) to the case of a population containing three alleles A_1, A_2, A_3,

$$(24) \qquad \frac{\partial \phi}{\partial t} = \frac{1}{4N} \frac{\partial^2}{\partial x_1^2} \{x_1(1 - x_1)\phi\} - \frac{1}{2N} \frac{\partial^2}{\partial x_1 \, \partial x_2} \{x_1 x_2 \phi\} + \frac{1}{4N} \frac{\partial^2}{\partial x_2^2} \{x_2(1 - x_2)\phi\} \qquad (0 < x_1 < x_1 + x_2 < 1),$$

with the initial condition

$$(25) \qquad \phi(p_1, p_2; x_1, x_2; 0) = \delta(x_1 - p_1)\,\delta(x_2 - p_2).$$

The solution in this case is (Kimura, 1956, 1964)

$$(26) \quad \phi(p_1, p_2; x_1, x_2; t) = \sum_{n=0}^{\infty} \sum_{j=0}^{\infty} C(n, j)(1 - x_3)^n T_n^1\left(\frac{x_1 - x_2}{1 - x_3}\right) \cdot J_j(2n + 5, 2n + 4, 1 - x_3) \exp\left[-\frac{(j + n + 2)(j + n + 3)t}{4N}\right]$$

$$= 5!\, p_1 p_2 p_3 \exp\left[-\frac{3t}{2N}\right] + \frac{7!}{2!} p_1 p_2 p_3 \left\{\sum_{i=1}^{3} \left(p_i - \frac{1}{3}\right) x_i\right\} \cdot \exp\left[-\frac{6t}{2N}\right] + \cdots,$$

where

$$(27)\quad C(n,j)=\frac{4(2j+2N+5)(j+2n+3)!(j+2N+4)!}{(n+1)(n+2)(2n+2)!(2n+3)!j!(j+1)!}p_1p_2p_3(1-p_3)^n \cdot T_n^1\left(\frac{p_1-p_2}{1-p_3}\right)J_j(2n+5,2n+4,1-p_3),$$

$$(28)\qquad x_3=1-x_1-x_2,$$

$$(29)\qquad p_3=1-p_1-p_2,$$

$T_n^1(z)$ denotes the Gegenbauer polynomials [cf. (12)], and $J_j(a,c,\rho)$ are the Jacobi polynomials, expressible in terms of ${}_2F_1$ as follows:

$$(30)\quad J_n(a,c,\rho)={}_2F_1(a+n,-n;c;\rho)=\binom{n+c-1}{n}^{-1}P_n^{(c-1,a-c)}(1-2\rho),$$

the last member in (30) giving the Jacobi polynomial in the alternative notation used, for example, in II.2(33), II.8(1)–II.8(4), Szegö (1975), and Erdélyi *et al.* (1953, Vol. II).

Allowing for migration between subpopulations in the case of two alleles A_1 and A_2, it is shown (Crow and Kimura, 1956) that

$$(31)\qquad \phi(p,x;t)=\sum_{i=0}^{\infty}X_i(x)\exp\left[-i\left\{\frac{m+(i-1)}{4N}\right\}t\right],$$

where

$$(32)\quad X_i(x)=x^{B-1}(1-x)^{A-B-1}\,{}_2F_1(A+i-1,-i;A-B;1-x) \cdot {}_2F_1(A+i-1,-i;A-B;1-p) \cdot\frac{\Gamma(A-B+i)\Gamma(A+2i)\Gamma(A+i-1)}{2!\{\Gamma(A-B)\}^2\Gamma(B+i)\Gamma(A+2i-1)}$$

$$(A=4Nm,B=4Nm\bar{x}),$$

m is the rate of migration in number of individuals exchanged per generation, and $\bar{x}$ is the frequency of A_1 in the immigrants. As $t\to\infty$, (31) converges to Wright's (1937) well-known formula for the steady-state gene frequency distribution with migration,

$$\phi(x)=\frac{\Gamma(4Nm)}{\Gamma(4Nm\bar{x})\Gamma(4Nm(1-\bar{x}))}x^{4Nm\bar{x}-1}(1-x)^{4Nm(1-\bar{x})-1},$$

which is a beta distribution [cf. 4.13(1)]. For the case of two loci each with a pair of alleles, A_1 and A_2 at the first locus, and B_1 and B_2 at the second, Wright's formula can be written

$$\phi(x,y)=C\bar{w}^{2N}x^{4N\nu_1-1}(1-x)^{4N\mu_1-1}y^{4N\nu_2-1}(1-y)^{4N\mu_2-1},$$

where x and y are, respectively, the frequencies of A_1 and B_1, and $\bar{w}$ is the average fitness of a population measured in selective values with respect to these two loci. Also, μ_1 is the mutation rate from A_1 to A_2, and v_1 is the rate in the reverse direction, while μ_2 and v_2 are the corresponding values for B_1 and B_2.

6.2 A Modified Bessel Function Model in Life Testing

Bhattacharya (1966) considers a lifetime model represented by the probability density function (p.d.f.)

(1) $$h(x;\beta) = \int_{\Omega} f(x;\theta)\,dG(\theta;\beta) \qquad (x > 0),$$

where

(2) $$f(x;\theta) = (1/\theta)e^{-x/\theta} \qquad (x > 0, \theta > 0),$$

and $G(\theta;\beta)$ is suitably chosen to represent a probability distribution function (P.D.F.) defined over the space $\theta \in \Omega$, such that

(3) $$G(0+;\beta) = 0,$$

and β is a single or vector-valued real parameter. The following two special cases are discussed:

Model I pertains to the case when

(4) $$dG(\theta;\beta) = (1/\beta)e^{-\theta/\beta}\,d\theta \qquad (\theta > 0, \beta > 0),$$

i.e., when the mean of (1) is subject to exponential variation over the positive real axis, and

Model II pertains to the case when

(5) $$dG(\theta;\alpha) = \frac{1}{\alpha B(p,q)}\left(\frac{\theta}{\alpha}\right)^{p-1}\left(1-\frac{\theta}{\alpha}\right)^{q-1} d\theta \qquad (0 \leqq \theta \leqq \alpha;\ p, q, \alpha > 0),$$

which means that the variations in the mean are restricted over a finite range $[0, \alpha]$, governed by a beta distribution of the first kind extended suitably over the range stated.

Now, for *Model I*, we have, from (1), (2), and (4), using Erdélyi *et al.* (1953, Vol. II, p. 82, Eq. (23)),

(6) $$h(x;\beta) = (2/\beta)K_0(2\sqrt{x/\beta}) \qquad (0 < x < \infty; \beta > 0),$$

where $K_\nu(z)$ is the modified Bessel function of the third kind

$$K_\nu(z) = \{\pi/(2\sin\nu\pi)\}\{I_{-\nu}(z) - I_\nu(z)\}, \tag{7}$$

$I_\nu(z)$ being the modified Bessel function of the first kind.

The Mellin transform of $K_\nu(z)$ is given by (Erdélyi *et al.*, 1954, Vol. I, p. 331, Eq. (26))

$$\int_0^\infty z^{s-1}K_\nu(z)\,dz = 2^{s-2}\Gamma\left(\frac{s-\nu}{2}\right)\Gamma\left(\frac{s+\nu}{2}\right) \qquad [\mathrm{Re}(s) > |\mathrm{Re}(\nu)|], \tag{8}$$

so that the rth moment of the distribution (6) is given by

$$\mu_r' = \frac{2}{\beta}\int_0^\infty x^r K_0(2\sqrt{x/\beta})\,dx = (r!)^2\beta^r, \tag{9}$$

from which we have for the mean and variance

$$E(x) = \beta, \tag{10}$$

$$\mathrm{Var}(x) = 3\beta^2. \tag{11}$$

The distribution function corresponding to (6) is given by

$$\begin{aligned} H(t;\beta) &= \int_0^t h(x;\beta)\,dx \\ &= \frac{1}{\beta}\int_0^\infty (1 - e^{-t/\theta})e^{-\theta/\beta}\,d\theta \\ &= 1 - (2/\beta)\sqrt{\beta t}\,K_1(2\sqrt{t/\beta}), \end{aligned} \tag{12}$$

using again Erdélyi *et al.* (1953, Vol. II, p. 82, Eq. (23)). Therefore, the conditional failure density, or the failure rate, is given by[1]

$$\begin{aligned} z(t;\beta) &= h(t;\beta)/[1 - H(t;\beta)] \\ &= K_0(2\sqrt{t/\beta})/[\sqrt{\beta t}\,K_1(2\sqrt{t/\beta})], \end{aligned} \tag{13}$$

which is shown to be monotonic decreasing in t, using differentiation.

For *Model II*, Bhattacharya uses the Riemann–Liouville fractional integral of order μ, defined for a function $\phi(x)$ as

$$R_\mu\{\phi(x); y\} = \frac{1}{\Gamma(\mu)}\int_0^y \phi(x)(y-x)^{\mu-1}\,dx, \tag{14}$$

which, for $\phi(x) = x^{\nu-1}e^{-\lambda/x}$, and for $\mathrm{Re}(\mu) > 0$, $\mathrm{Re}(\lambda/y) > 0$, turns out to be

[1] It may be noted that $z(t;\beta)$ corresponds to the function $\eta(x)$, which we have used in Chapters 4 and 5.

[see Erdélyi *et al.* (1954, Vol. II, p. 187, Eq. (18))]

(15) $$\lambda^{(\nu-1)/2} y^{\mu+(\nu-1)/2} \exp[-\lambda/(2y)] W_{(1/2)-\mu-(\nu/2),(\nu/2)}(\lambda/y),$$

where $W_{\kappa,\mu}(x)$ denotes the Whittaker function [cf. II.3(57) and II.5(13)]. From (2), (5), and (15), we obtain a Whittaker model of lifetime with the p.d.f.

(16) $$h(x;\alpha) = [\Gamma(p+q)/\{\alpha\Gamma(p)\}](x/\alpha)^{p/2-1} \exp[-x/(2\alpha)] \cdot W_{-q-(p-2)/2,(p-1)/2}(x/\alpha) \qquad (0 < x < \infty;\, p, q, \alpha > 0),$$

which, on using Erdélyi *et al.* (1953, Vol. I, p. 264, Eq. (2)), viz.,

(17) $$W_{\kappa,\mu}(x) = e^{-x/2} x^{c/2} \Psi(a, c; x) \qquad (a = \tfrac{1}{2} - \kappa + \mu,\, c = 2\mu + 1),$$

can be written

(18) $$h(x;\alpha) = \{\Gamma(p+q)/\alpha\Gamma(p)\}(x/\alpha)^{p-1} e^{-x/\alpha} \cdot \Psi(p+q-1, p; x/\alpha)$$
$$(0 < x < \infty;\, p, q, \alpha > 0;\, p + q > 1).$$

In the special case $p = q = 1$, (5) reduces to a uniform or rectangular distribution (cf. Section 4.13) over $[0, \alpha]$, and this is discussed in Bhattacharya and Holla (1965).

Using the result (Slater, 1960, p. 46; cf. also II.3(44)),

(19) $$\mathscr{L}[t^{b-1}\Psi(a, c; t)] = \{\Gamma(b)\Gamma(b - c + 1)/\Gamma(a + b - c + 1)\} \cdot {}_2F_1(a, b - c + 1; a + b - c + 1; 1 - s)$$
$$[\mathrm{Re}(b) > 0;\, \mathrm{Re}(c) < 1 + \mathrm{Re}(b);\, |1 - s| < 1],$$

the rth moment of (18) is given by

(20) $$\mu_r' = \frac{\alpha^r \Gamma(p+q)}{\Gamma(p)} \int_0^\infty z^{r+p-1} e^{-z} \Psi(p+q-1, p; z)\, dz = r!\alpha^r B(r+p, q)/B(p, q).$$

Also, the cumulative distribution corresponding to (18) is shown to be

(21) $$H(t;\alpha) = \int_0^t h(x;\alpha)\, dx = 1 - \{\Gamma(p+q)/\Gamma(p)\}(t/\alpha)^{(p-1)/2} \exp[-t/(2\alpha)] \cdot W_{-q-(p-1)/2,p/2}(t/\alpha),$$

and the failure rate to be

(22) $$z(t;\alpha) = \frac{W_{-q-(p-2)/2,(p-1)/2}(t/\alpha)}{\sqrt{\alpha t}\, W_{-q-(p-1)/2,p/2}(t/\alpha)} = \frac{\Psi(p+q-1, p; t/\alpha)}{t\Psi(p+q, p+1; t/\alpha)}.$$

Bhattacharya (1966) also considers the more general case, where

(23) $$f(x;\theta) = e^{-x/\theta}(x/\theta)^{p-1}/\{\theta\Gamma(p)\} \qquad (0 < x < \infty; p > 0),$$

(24) $$dG(\theta;\beta) = [e^{-\theta/\beta}(\theta/\beta)^{q-1}/\{\beta\Gamma(q)\}]\,d\theta \qquad (0 < \theta < \infty; q > 0),$$

which, on using (1), lead to

(25) $$h(x;\beta) = [2/\{\beta\Gamma(p)\Gamma(q)\}](x/\beta)^{(p+q)/2-1}K_{q-p}(2\sqrt{x/\beta})$$
$$(0 < x < \infty; p, q, \beta > 0).$$

The rth moment for this distribution, using (8), is given by

(26) $$\mu_r' = \beta^r\Gamma(r+p)\Gamma(r+q)/\{\Gamma(p)\Gamma(q)\}.$$

Also, using Erdélyi *et al.* (1953, Vol. II, p. 90, Eq. (7)), the corresponding distribution function is shown to be

(27) $$H(y;\beta) = \sqrt{y/\beta}\,\zeta(y;\beta)/\{2^{p+q-3}\Gamma(p)\Gamma(q)\},$$

where

(28) $$\zeta(y;\beta) = 2(q-1)K_{q-p}(2\sqrt{y/\beta})s_{p+q-2,q-p-1}(2\sqrt{y/\beta})$$
$$- K_{q-p-1}(2\sqrt{y/\beta})s_{p+q-1,q-p}(2\sqrt{y/\beta}),$$

where $2p-1$, $2q-1$ are not odd integers, and $s_{\mu,\nu}(z)$ is Lommel's function, defined by [see, for example, Erdélyi *et al.* (1953, Vol. II, p. 40, Eq. (69))]

(29) $$s_{\mu,\nu}(z) = [z^{\mu+1}/\{(\mu-\nu+1)(\mu+\nu+1)\}]$$
$$\cdot\,{}_1F_2[1;(\mu-\nu+3)/2,(\mu+\nu+3)/2;-z^2/4]$$
$$(\mu \pm \nu \text{ not an odd integer}),$$

and the failure rate in this case is given by

(30) $$z(t;\beta) = (2\sqrt{t/\beta})^{p+q-2}K_{q-p}(2\sqrt{t/\beta})$$
$$\cdot\,[2^{p+q-3}\Gamma(p)\Gamma(q)\beta - \sqrt{\beta t}\,\zeta(t;\beta)]^{-1}.$$

6.3 A Model for Spares Provisioning

Schweitzer (1967) studies the reliability of a system composed of N identical parts with an initial supply of m spares. The N parts in use are assumed to fail independently, each with a constant failure rate λ, while the spare parts fail independently, each with constant failure rate μ. Failed parts in use are immediately replaced by unfailed spares as long as these are available, and the system fails when no such spares are available. Spares that fail or are put into use are not replaced. The system is said to be in state k ($k = 0, 1, \ldots, m$) when there are exactly k unfailed spares and the system is operational, and in state F when it is failed. Define the transition probabilities

(1) $P_{mk}(t) = \text{Prob}\{\text{system is in state } k \text{ at time } t \mid \text{system is in state } m \text{ at time } 0\} \qquad (k = 0, 1, \ldots, m; m = 0, 1, 2, \ldots).$

The differential difference equations for this pure death process are

$$\frac{dP_{mk}(t)}{dt} = -\lambda_k P_{mk}(t) + (1 - \delta_{mk})\lambda_{k+1} P_{m,k+1}(t) \qquad (k = 0, 1, \ldots, m), \tag{2}$$

where

$$\lambda_k = N\lambda + k\mu. \tag{3}$$

The solution of (2) with the initial conditions

$$P_{mk}(0) = \delta_{mk} \tag{4}$$

is

$$P_{mk}(t) = (N\lambda/\mu + k + 1)_{m-k}[(1 - e^{-\mu t})^{m-k}/(m-k)!]e^{-\lambda_k t} \qquad (k = 0, 1, \ldots, m). \tag{5}$$

The system reliability $R_m(t)$ is defined by

(6) $R_m(t) = \text{Prob}\{\text{system operational at time } t \mid m \text{ unfailed spares at } 0\}$

$$= \sum_{k=0}^{m} P_{mk}(t)$$

$$= \sum_{k=0}^{m} \left(\frac{N\lambda}{\mu} + m - k + 1\right)_k \left(\frac{(1 - e^{-\mu t})^k}{k!}\right)(e^{-\mu t})^{(N\lambda/\mu + m - k)}.$$

Differentiating (6) with respect to t and using (1), we have

$$\frac{dR_m(t)}{dt} = -N\lambda P_{m0}(t) = -\mu\left(\frac{N\lambda}{\mu}\right)_{m+1}\left(\frac{(1 - e^{-\mu t})^m}{m!}\right)e^{-N\lambda t}. \tag{7}$$

The probability density of the system failure time is given by

$$f_m(t) = -\frac{dR_m(t)}{dt}. \tag{8}$$

The system reliability may be alternatively expressed as

$$R_m(t) = \int_t^\infty f_m(y)\,dy, \tag{9}$$

and on using (7) and (8), followed by the change of variables

$$x = e^{-\mu y}, \tag{10}$$

we have

$$R_m(t) = I_p(N\lambda/\mu, m+1) \qquad (p = e^{-\mu t}), \tag{11}$$

where

$$I_p(c,d) = \frac{\Gamma(c+d)}{\Gamma(c)\Gamma(d)} \int_0^p y^{c-1}(1-y)^{d-1}\, dy \qquad (c, d > 0) \tag{12}$$

is the incomplete beta function ratio and is related to the hypergeometric function by

$$I_p(c,d) = [\Gamma(c+d)/\{\Gamma(c+1)\Gamma(d)\}]p^c\, {}_2F_1(1-d, c; c+1; p). \tag{13}$$

From (11)–(13) it follows that

$$R_m(t) = (e^{-N\lambda t}/m!)(N\lambda/\mu + 1)_m\, {}_2F_1(-m, N\lambda/\mu; N\lambda/\mu + 1; e^{-\mu t}). \tag{14}$$

Schweitzer (1967) uses properties of the hypergeometric series for obtaining various asymptotic results on reliability. For large m and t, using (11), (12), and Stirling's approximation for the gamma function, it is shown that

$$R_m(t) \simeq \frac{m^{N\lambda/\mu}}{\Gamma(N\lambda/\mu)} \int_0^{e^{-\mu t}} x^{(N\lambda/\mu)-1}(1-x)^m\, dx. \tag{15}$$

Now, using the approximation

$$(1-x)^m \simeq e^{-mx}, \tag{16}$$

which is valid for $0 \leqq x < 1$, if $mx^2 \ll 1 - x$, we have

$$\begin{aligned} R_m(t) &= \frac{1}{\Gamma(N\lambda/\mu)} \int_0^{me^{-\mu t}} y^{N\lambda/\mu - 1} e^{-y}\, dy \\ &= \gamma(N\lambda/\mu, me^{-\mu t})/\Gamma(N\lambda/\mu). \end{aligned} \tag{17}$$

6.4 Chance-Constrained Linear Programming

Sengupta (1972) studies a chance-constrained linear programming (CCP) model with chi-square type deviates. Consider the linear programming (LP) problem

$$\text{maximize} \qquad z = c'x; \quad Ax \leqq b, \quad x \geqq 0, \tag{1}$$

where c and x are $n \times 1$ vectors, A is an $m \times n$ matrix, b is an $m \times 1$ vector, and the prime denotes the transpose. Let us take the chance-constrained relation

$$\text{(2)}\qquad \operatorname{Prob}\{a_i'x \leqq b_i\} \geqq u_i \qquad (i = 1, \ldots, m),$$

where a_i' is the ith row vector of A and u_i $(0 \leqq u_i \leqq 1)$ are preassigned. It is assumed that the elements c_j, b_i, and a_{ij} $(i = 1, \ldots, m; j = 1, \ldots, n)$ of the triple (c, b, A) are mutually independent chi-square deviates. Let us write

$$\text{(3)}\qquad P_i = a_i'x,$$

$$\text{(4)}\qquad Q_i = b_i - a_i'x,$$

$$\text{(5)}\qquad R_i = a_i'x/b_i = P_i/b_i \qquad (i = 1, \ldots, m).$$

Then the chance-constrained inequality (2) can be written

$$\operatorname{Prob}\{R_i \leqq 1\} \geqq u_i \qquad (i = 1, \ldots, m).$$

Now, P_i is distributed approximately like a noncentral chi-square variate, and b_i follows a central chi-square distribution. Hence, their ratio R_i follows a noncentral F-distribution, which is known to have a probability density function (p.d.f.)

$$\text{(6)}\qquad p(x) = \sum_{j=0}^{\infty} \left[\exp(-L/2) \frac{(L/2)^j}{\{j! B(\nu_1/2 + j, \nu_2/2)\}} \right] \left(\frac{\nu_1}{\nu_2} \right)^{\nu_1/2 + j} g(x),$$

where

$$\text{(7)}\qquad g(x) = x^{\nu_1/2 + j - 1}(1 + \nu_1 x/\nu_2)^{-(\nu_1 + \nu_2)/2 - j},$$

(8) $\nu_1, L =$ the degrees of freedom and the noncentrality parameter, respectively, of the numerator P_i,

(9) $\nu_2 =$ degrees of freedom of the denominator b_i,

and $B(m, n)$ is the beta function with arguments m, n.

Alternatively, when the distribution of $a_i'x$ can be approximated closely by a central chi-square distribution, Q_i is distributed like the difference of two independent central chi-square variates. This distribution has been explicitly derived by a number of authors and we may proceed as follows. If χ_m^2 and χ_n^2 are two independent central chi-square variates with characteristic functions $\phi_m(t)$ and $\phi_n(t)$, then the characteristic function (c.f.) $\phi(t)$ of

$$\text{(10)}\qquad X = \chi_n^2 - \chi_m^2$$

is given by

$$\text{(11)}\qquad \phi(t) = \phi_n(t)\phi_m(-t) = (1 - 2it)^{-n/2}(1 + 2it)^{-m/2}.$$

Therefore, using the inversion theorem on c.f.s, the p.d.f. of X is

(12) $$h_{n,m}(X) = (2\pi)^{-1}\int_{-\infty}^{\infty} \exp(-itx)(1-2it)^{-n/2}(1+2it)^{-m/2}\,dt,$$

whence we have[2] [see, e.g., Erdélyi *et al.* (1954, Vol. I, p. 119, Eq. (12))]

(13) $$h_{n,m}(X) = \begin{cases} \dfrac{\exp(-X/2)X^{(n+m)/2-1}}{2^{(n+m)/2}\Gamma(n/2)}\Psi\left(\dfrac{m}{2}, \dfrac{n+m}{2}; X\right) & (X \geqq 0) \\[2ex] \dfrac{\exp(X/2)(-X)^{(n+m)/2-1}}{2^{(n+m)/2}\Gamma(m/2)}\Psi\left(\dfrac{n}{2}, \dfrac{n+m}{2}; -X\right) & (X < 0) \end{cases}$$

[cf. also II.3(14)].

Using the exact distributions (6) and (10) and numerical tables, any desired tolerance level in the programming problem can be attained, as has been shown by Sengupta, who has tabulated numerical results. He also discusses the use of the exponential distribution in the CCP problem.

6.5 A Brand Choice Model

Jones (1973) proposes a brand choice model in the area of consumer behavior, which is composite of the Bernoulli, Markov, and linear learning models postulated earlier.

The Bernoulli model, which was originally proposed by Morrison [see Massy *et al.* (1970, Chapter 3)], studies the case of a binary choice between, say, a brand A, and all other brands collectively called brand B. The model assumes that the consumer's probability $p(t)$ of purchase of brand A at time t is constant over time and independent of the past purchase events. Thus,

(1) $$p(t) = p(0) = p,$$

say. The model can be made heterogeneous by allowing p to vary over the population of consumers. Let the cumulative probability distribution of p be

(2) $$F(p) = \begin{cases} w_1 & (p = 0) \\ w_1 + (1 - w_1)G(p) & (0 < p \leqq 1), \end{cases}$$

where

(3) $$0 \leqq w_1 \leqq 1$$

and $G(p)$ is a cumulative beta distribution, given by [cf. 6.3(12)]

(4) $$G(p) = \frac{\Gamma(a_1 + b_1)}{\Gamma(a_1)\Gamma(b_1)}\int_0^p x^{a_1-1}(1-x)^{b_1-1}\,dx.$$

[2] There are obvious misprints in the corresponding equations (6c) and (6d) of Sengupta (1972).

Thus, p is distributed as beta with a mass point at zero. The mean probability of purchase of brand A is

(5)
$$E(p) = (1 - w_1)\frac{\Gamma(a_1 + b_1)}{\Gamma(a_1)\Gamma(b_1)}\int_0^1 x^{a_1}(1 - x)^{b_1 - 1}\,dx$$
$$= (1 - w_1)a_1/(a_1 + b_1).$$

The Markov model allows for transitions in buying behavior from time t to $t + 1$. Morrison proposes that the transition matrix may be taken as

(6)
$$\Phi = \begin{array}{c} \quad A \qquad\quad B \\ \begin{bmatrix} \gamma + \delta & 1 - \gamma - \delta \\ \gamma & 1 - \gamma \end{bmatrix} \end{array},$$

where γ and δ are jointly distributed according to a Dirichlet distribution:

(7)
$$f(\gamma, \delta) = \begin{cases} [\Gamma(c + d + e)/\{\Gamma(c)\Gamma(d)\Gamma(e)\}] & \\ \cdot\, \gamma^{c-1}\delta^{d-1}(1 - \gamma - \delta)^{e-1} & (\gamma \geqq 0, \delta \geqq 0, \gamma + \delta \leqq 1) \\ 0 & \text{otherwise,} \end{cases}$$

which is independent of the value of p. It can be shown that the steady-state probability for state A corresponding to the transition matrix (6) is

(8)
$$\pi = \gamma/(1 - \delta),$$

and it can be derived from (7) that π is beta distributed with parameters c and e [see, e.g., Jones (1973, Appendix A)]. Hence, as in (5), the mean value of π is

(9)
$$E(\pi) = c/(c + e).$$

The linear learning model is again a binary choice model with

(10)
$$p(t + 1) = \begin{cases} \alpha + \beta + \lambda p(t) & \text{if brand } A \text{ was purchased at time } t \\ \alpha + \lambda p(t) & \text{if brand } B \text{ was purchased at time } t. \end{cases}$$

To account for heterogeneity of behavior, Jones proposes that the initial probability of purchase $p = p(0)$ may be taken to be beta distributed with a probability atom at zero as before and, independent of p, α, β, λ may be taken to be Dirichlet distributed with

(11)
$$f(\alpha, \beta, \lambda)$$
$$= \begin{cases} [\Gamma(j + k + m + n)/\{\Gamma(j)\Gamma(k)\Gamma(m)\Gamma(n)\}] & \\ \cdot\, \alpha^{j-1}\beta^{k-1}\lambda^{m-1}(1 - \alpha - \beta - \lambda)^{n-1} & \text{if } \alpha, \beta, \lambda \geqq 0; \alpha + \beta + \lambda \leqq 1 \\ 0 & \text{otherwise.} \end{cases}$$

It can be shown [see, e.g., Jones (1970)] that for given values of α, β and λ, the long-run mean probability of purchase μ for this model is independent of p and is given by

$$\mu = \alpha/\{1 - (\beta + \lambda)\}, \tag{12}$$

and, further, it can be shown as above that μ is beta distributed with parameters j and n. Hence

$$E(\mu) = j/(j + n). \tag{13}$$

6.6 Random Collision Processes

By analogy with statistical mechanics, consider a random collision process (Moran, 1961; Nishimura, 1974) with discrete time n and continuous states $x \in [0, \infty)$. Let $p_n(i)$ be the probability of state i at time n, and let $\pi(i, j|k, l)$ be the transition probability that two particles in states k and l are transformed by collision to states i and j, respectively. It is assumed that the collisions of the particles are random (independent) and that the number of particles is large. We have

$$p_{n+1}(i) = \sum_j \sum_{k,l} \pi(i, j|k, l) p_n(k) p_n(l). \tag{1}$$

Let the state parameter x be continuous and denote the kinetic energy of the particle, so that the sum of these energies is conserved; then we have

$$p_{n+1}(x) = \int_x^\infty \int_0^w \pi(x, w - x|y, w - y) p_n(y) p_n(w - y)\, dy\, dw. \tag{2}$$

Nishimura (1974) assumes that the transition probability densities satisfy the relation

$$w\pi(x, w - x|y, w - y) = \pi(u, 1 - u|v, 1 - v). \tag{3}$$

He also assumes that

$$\pi(x, w - x|y, w - y) = \pi(w - x, x|w - y, y), \tag{4}$$

and that $\pi(x, w - x|y, w - y)$ is independent of y $(0 \leqq y \leqq w)$. Define

$$g(u) = \pi(u, 1 - u|v, 1 - v). \tag{5}$$

As a special case, the following transition probability density is considered:

$$\pi(x, w - x|y, w - y) = \begin{cases} \{w^{2s-1} B(s, s)\}^{-1} x^{s-1} (w - x)^{s-1} \\ \qquad (x, w - x, y, w - y \geqq 0) \\ 0 \quad \text{otherwise,} \end{cases} \tag{6}$$

where

$$B(s,s) = \{\Gamma(s)\}^2/\Gamma(2s). \tag{7}$$

Then

$$g(u) = \frac{1}{B(s,s)} u^{s-1}(1-u)^{s-1} \tag{8}$$

is a symmetric beta distribution. Its kth moment is given by

$$v(k) = \int_0^1 u^k g(u)\,du = \frac{\Gamma(2s)\Gamma(s+k)}{\Gamma(s)\Gamma(2s+k)}. \tag{9}$$

Let

$$\mu_n(k) = \int_0^\infty x^k p_n(x)\,dx; \tag{10}$$

then it is shown (Nishimura, 1974, Theorem 2) that for every finite number K there exists the limit $\mu(k)$ $(k = 0, 1, \ldots, K)$ of the sequence $\{\mu_n(k)\}$, which is uniquely determined by

$$\mu(k) = v(k) \sum_{l=0}^{k} \binom{k}{l} \mu(l)\mu(k-l). \tag{11}$$

Now, it is easily seen that

$$\frac{\Gamma(s+k)}{\Gamma(s)} = \frac{\Gamma(2s)\Gamma(s+k)}{\Gamma(s)\Gamma(2s+k)} \cdot \sum_{l=0}^{k} \binom{k}{l} \frac{\Gamma(s+l)\Gamma(s+k-l)}{\Gamma(s)\Gamma(s)} \qquad (k = 0, 1, 2, \ldots). \tag{12}$$

From (9), (11), and (12), it follows that

$$\mu(k) = \alpha^{-k}\Gamma(s+k)/\Gamma(s), \tag{13}$$

which is the kth moment of the gamma distribution

$$p(x) = \begin{cases} \alpha^s x^{s-1} e^{-\alpha x}/\Gamma(s) & (x > 0) \\ 0 & \text{otherwise.} \end{cases} \tag{14}$$

For Kac's (1959) model in discrete time, we have $s = \frac{1}{2}$ and, therefore, the limiting probability distribution for kinetic energy given by

$$p(x) = \pi^{-1/2}\alpha^{1/2}x^{-1/2}e^{-\alpha x} \qquad (x > 0). \tag{15}$$

6.7 An Inventory Model with Limited Waiting Time

Das (1977) considers an $(S - 1, S)$ inventory model, where customers do not wait longer than a tolerable delay T. The system is assumed to start with an initial inventory of N items, and demands are taken to arrive in a Poisson stream at mean rate λ per unit time. Every time a demand occurs, a reorder is placed with a supply center, say a warehouse. Demands are filled immediately if there is stock on hand; otherwise they await replenishment (up to a maximum time T). Reorders are processed singly, the processing time being exponentially distributed with mean $1/\mu$. Let P_n be the steady-state probability that there are n reorders. Clearly,

(i) when $n \leqq N$, all reorders correspond to customer orders already filled and there are $N - n$ items on hand;

(ii) when $n > N$, the number of backorders is $n - N$.

The system described above behaves like an $M/M/1$ queue, in which the customers exhibit a reneging behavior only for $n > N$.

For a queuing system with reneging, the steady-state probability density that the age (or time since arrival) of the oldest customer in the system is t, conditional on there being a total of r customers in the system at that time, is given by [see Barrer (1957)]

(1) $$g_r(t) = \mu^r t^{r-1} e^{-\mu t}/\gamma(r, \mu T) \qquad (r \geqq 1; 0 < t \leqq T).$$

Therefore, the steady-state probability density that an order is lost, given that $n = N + r$, is

(2) $$\beta_r = g_r(T) = \mu^r T^{r-1} e^{-\mu T}/\gamma(r, \mu T).$$

By usual methods, we have the following equations for the system:

(3) $$-\lambda P_0 + \mu P_1 = 0;$$

(4) $$\lambda P_{n-1} - (\lambda + \mu)P_n + \mu P_{n+1} = 0 \qquad (1 \leqq n < N);$$

(5) $$\lambda P_{N-1} - (\lambda + \mu)P_N + (\mu + \beta_1)P_{N+1} = 0;$$

(6) $$\lambda P_{N+r-1} - (\lambda + \mu + \beta_r)P_{N+r} + (\mu + \beta_{r+1})P_{N+r+1} = 0 \qquad (r \geqq 1).$$

The solution of these equations is

(7) $$P_n = \rho^n P_0 \qquad (0 \leqq n \leqq N),$$

$$P_{N+r} = P_N \prod_{i=1}^{r} \{\lambda/(\mu + \beta_i)\} \qquad (r \geqq 1),$$

where it is assumed that

(8) $$\rho = \lambda/\mu < 1.$$

Now

$$(9)\qquad \prod_{i=1}^{r}\left\{\frac{\lambda}{\mu+\beta_i}\right\} = \rho^r \prod_{i=1}^{r}\left(1+\frac{\beta_i}{\mu}\right)^{-1}$$

$$= \rho^r \prod_{i=1}^{r}\left[\frac{\gamma(i,\alpha)}{\gamma(i,\alpha)+\alpha^{i-1}e^{-\alpha}}\right] \qquad (\alpha = \mu T)$$

on substitution from (2). As can be seen on integration, we have

$$(10)\qquad \gamma(1,\alpha) = 1 - e^{-\alpha},$$

$$(11)\qquad \gamma(i,\alpha) = (i-1)\gamma(i-1,\alpha) - \alpha^{i-1}e^{-\alpha} \qquad (i > 1)$$

Using (10) and (11) in (9) and simplifying, we find that

$$(12)\qquad \prod_{i=1}^{r}\left\{\frac{\lambda}{\mu+\beta_i}\right\} = \frac{\rho^r\gamma(r,\alpha)}{(r-1)!}.$$

Therefore

$$(13)\qquad \sum_{r=1}^{\infty}\prod_{i=1}^{r}\left\{\frac{\lambda}{\mu+\beta_i}\right\} = \sum_{r=1}^{\infty}\left\{\frac{\rho^r}{(r-1)!}\right\}\int_0^{\alpha} e^{-y}y^{r-1}\,dy$$

$$= \{\rho/(1-\rho)\}[1 - \exp\{-\alpha(1-\rho)\}],$$

and from the normalizing condition

$$(14)\qquad \sum_{n=0}^{\infty} P_n = 1,$$

we have

$$(15)\qquad P_0 = (1-\rho)/[1 - \rho^{N+1}\exp\{-\alpha(1-\rho)\}].$$

Das evaluates various measures of performance of the system and discusses the problem of optimization considering a linear cost structure.

6.8 Probability Distributions and Statistical Applications

Pearson (1924a) studied the hypergeometric series ${}_2F_1(\alpha, \beta; \gamma; x)$ to obtain the successive moments about the mean of the distribution ${}_2F_1(\alpha, \beta; \gamma; 1)$. He suggests the substitutions

$$(1)\qquad \alpha = -r, \quad \beta = -qn, \quad \gamma = pn - r + 1, \quad x = e^y \qquad (p + q = 1),$$

so that the series can be put in the form

$$(2)\qquad 1 + r\cdot\frac{qn}{pn-r+1}e^y + \frac{r(r-1)}{2!}\cdot\frac{qn(qn-1)}{(pn-r+1)(pn-r+2)}e^{2y} + \cdots,$$

and points out that if (2) is multiplied by the factor

$$\frac{M\cdot pn(pn-1)\cdots(pn-r+1)}{n(n-1)\cdots(n-r+1)}, \tag{3}$$

then, on putting $y = 0$, it represents the frequency distribution of unmarked individuals in M random samples of size r drawn from a population of n containing pn marked and qn unmarked individuals.[3] Pearson (1924b) discusses the generalization of this concept to the case of two variables corresponding to "a double hypergeometrical series." [Cf. Section II.6.]

Pearson and several other authors, who worked on his papers, have made a considerable use of the incomplete beta and gamma functions and the hypergeometric function in their studies of statistical frequency distributions. One such paper is a long one by Hansmann (1934) on the central nonnormal symmetrical frequency distribution. Hansmann started with Pearson's technique of finding a mathematical curve which has the same first four moments as any given frequency distribution and discussed the question of possible improvement by the consideration of higher moments.

Ayyangar (1934) considers the probability distribution obtained by multiplying (2) by (3) and dividing by M. It is clear that the mean of this probability distribution is rq. Let $\phi(t)$ be the coefficient of x^t in the distribution, i.e.,

$$\begin{aligned} \phi(t) = {} & \frac{pn(pn-1)\cdots(pn-r+1)}{n(n-1)\cdots(n-r+1)} \\ & \cdot \frac{r(r-1)\cdots(r-t+1)qn(qn-1)\cdots(qn-t+1)}{t!(pn-r+1)(pn-r+2)\cdots(pn-r+t)}. \end{aligned} \tag{4}$$

Ayyangar suggests a simple combinatorial method for obtaining the sth incomplete moment about the mean, defined by

$$v_s = \sum_{t=\rho}^{r} (t-rq)^s \phi(t). \tag{5}$$

Laha (1954) discusses some properties of the I-type Bessel function probability distributions given by

$$\begin{aligned} dF(x) = {} & (2/\beta)^{p-1} x^{(p-1)/2} \alpha^p \\ & \cdot \exp[-\beta^2/(4\alpha) - \alpha x]\, I_{p-1}(\beta x^{1/2})\, dx \qquad (0 \leqq x < \infty) \end{aligned} \tag{6}$$

through the use of characteristic functions. He proves that the sum

$$u = x_1 + x_2 + \cdots + x_n \tag{7}$$

[3] There seem to be some obvious misprints on p. 157 of Pearson (1924a).

of Bessel variates is a Bessel variate with density function

$$f(u) = (4u/\beta_0^2)^{(p_0-1)/2} \exp(-u - \beta_0^2/4)\, I_{p_0-1}(\beta_0 u^{1/2}), \tag{8}$$

where

$$p_0 = \sum_{j=1}^{n} p_j; \qquad \beta_0^2 = \sum_{j=1}^{n} \beta_j^2, \tag{9}$$

and the distribution of the x_j is given by

$$dF(x_j) = (2/\beta)^{p_j-1} x^{(p_j-1)/2} \cdot \exp(-x_j - \beta_j^2/4)\, I_{p_j-1}(\beta_j x_j^{1/2})\, dx_j \qquad (j = 1, \ldots, n). \tag{10}$$

Let $x_1, x_2, \ldots, x_n$ be n independent normal variates with means m_j $(j = 1, 2, \ldots, n)$ and common variance σ^2; then it is shown that the noncentral chi-square

$$\chi'^2 = \sum_{j=1}^{n} \frac{x_j^2}{\sigma^2} \tag{11}$$

has the density function

$$f(\chi'^2) = (\chi'^2/\lambda^2)^{(n-2)/4} \exp[-(\chi'^2 + \lambda^2)/2] I_{(1/2)n-1}((\chi'^2\lambda^2)^{1/2}), \tag{12}$$

$$\lambda^2 = \sum_{j=1}^{n} \frac{m_j^2}{\sigma^2}. \tag{13}$$

Further, Laha proves that if x_1, x_2 are two independent Bessel variates with distributions as given in (6) with parameters p_1, β_1 and p_2, β_2, respectively, then the ratio

$$w = x_1/x_2 \tag{14}$$

has the probability density function

$$f(w) = \exp[-(\beta_1^2 + \beta_2^2)/4] \sum_{r,s=0}^{\infty} \left(\frac{\beta_1}{2}\right)^{2r} \left(\frac{\beta_2}{2}\right)^{2s} \left(\frac{1}{\alpha^{r+s}}\right) \cdot \{r!s!B(p_1 + r, p_2 + s)\}^{-1} \frac{w^{p_1+r-1}}{(1+w)^{p_1+p_2+r+s}}. \tag{15}$$

It may be noted that we can write

$$f(w) = \exp[-(\beta_1^2 + \beta_2^2)/4]\{B(p_1, p_2)\}^{-1}\{w^{p_1-1}/(1+w)^{p_1+p_2}\} \cdot \Psi_2\left[p_1 + p_2; p_1, p_2; \frac{\beta_1^2 w}{4\alpha(1+w)}, \frac{\beta_2^2}{4\alpha(1+w)}\right], \tag{16}$$

where [cf. II.6(29) and II.6(30)]

(17) $$\Psi_2(\alpha;\gamma,\gamma';x,y) = \sum_{m,n=0}^{\infty} \frac{(\alpha)_{m+n}}{(\gamma)_m(\gamma')_n} \frac{x^m}{m!} \frac{y^n}{n!}$$

is a confluent hypergeometric function of two variables [Erdélyi *et al.* (1953, Vol. I, p. 225, Eq. (24))]. Further, it was shown by Laha and can be deduced from (16) that in the special case

(18) $$\beta_1 = \beta, \qquad \beta_2 \to 0,$$

the density function of

(19) $$v = w/(1 + w)$$

can be written

(20) $$f(v) = \exp(-\beta^2/4)\{B(p_1, p_2)\}^{-1} v^{p_1 - 1}(1 - v)^{p_2 - 1} \cdot {}_1F_1[p_1 + p_2; p_1; \beta^2 v/(4\alpha)].$$

Laha appends more results on the distributions of products and linear functions, etc., of independent Bessel variates.

Kabe (1958) discusses the use of Meijer's G function in statistics. He notes that the Mellin transform of a probability density function $f(x)$ $(0 < x < \infty)$ of a random variable X, viz.,

(21) $$g(s) = \int_0^\infty x^{s-1} f(x)\, dx$$

represents μ'_{s-1}, the $(s-1)$th moment of X about zero, and starts with the consideration of the distribution of the generalized correlation ratio U, whose $(s-1)$th moment is (Wilks, 1932, p. 484)

(22) $$g(s) = \prod_{j=1}^{n} \frac{\Gamma[\frac{1}{2}(N-j)]}{\Gamma[\frac{1}{2}(p-j)]} \prod_{j=1}^{n} \frac{\Gamma[\frac{1}{2}(p-j-2)+s]}{\Gamma[\frac{1}{2}(N-j-2)+s]}.$$

Using the inversion formula

(23) $$f(x) = \frac{1}{2\pi i} \int_{-i\infty}^{i\infty} x^{-s} g(s)\, ds,$$

we have the density function of U as

(24) $$f(U) = \prod_{j=1}^{n} \frac{\Gamma(\frac{1}{2}(N-j))}{\Gamma(\frac{1}{2}(p-j))} G_{n,n}^{n,0}\left(U \,\middle|\, \begin{matrix} a_1, \ldots, a_n \\ b_1, \ldots, b_n \end{matrix}\right),$$

where

(25) $$a_j = (N-2-j)/2, \qquad b_j = (p-2-j)/2, \qquad (j = 1, \ldots, n),$$

and

$$G^{m,n}_{p,q}\left(X\,\middle|\,\begin{matrix}a_1,\ldots,a_p\\ b_1,\ldots,b_q\end{matrix}\right) \tag{26}$$

is Meijer's G function [see, e.g., Section II.5 and Erdélyi *et al.* (1953, Vol. I, p. 207)]. In the special case $n = 1$, we have, from Erdélyi *et al.* (1953, Vol. I, p. 208, Eq. (5)),

$$\begin{aligned} G^{1,0}_{1,1}\left(U\,\middle|\,\begin{matrix}a_1\\ b_1\end{matrix}\right) &= \left\{\frac{U^{b_1}}{\Gamma(a_1-b_1)}\right\}{}_1F_0(1+b_1-a_1;-;U) \\ &= \{\Gamma(a_1-b_1)\}^{-1}U^{b_1}(1-U)^{a_1-b_1-1}, \end{aligned} \tag{27}$$

so that

$$\begin{aligned} f(u) = {} & [\Gamma((N-1)/2)/\{\Gamma((p-1)/2)\Gamma((N-p)/2)\}] \\ & \cdot U^{(p-3)/2}(1-U)^{(N-p-2)/2} \qquad (0 \leqq U \leqq 1), \end{aligned} \tag{28}$$

which is a beta distribution. Similarly, for $n = 2$, Kabe shows that

$$\begin{aligned} f(U) = {} & [\Gamma(N-2)/\{2\Gamma(p-2)\Gamma(N-p)\}] \\ & \cdot U^{(p-4)/2}(1-\sqrt{U})^{N-p-1} \qquad (0 \leqq U \leqq 1), \end{aligned} \tag{29}$$

which can be seen to be equivalent to Wilks's corresponding result, viz.,

$$\begin{aligned} f(U) = {} & \left[\frac{\Gamma[\frac{1}{2}(N-1)]\Gamma[\frac{1}{2}(N-2)]\,U^{(p-4)/2}(1-U)^{N-p-1}}{\Gamma[\frac{1}{2}(p-1)]\Gamma[\frac{1}{2}(p-2)]\Gamma(N-p)}\right] \\ & \cdot {}_2F_1\left(\frac{N-p-1}{2}, \frac{N-p}{2}; N-p; 1-U\right), \end{aligned} \tag{30}$$

on using Erdélyi *et al.* (1953, Vol. I, p. 101, Eq. (6)). It may be noted that (29) implies that for $n = 2$, $\sqrt{U}$ is beta distributed. Another problem discussed by Kabe (1958) is the following.

Let ρ_1 and ρ_2 be the canonical correlations between two sets of normal variables (X_1, X_2) and $(Y_1, \ldots, Y_p)$ $(p > 2)$, and let

$$Q = \rho_1\rho_2, \qquad Z = (1-\rho_1^2)(1-\rho_2^2). \tag{31}$$

Then, for the joint distribution $f(q, z)$ of Q and Z, we have (Girshick, 1939)

$$\begin{aligned} \mu'_{s-1,t-1} &= \int_0^\infty\int_0^\infty q^{s-1}z^{t-1}f(q,z)\,dq\,dz \\ &= \frac{\Gamma(n-1)\Gamma(p+s-2)\Gamma(n-p-3+2t)}{\Gamma(p-1)\Gamma(n-p-1)\Gamma(n+s-4+2t)} \\ &= g(s,t), \end{aligned} \tag{32}$$

say, which is the double Mellin transform of $f(q, z)$. The inverse transform can be expressed in terms of Meijer's G function and simplifies to

$$(33) \qquad f(q, z) = [\Gamma(n-1)/\{2\Gamma(p-1)(n-p-1)\}]z^{(n-p-3)/2}q^{p-2},$$

which agrees with the result given by Girshick (1939).

Kabe (1962) considers n independent gamma variates $X_1, X_2, \ldots, X_n$ with probability densities

$$(34) \quad f(x_j) = [x_j^{p_j-1}/\{2^{p_j}\Gamma(p_j)\}]e^{-x_j/2} \qquad (0 < x_j < \infty, j = 1, \ldots, n).$$

The characteristic function of the variable

$$(35) \qquad Z = \sum_{j=1}^{n} \frac{X_j}{\alpha_j} \qquad (\alpha_j > 0)$$

is given by

$$(36) \qquad \phi(t) = \prod_{j=1}^{n} \alpha_j^{p_j}(\alpha_j - 2it)^{-p_j},$$

whence, on inversion,

$$(37) \quad f(z) = \left[\alpha_1^{p_1}\alpha_2^{p_2}\cdots\alpha_n^{p_n}\Big/\left\{2^{\sum p_j}\Gamma\left(\sum p_j\right)\right\}\right]z^{\sum p_j - 1}\exp(-\alpha_1 z/2)$$

$$\cdot F_1^{(n-1)}\Big[p_2, p_3, \ldots, p_n; \sum p_j;$$

$$(\alpha_1 - \alpha_2)z/2; (\alpha_1 - \alpha_3)z/2, \ldots, (\alpha_1 - \alpha_n)z/2\Big],$$

where $F_1^{(n-1)}$ is a (confluent) hypergeometric series in $n - 1$ variables, defined by (Kabe, 1962, p. 1198, Eq. (2.4))

$$(37a) \quad F_1^{(n-1)}\Big[p_2, p_3, \ldots, p_n; \sum p_j; (\alpha_1 - \alpha_2)z/2, (\alpha_1 - \alpha_3)z/2, \ldots, (\alpha_1 - \alpha_n)z/2\Big]$$

$$= \sum_{m_2, \ldots, m_n = 0}^{\infty} \frac{(p_2)_{m_2}\cdots(p_n)_{m_n}}{(\sum p_j)_{m_2+\cdots+m_n}} \frac{[(\alpha_1 - \alpha_2)z/2]^{m_2}}{m_2!}$$

$$\cdots \frac{[(\alpha_1 - \alpha_n)z/2]^{m_n}}{m_n!}.$$

As a matter of fact, in terms of the well-known confluent hypergeometric $\Psi_2^{(n)}$ function of n variables (Erdélyi *et al.*, 1954, Vol. I, p. 385), we have (see also

Section II.7)

(37b)
$$F_1^{(n)}[\beta_1, \ldots, \beta_n; \gamma; z_1, \ldots, z_n] \equiv \Psi_2^{(n)}[\beta_1, \ldots, \beta_n; \gamma; z_1, \ldots, z_n] = \lim_{\alpha \to \infty} F_D^{(n)}[\alpha, \beta_1, \ldots, \beta_n; \gamma; z_1/\alpha, \ldots, z_n/\alpha],$$

where [see II.7(4)]

(37c)
$$F_D^{(n)}[\alpha, \beta_1, \ldots, \beta_n; \gamma; z_1, \ldots, z_n] = \sum_{m_1, \ldots, m_n = 0}^{\infty} \frac{(\alpha)_{m_1 + \cdots + m_n}(\beta_1)_{m_1} \cdots (\beta_n)_{m_n}}{(\gamma)_{m_1 + \cdots + m_n}} \frac{z_1^{m_1}}{m_1!} \cdots \frac{z_n^{m_n}}{m_n!}$$

is the relatively more familiar Lauricella's hypergeometric function of the fourth type in n variables (Lauricella, 1893; Exton, 1976). Equation (37) is used by Kabe to find the exact distribution of the criterion L employed in the simultaneous testing of the sample means in multivariate analysis.

Ruben (1960) discusses spherical normal distributions in Euclidean N-space. Defining χ_N^2 and $\chi_{N,k}^2$ to be variates distributed as χ^2 with N degrees of freedom (d.f.) and a noncentral χ^2 variate with N d.f. and noncentrality parameter k, respectively, so that

(38)
$$\chi_{N,k}^2 = \sum_{i=1}^{N} (x_i - k_i)^2; \qquad k^2 = \sum_{i=1}^{N} k_i^2,$$

where x_i are independent $N(0, 1)$ variates[4] and

(39)
$$\chi_{N,0}^2 \equiv \chi_N^2,$$

he has shown that the distribution function of χ_N^2 is

(40)
$$F_N(a^2) = \text{Prob}\{\chi_N^2 \leqq a^2\} = \{2^{N/2}\Gamma(N/2)\}^{-1}\gamma(N/2, a^2),$$

where $\gamma(\alpha, x)$ is the incomplete gamma function, the corresponding density function being

(41)
$$f_N(a^2) = \{2^{N/2}\Gamma(N/2)\}^{-1} \exp(-a^2/2)(a^2)^{N/2-1},$$

and the density function for $\chi_{N,k}^2$ is

(42)
$$g_{N,k}(a^2) = \tfrac{1}{2}k^{-(N/2-1)}a^{N/2-1} \exp[-(k^2 + a^2)/2]\, I_{N/2-1}(ka),$$

where $I_n(z)$ is the modified Bessel function of the first kind and order n. Ruben further proves that the variate

(43)
$$t_{N-1} = x_1 \Bigg/ \left[\sum_{i=2}^{N} \frac{x_i^2}{n-1}\right]^{1/2}$$

[4] $N(0, 1)$ variate means a normal variate with mean zero and variance one.

has the density function

$$(44)\qquad g_{N-1}(c) = \left[(N-1)^{1/2} B\left(\frac{N-1}{2}, \frac{1}{2}\right)\right]^{-1} \left(1 + \frac{c^2}{N-1}\right)^{-N/2},$$

where $B(\alpha, \beta)$ is the well-known beta function.

With x_i defined as above, let

$$(45)\qquad L \equiv L(\mathbf{x}) = \sum_{i=1}^{N} p_i x_i,$$

$$(46)\qquad Q \equiv Q(\mathbf{x}) = \sum_{i=1}^{N} q_i x_i^2 \qquad (p_i, q_i \text{ real});$$

then it is shown by Ruben (1962) that

$$(47)\qquad \left|\frac{E\{Q^j H_{2j}(L/Q^{1/2})\}}{(2j)!}\right| \leqq \left(\frac{n}{2}\right)_j \left(q_0 + \frac{1}{2}\sum_i p_i^2\right)^j \Big/ j!,$$

where

$$(48)\qquad q_0 = \max_i |q_i|,$$

and $H_k(t)$ is the Hermite polynomial of degree k [see II.3(64)].

Stacy (1962) suggests, for a random variable X, the probability density function

$$(49)\qquad f(x; a, d, p) = pa^{-d}x^{d-1} \exp[-(x/a)^p]/\Gamma(d/p) \qquad (x \geqq 0; a, d, p > 0),$$

the corresponding cumulative distribution function being

$$(50)\qquad F(x; a, d, p) = \gamma(d/p, (x/a)^p)/\Gamma(d/p),$$

and the moment generating function being

$$(51)\qquad M(\theta) = E(e^{\theta X}) = \sum_{r=0}^{\infty} \int_0^{\infty} \left\{\frac{(\theta x)^r}{r!}\right\} f(x; a, d, p)\, dx$$

$$= \sum_{r=0}^{\infty} \left\{\frac{(\theta a)^r}{r!}\right\} \frac{\Gamma((d+r)/p)}{\Gamma(d/p)}.$$

Further, Stacy points out that if $X_1, \ldots, X_n$ are n independent random variables, X_i having the density function

$$(52)\qquad f(x_i; a, d, p) \qquad (i = 1, \ldots, n),$$

then

$$(53)\qquad Z_n = \sum_{i=1}^{n} \left(\frac{X_i}{a_i}\right)^{p_i}$$

will have the density function

$$f(z_n) = f\left(z_n; 1, \sum_{i=1}^{n} \left(\frac{d_i}{p_i}\right), 1\right). \tag{54}$$

The rest of the paper is devoted to the discussion of the distribution of the sum

$$Y = \sum_{i=1}^{n} X_i. \tag{55}$$

Kemp and Kemp (1965) define the Hermite distribution, whose probability generating function (p.g.f.) is of the form

$$P(s) = \exp[a_1(s-1) + a_2(s^2-1)]. \tag{56}$$

Setting

$$a_1 = \alpha\beta, \qquad a_2 = \alpha^2/2 \tag{57}$$

in (56) yields

$$\begin{aligned} P(s) &= \exp\{-(\alpha\beta + \alpha^2/2)\}\exp(\alpha\beta s + \alpha^2 s^2/2) \\ &= \exp\{-(\alpha\beta + \alpha^2/2)\}\sum_{n=0}^{\infty} \frac{H_n^*(\beta)\alpha^n s^n}{n!}, \end{aligned} \tag{58}$$

so that

$$\begin{aligned} P_0 &= \exp\{-(\alpha\beta + \alpha^2/2)\}, \\ P_n &= P_0 \alpha^n H_n^*(\beta)/n!, \end{aligned} \tag{59}$$

where $H_n^*(x)$ is the "modified" Hermite polynomial

$$H_n^*(x) = i^{-n} He_n(ix), \tag{60}$$

$He_n(x)$ being the Hermite polynomials, defined by [see, e.g., II.3(65) and Kendall and Stuart (1963, p. 155)]

$$\exp\left(tx - \frac{t^2}{2}\right) = \sum_{n=0}^{\infty} He_n(x)\frac{t^n}{n!}. \tag{61}$$

It may be noted here that another definition, with notation $H_n(x)$, is given, e.g., in Erdélyi *et al.* (1953, Vol. II) as

$$\exp(2xz - z^2) = \sum_{n=0}^{\infty} H_n(x)\frac{z^n}{n!}, \tag{62}$$

so that $He_n(x) = 2^{-n/2} H_n(x/\sqrt{2})$, $n \geqq 0$ [cf. II.3(65) and Magnus *et al.* (1966, p. 250)]. From (58), we can write

$$P_n = \exp\{-(a_1 + a_2)\}\sum_{j=0}^{[n/2]} \frac{a_1^{n-2j} a_2^j}{(n-2j)!\,j!}, \tag{63}$$

whence

$$(64)\qquad P_{2r} = \exp\{-(a_1 + a_2)\}(a_2^r/r!)\,{}_1F_1[-r; \tfrac{1}{2}; -a_1^2/(4a_2)],$$

$$(65)\qquad P_{2r+1} = \exp\{-(a_1 + a_2)\}(a_2^r/r!)a_1\,{}_1F_1[-r; \tfrac{3}{2}; -a_1^2/(4a_2)].$$

Mathai and Saxena (1966) propose a generalized hypergeometric probability distribution with density function

$$(66)\qquad p(x) = \begin{cases} Kx^{c-1}\,{}_2F_1(\alpha, \beta; \gamma; -ax^d) \\ \qquad (x > 0, c > 0, \alpha - c/d > 0, \beta - c/d > 0) \\ 0 \qquad (\text{elsewhere}), \end{cases}$$

where

$$(67)\quad K = da^{c/d}\Gamma(\alpha)\Gamma(\beta)\Gamma(\gamma - c/d)/\{\Gamma(c/d)\Gamma(\gamma)\Gamma(\alpha - c/d)\Gamma(\beta - c/d)\}.$$

Replacing a by a/α and letting $\alpha \to \infty$, they obtain another distribution with density function

$$(68)\qquad f(x) = [da^{c/d}\Gamma(\beta)\Gamma(\gamma - c/d)/\{\Gamma(c/d)\Gamma(\gamma)\Gamma(\beta - c/d)\}] \cdot x^{c-1}\,{}_1F_1(\beta; \gamma; -ax^d),$$

and they show how various well-known distributions like beta, gamma, exponential, Weibull, etc., can be obtained from (68) as special cases as well as the distribution proposed by Stacy (1962) [see (49)]. The characteristic function of the distribution is shown to be

$$(69)\quad \phi(t) = \int_0^\infty e^{itx} p(x)\,dx$$
$$= [da^{c/d}\Gamma(\gamma - c/d)(-it)^{-c}/\{\Gamma(c/d)\Gamma(\alpha - c/d)\Gamma(\beta - c/d)\}] \cdot H_{3,2}^{1,3}\left[a(-it)^{-d}\,\middle|\,\begin{matrix}(1-c, d), (1-\alpha, 1), (1-\beta, 1) \\ (0, 1), (1-\gamma, 1)\end{matrix}\right]$$

in terms of Fox's H function. Also, using Erdélyi *et al.* (1954, Vol. II, p. 399, Eq. (7)), we have the distribution function

$$(70)\qquad F(x) = \int_0^x p(y)\,dy = (Kx^c/c)\,{}_3F_2(\alpha, \beta, c/d; \gamma, c/d + 1; -ax^d).^5$$

In a special case of the distribution (66), Mathai and Saxena have obtained the density function for the largest-order statistic in terms of the generalized

[5] The corresponding result of Mathai and Saxena (1966, p. 130, Eq. (11)) seems to be in error.

hypergeometric function ${}_3F_2$, and the density function for the ratio of two independent random variables distributed as in (66) in terms of the H function.

Mathai and Saxena (1968) consider various distributions with density functions of the form

$$p(x) = Ke^{-\alpha x}x^{\beta}f(x), \tag{71}$$

where K is a constant and $f(x)$ a modified Bessel function, parabolic cylinder function, or the hypergeometric function ${}_1F_1$, ${}_1F_2$, or ${}_1F_3$, and show that all these distributions possess the "gamma property," viz., that the mean of a random sample from any of these distributions will follow a distribution of the same form with modified parameters.

Manocha (1968) proposes a "generalized Laguerre" distribution with density function

$$p_n(\alpha, x) = \begin{cases} Kx^{\alpha}e^{-x}L_n^{(\alpha)}(2x) & (x \geq 0; n = 0, 1, 2, \ldots) \\ 0 & (\text{elsewhere}), \end{cases} \tag{72}$$

where

$$K = (-1)^n n!/\Gamma(\alpha + n + 1) \tag{73}$$

and

$$L_n^{(\alpha)}(x) = \binom{\alpha + n}{n} {}_1F_1(-n; \alpha + 1; x) \tag{74}$$

is the Laguerre polynomial. He proves that

$$E(X^r) = (1 + \alpha)_r \, {}_2F_1(-n, -r; \alpha + 1; 2), \tag{75}$$

and obtains expressions for the distribution function and its characteristic function, as well as the distribution of the largest-order statistic for the sample size $n = 2$.

McNolty (1968) proposes the following "radial" probability density function for the distribution of the distance from a typical impact point to a point target [cf. (8)]:

$$g(S) = \{(\lambda S)^Q/\gamma^{Q-1}\} \exp[-(\lambda S^2 + \gamma^2/\lambda)/2] \, I_{Q-1}(\gamma S) \quad (\gamma, Q > 0), \tag{76}$$

where $I_\nu(z)$ stands for the modified Bessel function of the first kind. From (75), using the change of variables

$$X = S\cos\theta, \qquad Y = S\sin\theta, \qquad S^2 = X^2 + Y^2, \tag{77}$$

he obtains the density function for the component X or Y:

$$f(x) = 2^{(Q/4)-(3/2)}\{|x|^{Q/2}/\gamma^{(Q/2)-1}\}\lambda^{Q/2} \exp[-\{\lambda x^2 + \gamma^2/(2\lambda)\}] \cdot I_{(Q/2)-1}(\gamma|x|/\sqrt{2}) \qquad (-\infty < x < \infty), \tag{78}$$

and for the phase angle θ[6]:

(79) $$h(\theta) = |(\sin 2\theta)/2|^{Q-1} \exp\{-\gamma^2/(2\lambda)\} \cdot \sum_{i=0}^{\infty} \sum_{j=0}^{\infty} \left(\frac{\gamma}{2}\right)^{2i+2j} \Gamma(Q+i+j) \sin^{2i}\theta \sin^{2j}\theta \div \left[2\lambda^{i+j} i!j!\Gamma\left(\frac{Q}{2}+i\right)\Gamma\left(\frac{Q}{2}+j\right)\right] \qquad (0 \leqq \theta \leqq 2\pi).$$

He shows this by considering particular values of the parameters in (76), for radial distributions like the Rayleigh, the Maxwell–Boltzmann, chi, noncentral chi, and one-sided Gaussian, and finds the corresponding component and phase distributions, which in the first case (Rayleigh) are, respectively, Gaussian and uniform distributions.

Srivastava and Singhal (1972) propose a more general form of probability density function:

(80) $$p(x) = \begin{cases} U[\alpha, \beta, \gamma: (a_p, A_p)_n; (b_q, B_q)_m] x^{\beta-1} H_{p,q}^{m,n}[a x^{\gamma}] \\ \quad \left(x > 0, \gamma > 0; -\min\limits_{1 \leqq j \leqq m} \{b_j/B_j\} < \beta/\gamma < -\max\limits_{1 \leqq j \leqq n} \{(a_j - 1)/A_j\}\right) \\ 0 \qquad \text{(elsewhere)}, \end{cases}$$

where (cf. also Section II.5)

(81) $$U \equiv U[\alpha, \beta, \gamma: (a_p, A_p)_n; (b_q, B_q)_m] = \gamma \alpha^{\beta/\gamma} \prod_{j=m+1}^{q} \Gamma\left[1 - b_j - \left(\frac{\beta}{\gamma}\right) B_j\right] \prod_{j=n+1}^{p} \Gamma\left[a_j + \left(\frac{\beta}{\gamma}\right) A_j\right] \cdot \left\{\prod_{j=1}^{m} \Gamma\left[b_j + \left(\frac{\beta}{\gamma}\right) B_j\right] \prod_{j=1}^{n} \Gamma\left[1 - a_j - \left(\frac{\beta}{\gamma}\right) A_j\right]\right\}^{-1},$$

(82) $$H_{p,q}^{m,n}(z) = H_{p,q}^{m,n}\left[z \middle| \begin{matrix} (a_1, A_1), \ldots, (a_p, A_p) \\ (b_1, B_1), \ldots, (b_q, B_q) \end{matrix}\right]$$

is the H function, and it is assumed that the parameters of the function are so restricted that $p(x)$ remains nonnegative. They point out that the H function can be specialized to Meijer's G function, from which the hypergeometric functions ${}_2F_1$ and ${}_1F_1$ and, therefore, the distributions (66) and (68), proposed by Mathai and Saxena (1966), can be obtained as special cases by specializing the parameters without having recourse to a limiting process. They show that

[6] Incidentally, the double series in (79) can readily be rewritten in terms of the confluent hypergeometric Ψ_2 function defined by II.6(30).

the corresponding distribution function is given by

$$(83)\qquad F(x) = \int_0^x p(y)\,dy$$

$$= Ux^{\beta} H^{m,n+1}_{p+1,q+1}\left[\alpha x^{\gamma}\middle|\begin{matrix}(1-\beta,\gamma),(a_1,A_1),\ldots,(a_p,A_p)\\(b_1,B_1),\ldots,(b_q,B_q),(-\beta,\gamma)\end{matrix}\right],$$

and the characteristic function is

$$(84)\qquad \phi(t) = \int_0^{\infty} e^{itx} p(x)\,dx$$

$$= U(-it)^{-\beta} H^{m,n+1}_{p+1,q}\left[\alpha(-it)^{-\gamma}\middle|\begin{matrix}(1-\beta,\gamma),(a_1,A_1),\ldots,(a_p,A_p)\\(b_1,B_1),\ldots,(b_q,B_q)\end{matrix}\right],$$

where U is given by (81). They also find the distribution of the largest-order statistic and of the ratio of two independent variates distributed as in (80) with different parameters and simplify the results in terms of the H function.

Gupta and Jain (1974) define a generalized Hermite (GH) distribution as one having the probability generating function

$$(85)\qquad P(s) = \exp\{-(\alpha+\beta)\}\exp(\alpha s + \beta s^m)$$

$$= \sum_{n=0}^{\infty} P_{n,m}(\alpha,\beta)s^n \qquad [P_{n,m}(\alpha,\beta) \geqq 0],$$

so that the probability that a GH variate has the value n is

$$(86)\qquad P_{n,m}(\alpha,\beta) = \exp\{-(\alpha+\beta)\}\, H_{n,m}(\alpha,\beta)/n! \qquad (n = 0,1,2,\ldots),$$

where $H_{n,m}(\alpha,\beta)$ are the "generalized" Hermite polynomials defined by them as [see also II.9(7) defining the *same* polynomials $\{g_n^m(\alpha,\beta)\}$, which were, in fact, introduced and studied earlier by Gould and Hopper (1962)]

$$(87)\qquad \exp(\alpha s + \beta s^m) = \sum_{n=0}^{\infty} H_{n,m}(\alpha,\beta)\frac{s^n}{n!}.$$

It follows that

$$(88)\qquad P_{j,m} = K\alpha^j/j! \qquad (j = 0,1,\ldots,m-1),$$

$$(89)\qquad P_{m,m} = (K\alpha^m + \beta m!)/m!,$$

where

$$(90)\qquad K = \exp\{-(\alpha+\beta)\}.$$

Also, they show that in terms of the generalized hypergeometric functions, we can write

(91) $$P_{n,m}(\alpha, \beta) = (K\alpha^n/n!) \cdot {}_mF_0\left(\frac{-n}{m}, \frac{-n+1}{m}, \ldots, \frac{-n+m-1}{m}; -; -\frac{\beta m^m}{\alpha^m}\right),$$

and that the sum of two independent GH variates with parameters (α_1, β_1, m) and (α_2, β_2, m) is a GH variate with parameters $(\alpha_1 + \alpha_2, \beta_1 + \beta_2, m)$.

6.9 Generalized Birth and Death Processes

General birth and death processes have been discussed in such standard works as those by Kendall (1948), Bartlett (1955), Reuter (1957), Bharucha-Reid (1960), Bailey (1964), Feller (1968), and Iosikscu and Taut (1973). Recently, Getz (1975, 1976) considered the most general case when there are births, deaths, and immigration and emigration, and obtained expressions for the mean $\bar{n}(t)$ and the variance $\sigma^2(t)$ of the size of the population at time t. Unfortunately, as Kapur (1978a) observes, Getz's expressions for $\bar{n}(t)$ and $\sigma^2(t)$ are in error; their corrected versions, however, become extremely complicated when all the four processes are involved. This might account for the fact that the earlier discussions on the subject are confined to only three processes, viz., birth, death, and immigration.

Kapur (1978a) initiates his study of the most general case in its steady state and finds expressions for the probability distribution, probability generating function, and moments in terms of generalized hypergeometric functions. He also uses many known relations between hypergeometric functions to give their interesting interpretations in terms of probabilities of extinctions for these generalized birth and death processes. To describe his work more fully, let the probabilities of birth and death in the time-interval $(t, t + \Delta t)$ be

$$\lambda_n \Delta t + o(\Delta t) \qquad \text{and} \qquad \mu_n \Delta t + o(\Delta t),$$

respectively, the population at time t being n. Then we readily obtain

(1) $$p_n(t + \Delta t) = p_{n+1}(t)\mu_{n+1} \Delta t + p_{n-1}(t)\lambda_{n-1} \Delta t + p_n(t)[1 - \lambda_n \Delta t - \mu_n \Delta t] + o(\Delta t), \qquad n = 1, 2, 3, \ldots,$$

(2) $$p_0(t + \Delta t) = p_1(t)\mu_1 \Delta t + p_0(t)[1 - \lambda_0 \Delta t] + o(\Delta t),$$

which lead to the system of the differential difference equations

(3) $$\frac{d}{dt}\{p_n(t)\} = p_{n+1}(t)\mu_{n+1} + p_{n-1}(t)\lambda_{n-1} - p_n(t)[\lambda_n + \mu_n], \qquad n \geqq 1,$$

(4) $$\frac{d}{dt}\{p_0(t)\} = p_1(t)\mu_1 - p_0(t)\lambda_0.$$

For the steady state, we put

$$\frac{d}{dt}\{p_n(t)\} = 0, \qquad n = 0, 1, 2, \ldots,$$

and solve the difference equations resulting from (3) and (4) to get

(5) $$p_n = p_0 \frac{\lambda_0 \lambda_1 \cdots \lambda_{n-1}}{\mu_1 \mu_2 \cdots \mu_n}, \qquad n = 1, 2, 3, \ldots.$$

Thus the probability p_0 of ultimate extinction is given by

(6) $$p_0 \left\{ 1 + \sum_{n=1}^{\infty} \frac{\lambda_0 \lambda_1 \cdots \lambda_{n-1}}{\mu_1 \mu_2 \cdots \mu_n} \right\} = 1,$$

provided that the infinite series converges.

Kapur (1978a) sets

(7) $$\lambda_n = (na_1 + b_1)(na_2 + b_2) \cdots (na_p + b_p),$$
$$\mu_n = (nc_1 + d_1)(nc_2 + d_2) \cdots (nc_q + d_q),$$

so that (5) yields

(8) $$p_n = p_0 \frac{(\alpha_1)_n (\alpha_2)_n \cdots (\alpha_p)_n}{(\beta_1)_n (\beta_2)_n \cdots (\beta_q)_n} x^n,$$

where, and in what follows,

(9) $$(\alpha)_n = \frac{\Gamma(\alpha + n)}{\Gamma(\alpha)} = \begin{cases} 1, & n = 0, \\ \alpha(\alpha + 1) \cdots (\alpha + n - 1), & n = 1, 2, 3, \ldots, \end{cases}$$

(10) $$\alpha_i = \frac{b_i}{a_i}, \qquad i = 1, 2, \ldots, p,$$

(11) $$\beta_j = \frac{d_j}{c_j}, \qquad j = 1, 2, \ldots, q,$$

and

(12) $$x = a_1 a_2 \cdots a_p / c_1 c_2 \cdots c_q,$$

it being assumed that

(13) $a_i, b_i \neq 0 \quad (i = 1, 2, \ldots, p)$ and $c_j, d_j \neq 0 \quad (j = 1, 2, \ldots, q)$.

For determining p_0, we have from (6)

(14) $$p_0 \sum_{n=0}^{\infty} \frac{(\alpha_1)_n (\alpha_2)_n \cdots (\alpha_p)_n}{(\beta_1)_n (\beta_2)_n \cdots (\beta_q)_n} x^n = 1,$$

or, equivalently,

$$p_0 \; {}_{p+1}F_q\left[\begin{matrix}1, \alpha_1, \alpha_2, \ldots, \alpha_p; \\ \beta_1, \beta_2, \ldots, \beta_q;\end{matrix} \; x\right] = 1, \tag{15}$$

in terms of the generalized hypergeometric series, which, incidentally, converges for all finite x if $p < q$ and diverges for all $x \neq 0$ if $p > q$; if $p = q$, this series converges for $x < 1$, and also when $x = 1$, provided that

$$\sum_{j=1}^{q} \beta_j - \sum_{j=1}^{\infty} \alpha_j > 1. \tag{16}$$

The *moment generating function* is given by

$$\phi(\theta) = \sum_{n=0}^{\infty} p_n e^{n\theta} = {}_{p+1}F_q\left[\begin{matrix}1, \alpha_1, \alpha_2, \ldots, \alpha_p; \\ \beta_1, \beta_2, \ldots, \beta_q;\end{matrix} \; xe^{\theta}\right] \cdot \left\{{}_{p+1}F_q\left[\begin{matrix}1, \alpha_1, \alpha_2, \ldots, \alpha_p; \\ \beta_1, \beta_2, \ldots, \beta_q;\end{matrix} \; x\right]\right\}^{-1}, \tag{17}$$

which, on replacing e^{θ} by s, yields the *probability generating function* $\phi(\ln s)$.

The factorial moment of order r is given by

$$\begin{aligned} \mu'_{(r)} &= \sum_{n=r}^{\infty} n(n-1)(n-2)\cdots(n-r+1)p_n \\ &= \sum_{n=0}^{\infty} \frac{(n+r)!}{n!} p_{n+r} \\ &= r! \frac{(\alpha_1)_r(\alpha_2)_r\cdots(\alpha_p)_r}{(\beta_1)_r(\beta_2)_r\cdots(\beta_q)_r} x^r \\ &\quad \cdot {}_{p+1}F_q\left[\begin{matrix}1+r, \alpha_1+r, \alpha_2+r, \ldots, \alpha_p+r; \\ \beta_1+r, \beta_2+r, \ldots, \beta_q+r;\end{matrix} \; x\right] \\ &\quad \cdot \left\{{}_{p+1}F_q\left[\begin{matrix}1, \alpha_1, \alpha_2, \ldots, \alpha_p; \\ \beta_1, \beta_2, \ldots, \beta_q;\end{matrix} \; x\right]\right\}^{-1}. \end{aligned} \tag{18}$$

From the aforementioned conditions of convergence of the generalized hypergeometric series, it thus follows that the moments of all orders exist if $p < q$ or if $p = q$ and

$$a_1 a_2 \cdots a_p < c_1 c_2 \cdots c_q;$$

if, however, $p = q$ and

$$a_1 a_2 \cdots a_p = c_1 c_2 \cdots c_q,$$

the moment of the rth order will exist whenever

$$\sum_{j=1}^{q} \beta_j - \sum_{j=1}^{p} \alpha_j > 1 + r, \qquad r = 1, 2, 3, \ldots . \tag{19}$$

In another general case considered by Kapur (1978a), he sets in (5) and (6)

$$\lambda_n = (na_1 + b_1)(na_2 + b_2) \cdots (na_p + b_p),$$
$$\mu_n = n(nc_1 + d_1)(nc_2 + d_2) \cdots (nc_q + d_q), \tag{20}$$

so that, using the notation of this section,

$$p_0 \, {}_pF_q\left[\begin{matrix}\alpha_1, \ldots, \alpha_p; \\ \beta_1, \ldots, \beta_q;\end{matrix} \; x\right] = 1, \tag{21}$$

$$\phi(\theta) = {}_pF_q\left[\begin{matrix}\alpha_1, \ldots, \alpha_p; \\ \beta_1, \ldots, \beta_q;\end{matrix} \; xe^{\theta}\right] \cdot \left\{ {}_pF_q\left[\begin{matrix}\alpha_1, \ldots, \alpha_p; \\ \beta_1, \ldots, \beta_q;\end{matrix} \; x\right]\right\}^{-1}, \tag{22}$$

and

$$\mu'_{(r)} = \frac{(\alpha_1)_r \cdots (\alpha_p)_r}{(\beta_1)_r \cdots (\beta_q)_r} x^r \, {}_pF_q\left[\begin{matrix}\alpha_1 + r, \ldots, \alpha_p + r; \\ \beta_1 + r, \ldots, \beta_q + r;\end{matrix} \; x\right] \cdot \left\{ {}_pF_q\left[\begin{matrix}\alpha_1, \ldots, \alpha_p; \\ \beta_1, \ldots, \beta_q;\end{matrix} \; x\right]\right\}^{-1}. \tag{23}$$

Making use of certain known theorems on the sums of generalized hypergeometric series for particular values of p, q, α, β, and x, Kapur (1978a) discusses a number of special cases of his choices (7) and (20). On the other hand, since contiguous hypergeometric functions are known to be related, Kapur (1978a,b) uses these relations to derive several interesting relations between probabilities of ultimate extinction of contiguous birth and death processes. Kapur (1978b) also derives a linear relation between consecutive moments, again by using a known property of the generalized hypergeometric function (Rainville, 1960, Chapter 5, Section 48).

Kapur and Kumar (1978) study an interesting modification of the system of differential difference equations (3) and (4) in order to account for situations with *twin* births, while Kapur (1979) uses some well-known special summation theorems for the Gaussian hypergeometric function to determine the moments of all orders for the probability distribution of the ultimate size of the population in twelve different birth and death processes that he obtains by further specializing his choices in (7).

Bibliography

Abramowitz, M., and Stegun, I. A. (1964). "Handbook of Mathematical Functions." National Bureau of Standards, Washington, D.C.

Adiri, I., and Avi-Itzhak, B. (1969). A time-sharing queue. *Management Sci.* **15**, 639–657.

Aggarwal, O. P. (1955). Some minimax invariant procedures for estimating a cumulative distribution function. *Ann. Math. Statist.* **26**, 450–463.

Airey, J. R., and Webb, H. A. (1918). The practical importance of the confluent hypergeometric function. *Philos. Mag.* **36**, 129–141.

Alagaraja, K. (1970). A queue with particular bulk service having dependent service times. *Yokohama Math. J.* **18**, 15–22.

Ancker, C. J., Jr. (1964). Stochastic duels with limited ammunition supply. *Oper. Res.* **12**, 38–50.

Ancker, C. J., Jr., and Gafarian, A. V. (1962). Queuing with impatient customers who leave at random. *J. Ind. Eng.* **13**, 86–87.

Ancker, C. J., Jr., and Gafarian, A. V. (1963a). Queuing with reneging and multiple heterogeneous servers. *Naval Res. Logist. Quart.* **10**, 125–149.

Ancker, C. J., Jr., and Gafarian, A. V. (1963b). Some queuing problems with balking and reneging. I. *Oper. Res.* **11**, 88–100.

Ancker, C. J., Jr., and Gafarian, A. V. (1963c). Some queuing problems with balking and reneging. II. *Oper. Res.* **11**, 928–937.

Ansell, J., Bendell, A., and Humble, S. (1980). Nested renewal processes. *Adv. in Appl. Probab.* **12**, 880–892.

Appell, P., and Kampé de Fériet, J. (1926). "Fonctions Hypergéométriques et Hypersphériques: Polynômes d'Hermite." Gauthier-Villars, Paris.

Arora, K. L. (1964). Two-server bulk-service queuing process. *Oper. Res.* **12**, 286–294.

Askey, R. (1975). "Orthogonal Polynomials and Special Functions." (*Regional Conf. Ser. in Appl. Math.* **21**). Soc. Industr. Appl. Math., Philadelphia, Pennsylvania.

Askey, R. (1978). Jacobi's generating function for Jacobi polynomials. *Proc. Amer. Math. Soc.* **71**, 243–246.

Avi-Itzhak, B., and Ben-Tuvia, S. (1963). A problem of optimizing a collecting reservoir system. *Oper. Res.* **11**, 122–136.

Ayyangar, A. A. K. (1934). A note on the incomplete moments of the hypergeometrical series. *Biometrika* **26**, 264–265.

Bailey, N. T. J. (1954). A continuous time treatment of a simple queue using generating functions. *J. Roy. Statist. Soc. Ser. B* **16**, 288–291.

Bailey, N. T. J. (1957). Some further results in the non-equilibrium theory of a simple queue. *J. Roy. Statist. Soc. Ser. B* **19**, 326–333.

Bailey, N. T. J. (1964). "The Elements of Stochastic Processes." Wiley, New York.

Bailey, N. T. J. (1975). "The Mathematical Theory of Infectious Diseases." Hafner, New York.

Bailey, W. N. (1935). "Generalized Hypergeometric Series," Cambridge Math. Tract No. 32. Cambridge Univ. Press, London and New York. Reprinted in 1964 by Stechert-Hafner, New York.

Banerjee, D. P. (1958). On the exact distribution of a test in multivariate analysis. *J. Roy. Statist. Soc. Ser. B* **20**, 108–110.

Barlow, R. E., and Proschan, F. (1975). "Statistical Theory of Reliability and Life Testing." Holt, New York.

Barrer, D. Y. (1957). Queuing with impatient customers and ordered service. *Oper. Res.* **5**, 650–656.

Bartlett, M. S. (1955). "An Introduction to Stochastic Processes, with Special Reference to Methods and Applications." Cambridge Univ. Press, London and New York.

Bateman, H. (1931). The k-function, a particular case of the confluent hypergeometric function. *Trans. Amer. Math. Soc.* **33**, 817–831.

Beckmann, P. (1973). "Orthogonal Polynomials for Engineers and Physicists." Golem Press, Boulder, Colorado.

Beightler, C. S., Mitten, L. G., and Nemhauser, G. L. (1961). A short table of Z-transforms and generating functions. *Oper. Res.* **9**, 574–578.

Belkin, B., Stone, L. D., and Snyder, M. A. (1972). Crossing problems for non-constant thresholds and certain non-Markov processes. *J. Appl. Probab.* **9**, 769–788.

Ben-Israel, A., and Naor, P. (1960a). A problem of delayed service. I. *J. Roy. Statist. Soc. Ser. B* **22**, 245–269.

Ben-Israel, A., and Naor, P. (1960b). A problem of delayed service. II. *J. Roy. Statist. Soc. Ser. B* **22**, 270–276.

Benjamin, E. (1948). Some applications of the Mellin transform in statistics. *Ann. Math. Statist.* **19**, 370–379.

Bharucha-Reid, A. T. (1960). "Elements of the Theory of Markov Processes and their Applications." McGraw-Hill, New York.

Bhat, B. R. (1969). Used item replacement policy. *J. Appl. Probab.* **6**, 309–318.

Bhat, U. N. (1968). "A Study of Queueing Systems $M/G/1$ and $GI/M/1$." Springer-Verlag, Berlin and New York.

Bhat, U. N. (1972). "Elements of Applied Stochastic Processes." Wiley, New York.

Bhattacharya, S. K. (1966). A modified Bessel function model in life testing. *Metrika* **11**, 133–144.

Bhattacharya, S. K. (1966a). Confluent hypergeometric distributions of discrete and continuous type with application in accident proneness. *Calcutta Statist. Assoc. Bull.* **15**, 20–31.

Bhattacharya, S. K., and Holla, M. S. (1965). On a life test distribution with stochastic deviations in the mean. *Ann. Inst. Statist. Math.* **17**, 97–104.

Boersma, J. (1962). On a function which is a special case of Meijer's G-function. *Compositio Math.* **15**, 34–63.

Borovkov, A. A. (1976). "Stochastic Processes in Queueing Theory" (Translated from the Russian by K. Wickwire). Springer-Verlag, Berlin and New York.

Borthakur, A. (1971). A Poisson queue with a general bulk service rule. *J. Assam Sci. Soc.* **14**, 162–167.

Borthakur, A., and Medhi, J. (1974). A queuing system with arrivals and service in batches of variable size. *Cahiers Centre Études Rech. Opér.* **16**, 117–126.

Braaksma, B. L. J. (1964). Asymptotic expansions and analytic continuations for a class of Barnes-integrals. *Compositio Math.* **15**, 239–341.

Brafman, F. (1957). Some generating functions of Laguerre and Hermite polynomials. *Canad. J. Math.* **9**, 180–187.

Brigham, G. (1955). On a congestion problem in an aircraft factory. *Oper. Res.* **3**, 412–428.

Brill, P. H. (1979). An embedded level crossing technique for dams and queues. *J. Appl. Probab.* **16**, 174–186.

Brockmeyer, E., Halstrøm, H. L., and Jensen, A. (1948). "The Life and Works of A. K. Erlang." Copenhagen Telephone Co., Copenhagen.

Brockwell, P. J. (1964). An asymptotic expansion for the tail of a binomial distribution and its application to queuing theory. *J. Appl. Probab.* **1**, 161–167.

Brooks, R. J. (1980). On the relative efficiency of two paired-data experiments. *J. Roy. Statist. Soc. Ser. B* **42**, 186–191.

Buchholz, H. (1969). "The Confluent Hypergeometric Function" (Translated from the German by H. Lichtblau and K. Wetzel). Springer-Verlag, Berlin and New York.

Burchnall, J. L. (1942). Differential equations associated with hypergeometric functions. *Quart. J. Math. Oxford Ser.* **13**, 90–106.

Buschman, R. G. (1972). Contiguous relations and related formulas for the *H*-function of Fox. *Jñānabha Sect. A* **2**, 39–47.

Buschman, R. G. (1977). *H*-functions of two variables. II. *Jñānabha* **7**, 107–118.

Buschman, R. G. (1978). *H*-function of two variables. I. *Indian J. Math.* **20**, 139–153.

Buschman, R. G. (1979). *H*-functions of two variables. III. *Pure Appl. Math. Sci.* **9**, 13–18.

Buschman, R. G. (1979a). *H*-functions of *N* variables. *Ranchi Univ. Math. J.* **10**, 81–88.

Buschman, R. G., and Srivastava, H. M. (1975). Inversion formulas for the integral transformation with the *H*-function as kernel. *Indian J. Pure Appl. Math.* **6**, 583–590.

Campbell, L. L. (1966). Exponential entropy as a measure of extent of a distribution. *Z. Wahrsch. Verw. Gebiete* **5**, 217–225.

Carlitz, L., and Srivastava, H. M. (1976). Some hypergeometric polynomials associated with the Lauricella function F_D of several variables. I and II. *Mat. Vesnik* **13 (28)**, 41–47, 143–152.

Carlson, B. C. (1963). Lauricella's hypergeometric function F_D. *J. Math. Anal. Appl.* **7**, 452–470.

Carlson, B. C. (1977). "Special Functions of Applied Mathematics." Academic Press, New York.

Champernowne, D. G. (1956). An elementary method of the solution of the queuing problem with a single server and a constant parameter. *J. Roy. Statist. Soc. Ser. B* **18**, 125–128.

Chan, J., and Conolly, B. (1978). Comparative effectiveness of certain queueing systems with adaptive demand and service mechanisms. *Comput. Oper. Res.* **5**, 187–196.

Chandrasekhar, S. (1943). Stochastic problems in physics and astronomy. *Rev. Modern Phys.* **15**, 1–89. Reprinted in "Selected Papers on Noise and Stochastic Processes" (N. Wax, ed.). Dover, New York, 1954.

Chaudhry, M. L., and Templeton, J. G. C. (1973). A note on the distribution of a busy period for $M/M/c$ queuing system. *Math. Operationsforsch. Statist.* **4**, 75–79.

Chen, W. (1980). On the weak form of Zipf's law. *J. Appl. Probab.* **17**, 611–622.

Chia, A. B., and Watterson, G. A. (1969). Demographic effects on the rate of genetic evolution. I. Constant size populations with two genotypes. *J. Appl. Probab.* **6**, 231–248.

Chihara, T. S. (1978). "An Introduction to Orthogonal Polynomials." Gordon and Breach, New York.

Choo, Q. H., and Conolly, B. (1980). Waiting time analysis for a tandem queue with correlated service. *European J. Oper. Res.* **4**, 337–345.

Clapham, J. C., and Lambe, T. A. (1963). The sorting of a sawmill lumber. *Oper. Res.* **11**, 502–518.

Clarke, A. B. (1953). On time-dependent waiting line processes (Abst.). *Ann. Math. Statist.* **24**, 491–492.

Clarke, A. B. (1956). A waiting time process of Markov type. *Ann. Math. Statist.* **27**, 452–459.

Cohen, J. W. (1969). "The Single Server Queue." North-Holland Publ., Amsterdam.

Conolly, B. W. (1958a). A difference equation technique applied to the simple queue. *J. Roy. Statist. Soc. Ser. B* **20**, 165–167.

Conolly, B. W. (1958b). A difference equation technique applied to the simple queue with arbitrary arrival interval distribution. *J. Roy. Statist. Soc. Ser. B* **20**, 167–175.

Conolly, B. W. (1959). The busy period in relation to the queuing process $GI/M/1$. *Biometrika* **46**, 246–251.

Conolly, B. W. (1965). Marche aléatoire dont la repartition de la longeur des étapes suit une loi exponentielle negative. *Ann. Inst. H. Poincaré Sect. B* **2**, 173–184.

Conolly, B. W. (1971a). On randomized random walks. *SIAM Rev.* **13**, 81–99.

Conolly, B. W. (1971b). Some applications of the theory of infinite capacity service systems to a single server system with linearly state dependent service. *J. Appl. Probab.* **8**, 202–207.

Conolly, B. W. (1974). The generalized state-dependent Erlangian queue: the busy period. *J. Appl. Probab.* **11**, 618–623.

Conolly, B. (1975). "Lecture Notes on Queueing Systems." Wiley (Halsted), New York.

Conolly, B. (1980). "Techniques in Operational Research," Vol. 2, Models, Search and Randomization. Wiley (Halsted), New York.

Conolly, B. (1981). "Techniques in Operational Research," Vol. 1, Queueing Systems. Wiley (Halsted), New York (to be published).

Conolly, B. W., and Chan, J. (1977). Generalized birth and death queueing processes: recent results. *Adv. in Appl. Probab.* **9**, 125–140.

Conolly, B. W., and Choo, Q. H. (1979). The waiting time process for a generalized correlated queue with exponential demand and service. *SIAM J. Appl. Math.* **37**, 263–275.

Consul, P. C. (1969). On the exact distributions of Votaw's criteria for testing compound symmetry of a covariance matrix. *Ann. Math. Statist.* **40**, 836–843.

Consul, P. C. (1970a). On generalization of binomial distribution (Abst.). *Ann. Math. Statist.* **41**, 1386.

Consul, P. C. (1970b). On generalization of Poisson distribution. (Abst.). *Ann. Math. Statist.* **41**, 1387.

Conway, R. W. (1963). Some tactical problems in digital simulation. *Management Sci.* **10**, 47–61.

Cooper, R. D., Hoare, M. R., and Rahman, M. (1977). Stochastic processes and special functions: on the probabilistic origin of some positive kernels associated with classical orthogonal polynomials. *J. Math. Anal. Appl.* **61**, 261–292.

Copson, E. T. (1935). "Theory of Functions of a Complex Variable." Oxford Univ. Press, London and New York.

Courant, R., and Hilbert, D. (1950). "Methoden der Mathematischen Physik," Rev. ed. Springer-Verlag, Berlin and New York.

Cox, D. R. (1955). The anylysis of non-Markovian stochastic processes by the inclusion of supplementary variables. *Proc. Cambridge Philos. Soc.* **51**, 433–441.

Cox, D. R. (1961). A simple congestion system with incomplete service. *J. Roy. Statist. Soc. Ser. B* **23**, 215–222.

Cox, D. R. (1962). "Renewal Theory." Methuen, London.

Cox, D. R., and Miller, H. D. (1965). "The Theory of Stochastic Processes." Methuen, London.

Cox, D. R., and Smith, W. L. (1961). "Queues." Methuen, London.

Cromie, M. V., and Chaudhry, M. L. (1976). Analytically explicit results for the queueing system $M/M^x/c$ with charts and tables for certain measures of efficiency. *Oper. Res. Quart.* **27**, 733–745.

Cromie, M. V., Chaudhry, M. L., and Grassmann, W. K. (1979). Further results for the queuing system $M^x/M/c$. *J. Oper. Res. Soc.* **30**, 755–763.

Crow, J. F., and Kimura, M. (1956). Some genetic problems in natural populations. *Proc. Berkeley Symp. Math. Statist. Probab., 3rd* **4**, 1–22.

Daley, D. J. (1969). The total waiting time in a busy period of a stable single-server queue. I. *J. Appl. Probab.* **6**, 550–564.

Daley, D. J. (1971). The definition of a multi-dimensional generalization of shot noise. *J. Appl. Probab.* **8**, 128–135.

Daley, D. J., and Jacobs, D. R., Jr. (1969). The total waiting time in a busy period of a stable single-server queue. II. *J. Appl. Probab.* **6**, 565–572.

Danese, A. E. (1976–1977). Present status and current trends in the theory of orthogonal polynomials and special functions. *Rend. Sem. Mat. Univ. Politec. Torino* **35**, 5–20.

Daniels, H. E. (1969). The minimum of a stationary Markov process superimposed on a *U*-shaped trend. *J. Appl. Probab.* **6**, 399–408.

Das, C. (1977). The $(S-1, S)$ inventory model under time limit on backorders. *Oper. Res.* **25**, 835–850.

Degerine, S. (1979). Lois de von Mises et lois liées. *Ann. Inst. H. Poincaré Sect. B* **15**, 63–77.

De Meyer, A., and Teugels, J. L. (1980). On the asymptotic behaviour of the distributions of the busy period and service time in $M/G/1$. *J. Appl. Probab.* **17**, 802–813.

Dick, R. S. (1970). Some theorems on single-server queue with balking. *Oper. Res.* **18**, 1193–1206.

Dill, W. R., Gaver, D. P., and Weber, W. L. (1966). Models and modelling for manpower planning. *Management Sci.* **13**, B142–B167.

Dobbie, J. M. (1961). A double-ended queuing system of Kendall. *Oper. Res.* **9**, 755–757.

Dobbie, J. M. (1964). Surveillance of a region by detection and tracking operations. *Oper. Res.* **12**, 379–394.

Dobrusin, R. L. (1953). Limit theorems for a Markov chain of two states. *Izv. Akad. Nauk SSSR Ser. Mat.* **17**, 291–330.

Doetsch, G. (1961). "Guide to the Applications of Laplace Transforms." Van Nostrand, London.

Donelson, J., III, and Maltz, F. (1972). A comparison of linear versus non-linear prediction for polynomial functions of the Ornstein–Uhlenbeck process. *J. Appl. Probab.* **9**, 725–744.

Dua, S., Khadilkar, S., and Sen, K. (1976). A modified random walk in the presence of partially reflecting barriers. *J. Appl. Probab.* **13**, 169–175.

Dyer, M. E., and Proll, L. G. (1977). On the validity of marginal analysis for allocating servers in $M/M/c$ queues. *Management Sci.* **23**, 1019–1022.

Edie, L. C. (1954). Traffic delays at toll booths. *Oper. Res.* **2**, 107–138.

Ehlers, P. F., and Enns, E. G. (1981). Random secants of a convex body generated by surface randomness. *J. Appl. Probab.* **18**, 157–166.

Engel, J., and Zijlstra, M. (1980). A characterization of the gamma distribution by the negative binomial distribution. *J. Appl. Probab.* **17**, 1138–1144.

Erdélyi, A. (1936). Über einige bestimmte Integrale in denen die Whittakerschen $M_{k,m}$-Funktionen auftreten. *Math. Z.* **40**, 693–702.

Erdélyi, A. (1956). "Asymptotic Expansions." Dover, New York.

Erdélyi, A., Magnus, W., Oberhettinger, F., and Tricomi, F. G. (1953). "Higher Transcendental Functions," Vols. I and II. McGraw-Hill, New York.

Erdélyi, A., Magnus, W., Oberhettinger, F., and Tricomi, F. G. (1954). "Tables of Integral Transforms," Vols. I and II. McGraw-Hill, New York.

Erdélyi, A., Magnus, W., Oberhettinger, F., and Tricomi, F. G. (1955). "Higher Transcendental Functions," Vol. III. McGraw-Hill, New York.

Exton, H. (1973). Contribution to the study of Lauricella's functions $F_D^{(n)}$. *Nanta Math.* **6**, 65–70.

Exton, H. (1976). "Multiple Hypergeometric Functions and Applications." Wiley (Halsted), New York.

Exton, H. (1978). "Handbook of Hypergeometric Integrals. Theory, Applications, Tables, Computer Programs." Wiley (Halsted), New York.

Fama, E. F. (1965). Behavior of stock market prices. *J. Business* **38**, 34–105.

Feller, W. (1966). "An Introduction to Probability Theory and Its Applications," Vol. II. Wiley, New York.

Feller, W. (1968). "An Introduction to Probability Theory and Its Applications," Vol. I, 3rd ed. Wiley, New York.

Finch, P. D. (1969). A note on the queuing system $GI/E_k/1$. *J. Appl. Probab.* **6**, 708–710.

Foster, F. G. (1959). A unified theory for stock, storage and queue control. *Oper. Res. Quart.* **10**, 121–130.

Fox, C. (1961). The G and H functions as symmetrical Fourier kernels. *Trans. Amer. Math. Soc.* **98**, 395–429.

Fox, M. (1965). A simplified model for the formation, movement, and dissipation of fair weather cumulus clouds. *J. Appl. Probab.* **2**, 178–185.

Fujisawa, T. (1962). On a waiting-line process with a particular bulk-service. *Yokohama Math. J.* **10**, 45–51.
Gart, J. J. (1965). Some stochastic models relating time and dosage in response curves. *Biometrics* **21**, 583–599.
Gaur, K. N. (1972). Mathematical Analysis of Some Queuing Models. Ph.D. Thesis, Jodhpur Univ., India.
Gaur, K. N., and Kashyap, B. R. K. (1975). The busy period in a priority queue with limited waiting space. *J. Math. Sci.* **10**, 46–53.
Gaur, K. N., and Kashyap, B. R. K. (1977). The number served in a head-of-the-line priority queue with general service. *Defence Sci. J.* **27**, 67–74.
Gaver, D. P. (1954). The influence of servicing times in queuing processes. *Oper. Res.* **2**, 139–149.
Gaver, D. P., Jr. (1963). Accommodation of second-class traffic. *Oper. Res.* **11**, 72–87.
Gaver, D. P., Jr. (1968). Diffusion approximations and models for certain congestion problems. *J. Appl. Probab.* **5**, 607–623.
Gaver, D. P., Jr. (1969). Highway delays resulting from flow-stopping incidents. *J. Appl. Probab.* **6**, 137–153.
Gaver, D. P., and Lewis, P. A. (1980). First-order autoregressive gamma sequences and point processes. *Adv. in Appl. Probab.* **12**, 727–745.
Gavish, B., and Schweitzer, P. J. (1977). The Markovian queue with bounded waiting time. *Management Sci.* **23**, 1349–1357.
Getz, W. M. (1974). Optimal control of a birth and death population model. *Math. Biosci.* **23**, 87–111.
Getz, W. M. (1976). Modelling and control of birth and death processes. Report No. WISK-196. National Institute of Mathematical Sciences, Pretoria.
Ghare, P. M. (1968). Multichannel queuing system with bulk service. *Oper. Res.* **16**, 189–192.
Ghosal, A. (1970). "Some Aspects of Queuing and Storage Systens." Springer-Verlag, Berlin and New York.
Ghurye, S. G., and Olkin, I. (1962). A characterization of the multivariate normal distribution. *Ann. Math. Statist.* **33**, 533–541.
Gibson, A. E. (1968). Some Aspects of Time-Dependent One Dimensional Random Walks. Ph.D. Thesis, Virginia Polytechnic Institute.
Gibson, A. E., and Conolly, B. W. (1971). On a three-state sojourn time problem. *J. Appl. Probab.* **8**, 716–723,
Gilchrist, W. G. (1961). Some sequential tests using range. *J. Roy. Statist. Soc. Ser. B* **23**, 335–342.
Gilchrist, R. (1976). The inter-arrival times of accepted customers in an $M/G/1$ queue with finite capacity. *J. Appl. Probab.* **13**, 633–638.
Girshick, M. A. (1939). On the sampling theory of the roots of determinantal equations. *Ann. Math. Statist.* **10**, 203–224.
Gnedenco, B. V., and Kovalenco, I. N. (1968). "Introduction to Queueing Theory" (Translated from the Russian by E. R. Kondor). Israel Program for Scientific Translations, Jerusalem.
Göbel, F. (1976). "Queueing Models Involving Buffers," Mathematical Centre Tracts No. 60. Mathematisch Centrum, Amsterdam.
Goldstein, S. (1953). On the mathematics of exchange processes in fixed columns. I. *Proc. Roy. Soc. London Ser. A* **219**, 151–171.
Good, I. J. (1961). The frequency count of a Markov chain and the transition to continuous time. *Ann. Math. Statist.* **32**, 41–48.
Gould, H. W., and Hopper, A. T. (1962). Operational formulas connected with two generalizations of Hermite polynomials. *Duke Math. J.* **29**, 51–63.
Greenberg, I. (1964). The distribution of busy time for a simple queue. *Oper. Res.* **12**, 503–504.
Greenberg, I. (1973). Distribution-free analysis of $M/G/1$ and $G/M/1$ queues. *Oper. Res.* **21**, 629–635.
Greenberg, I. (1980). The moments of coverage of a linear set. *J. Appl. Probab.* **17**, 865–868.

Greenwood, J. A., and Durand, D. (1955). The distribution of length and components of the sum of n random unit vectors. *Ann. Math. Statist.* **26**, 233–246.

Gross, D., and Harris, C. M. (1974). "Fundamentals of Queueing Theory." Wiley, New York.

Grosswald, E. (1978). "Bessel Polynomials." Springer-Verlag, Berlin and New York.

Grubbs, F. E. (1964). Approximate circular and noncircular offset probabilities of hitting. *Oper. Res.* **12**, 51–62.

Grubbs, F. E. (1968). Expected target damage for a salvo of rounds with elliptical normal delivery and damage functions. *Oper. Res.* **16**, 1021–1026.

Gumbel, E. J. (1958). "Statistics of Extremes." Columbia Univ. Press, New York.

Gumbel, E. J., Greenwood, J. A., and Durand, D. (1953). The circular normal distribution: theory and tables. *J. Amer. Statist. Assoc.* **48**, 131–152.

Gupta, K. C., and Jain, U. C. (1966). The H-function. II. *Proc. Nat. Acad. Sci. India Sect. A* **36**, 594–609.

Gupta, R. P., and Jain, G. C. (1974). A generalized Hermite distribution and its properties. *SIAM J. Appl. Math.* **27**, 359–363.

Gurland, J. (1955). Distribution of the maximum of the arithmetic mean of correlated random variables. *Ann. Math. Statist.* **26**, 294–300.

Hadidi, N. (1969). On the service time distribution and the waiting time process of a potentially infinite capacity queuing system. *J. Appl. Probab.* **6**, 594–603.

Hadidi, N. (1972). On the "output" process of a state dependent queue. *Skand. Aktuarietidskr.* **10**, 182–186.

Hadidi, N. (1974). Busy period of queues with state dependent arrival and service rates. *J. Appl. Probab.* **11**, 842–848.

Hadidi, N. (1976). Appointment scheme as compared to random service and arrival times from specific distributions. (Abst.). *Adv. in Appl. Probab.* **8**, 250–251.

Hadidi, N., and Conolly, B. W. (1969a). On the improvement of the operational characteristics of single-server queues by the use of a queue-length-dependent service mechanism. *Appl. Statist.* **18**, 229–240.

Hadidi, N., and Conolly, B. W. (1969b). On the reduction of congestion. Statistic. Res. Report No. 6, Institute of Mathematics, Univ. of Oslo.

Hansmann, G. H. (1934). On certain non-normal symmetrical frequency distributions. *Biometrika* **26**, 129–195.

Harris, C. M. (1967a). Queues with stochastic service rates. *Naval Res. Logist. Quart.* **14**, 219–230.

Harris, C. M. (1967b). Queues with state-dependent stochastic service rates. *Oper. Res.* **15**, 117–130.

Harris, C. M. (1968a). Correction to "Queues with state-dependent stochastic service rates". *Oper. Res.* **16**, 885–886.

Harris, C. M. (1968b). The Pareto distribution as a queue service discipline. *Oper. Res.* **16**, 307–313.

Hasija, R. L. (1972). A two warehouse inventory problem. *Metrika* **18**, 133–136.

Hasofer, A. M. (1964). On the single-server queue with non-homogeneous Poisson input and general service time. *J. Appl. Probab.* **1**, 369–384.

Hasofer, A. M. (1965). Some perturbation results for the single-server queue with Poisson input. *J. Appl. Probab.* **2**, 462–466.

Heathcote, C. R. (1959). The time dependent problem for a queue with preemptive priorities. *Oper. Res.* **7**, 670–680.

Heathcote, C. R. (1960). A single queue with several preemptive priority classes. *Oper. Res.* **8**, 630–638.

Heathcote, C. R., and Moyal, J. E. (1959). The random walk (in continuous time) and its application to the theory of queues. *Biometrika* **46**, 400–411.

Heyman, D. P. (1974). An approximation for the busy period of the $M/G/1$ queue using a diffusion model. *J. Appl. Probab.* **11**, 159–169.

Holla, M. S. (1966). On a Poisson-inverse Gaussian distribution. *Metrika* **11**, 115–121.
Holla, M. S. (1970). On a noncentral chi-square distribution in the analysis of weapon system effectiveness. *Metrika* **15**, 9–14.
Holmes, P. T., and Strawderman, W. E. (1969). A note on the waiting times between record observations. *J. Appl. Probab.* **6**, 711–714.
Holst, L. (1980). On the lengths of the pieces of a stick broken at random. *J. Appl. Probab.* **17**, 623–634.
Howard, R. A. (1965). Dynamic inference. *Oper. Res.* **13**, 712–733.
Humbert, P. (1920). La fonction $W_{k,\mu_1,\mu_2,\ldots,\mu_n}(x_1, x_2, \ldots, x_n)$. *C. R. Acad. Sci. Paris* **171**, 428–430.
Humbert, P. (1921a). Sur les polynomes hypergéométriques. *C. R. Acad. Sci. Paris* **172**, 1282–1283.
Humbert, P. (1921b). Formule de multiplication pour la fonction de Kummer, $\Phi(\alpha, \gamma, x)$. *C. R. Acad. Sci. Paris* **173**, 217–219.
Hunt, G. C. (1956). Sequential arrays of waiting lines. *Oper. Res.* **4**, 674–683.
Iglehart, D. L. (1965). Limiting diffusion approximations for the many server queue and the repairman problem. *J. Appl. Probab.* **2**, 429–441.
Iosikscu, M., and Taut, P. (1973). "Stochastic Processes and Applications in Biology and Medicine." Springer-Verlag, Berlin and New York.
Ismail, M. E. H., and Kelker, D. H. (1979). Special functions, Stieltjes transforms and infinite divisibility. *SIAM J. Math. Anal.* **10**, 884–901.
Ismail, M. E. H., and May, C. P. (1979). Special functions, infinite divisibility and transcendental equations. *Math. Proc. Cambridge Philos. Soc.* **85**, 453–464.
Ismail, M. E. H., and Tamhankar, M. V. (1979). A combinatorial approach to some positivity problems. *SIAM J. Math. Anal.* **10**, 478–485.
Ivan, C. (1981). Weak convergence of the simple birth-and-death process. *J. Appl. Probab.* **18**, 245–252.
Jackson, R. R. P., and Aspden, P. (1980). A transient solution to the multistage Poisson queuing system with infinite servers. *Oper. Res.* **28**, 618–622.
Jackson, R. R. P., and Henderson, J. C. (1966). The time-dependent solution to the many server Poisson queue. *Oper. Res.* **14**, 720–723.
Jaiswal, N. K. (1960). Time-dependent solution of the bulk-service queuing problem. *Oper. Res.* **8**, 773–781.
Jaiswal, N. K. (1968). "Priority Queues." Academic Press, New York.
Jensen, L. (1974). Solving a singular diffusion equation occurring in population genetics. *J. Appl. Probab.* **11**, 1–15.
Jensen, L., and Pollak, E. (1969). Random selective advantages of a gene in a finite population. *J. Appl. Probab.* **6**, 19–37.
Jones, J. M. (1970). A dual-effects model of brand choice. *J. Marketing Res.* **7**, 458–464.
Jones, J. M. (1973). A composite heterogeneous model for brand choice behavior. *Management Sci.* **19**, 499–509.
Jorgenson, D. W., and McCall, J. J. (1963). Optimal scheduling of replacement and inspection. *Oper. Res.* **11**, 732–746.
Jupp, P. E., and Mardia, K. V. (1980). A general correlation coefficient for directional data and related regression problems. *Biometrika* **67**, 163–173.
Kabak, I. W. (1968). Blocking and delays in $M^{(n)}/M/c$ bulk queuing systems. *Oper. Res.* **16**, 830–840.
Kabak, I. W. (1970). Blocking and delays in $M^{(x)}/M/c$ bulk arrival queuing systems. *Management Sci.* **17**, 112–115.
Kabak, I. W., and Owen, J. (1970). Random variables, the time value of money and capital expenditures. *Management Sci.* **17**, 142–145.
Kabe, D. G. (1958). Some applications of the Meijer *G*-functions to distribution problems in statistics. *Biometrika* **45**, 578–580.

Kabe, D. G. (1962). On the exact distribution of a class of multivariate test criteria. *Ann. Math. Statist.* **33**, 1197–1200.

Kac, M. (1959). "Probability and Related Topics in Physical Sciences." International Publ., London.

Kampé de Fériet, J. (1921). Les fonctions hypergéométriques d'ordre superieur à deux variables. *C. R. Acad. Sci. Paris* **173**, 401–404.

Kampé de Fériet, J. (1937). "La fonction hypergéométrique" (Mémor. Sci. Math. Fasc. **85**). Gauthier-Villars, Paris.

Kampé de Fériet, J. (1957). "Fonctions de la physique mathématique." Centre National de la Recherche Scientifique, Paris.

Kaplan, M. (1980). A two-fold tandem net with deterministic links and source interference. *Oper. Res.* **28**, 512–526.

Kapur, J. N. (1978a). On generalised birth and death processes and generalised hypergeometric functions. *Indian J. Math.* **20**, 57–69.

Kapur, J. N. (1978b). Application of generalized hypergeometric functions to generalized birth and death processes. *Indian J. Pure Appl. Math.* **9**, 1059–1069.

Kapur, J. N. (1978c). Some new results in population dynamics. *Nat. Acad. Sci. India Lett.* **1**, 329–331.

Kapur, J. N. (1979). Moments for some general birth and death processes. *J. Indian Acad. Math.* **1** (1), 10–17.

Kapur, J. N (1979a). Probabilities of ultimate extinction for general birth and death process. *Pure Appl. Math. Sci.* **10**, 105–108.

Kapur, J. N., and Kumar, U. (1978). Generalised birth and death processes with twin births. *Nat. Acad. Sci. India Lett.* **1**, 30–32.

Karlin, S., and McGregor, J. L. (1957a). The differential equations of birth-and-death processes, and the Stieltjes moment problem. *Trans. Amer. Math. Soc.* **85**, 489–546.

Karlin, S., and McGregor, J. L. (1957b). The classification of birth and death processes. *Trans. Amer. Math. Soc.* **86**, 366–400.

Karlin, S., and McGregor, J. L. (1958). Many server queuing processes with Poisson input and exponential service times. *Pacific J. Math.* **8**, 87–118.

Karlin, S., and McGregor, J. L. (1960). Classical diffusion processes and total positivity. *J. Math. Anal. Appl.* **1**, 163–183.

Karlin, S., and McGregor, J. L. (1965). Ehrenfest urn models. *J. Appl. Probab.* **2**, 352–376.

Karlsson, P. W. (1973). Reduction of certain generalized Kampé de Fériet functions. *Math. Scand.* **32**, 265–268.

Kashyap, B. R. K. (1965a). The random walk with partially reflecting barriers with application to queuing theory. *Proc. Nat. Inst. Sci. India Part A* **31**, 527–535.

Kashyap, B. R. K. (1965b). A double-ended queuing system with limited waiting space. *Proc. Nat. Inst. Sci. India Part A* **31**, 559–571.

Kashyap, B. R. K. (1965c). Some Problems in Queuing Theory. Ph.D. Thesis, Jodhpur Univ., India.

Kashyap, B. R. K. (1966). The double-ended queue with bulk service and limited waiting space. *Oper. Res.* **14**, 822–834.

Kashyap, B. R. K. (1967). Further results for the double-ended queue. *Metrika* **11**, 168–186.

Kashyap, B. R. K. (1970a). Busy period distributions for the double-ended queue. Proc. Symposium Operations Res., *Indian Nat. Sci. Acad.* **42**, 316–323.

Kashyap, B. R. K. (1970b). Busy periods for the double-ended queue with general arrivals. *J. Math. Sci.* **5**, 11–17.

Kashyap, B. R. K., and Mathur, K. C. (1970). Information generating function for the r-power distribution. *Defence Sci. J.* **20**, 177–182.

Kashyap, B. R. K., and Srivastava, H. M. (1974). On distribution-free analysis of $M/G/1$ and $G/M/1$ queues. Working Paper Ser. No. **9**, Indian Institute of Management, Calcutta.

Katti, S. K. (1960). Moments of the absolute difference and absolute deviation of discrete distributions. *Ann. Math. Statist.* **31**, 78–85.

Keilson, J. (1964). A review of transient behavior in regular diffusion and birth-death processes. *J. Appl. Probab.* **1**, 247–266.

Keilson, J. (1965) "Green's Function Methods in Probability Theory." Hafner, New York.

Keilson, J. (1966). The ergodic queue length distribution for queuing systems with finite capacity. *J. Roy. Statist. Soc. Ser. B* **28**, 190–201.

Keilson, J., and Kooharian, A. (1960). On time dependent queuing processes. *Ann. Math. Statist.* **31**, 104–112.

Kemp, A. W., and Kemp, C. D. (1969). Branching and clustering models associated with the "lost-games distribution." *J. Appl. Probab.* **6**, 700–703.

Kemp, C. D., and Kemp, A. W. (1965). Some properties of the "Hermite" distribution. *Biometrika* **52**, 381–394.

Kendall, D. G. (1948). On the generalized birth and death processes. *Ann. Math. Statist.* **10**, 1–18.

Kendall, D. G. (1949). Stochastic processes and population growth. *J. Roy. Statist. Soc. Ser. B* **11**, 230–264.

Kendall, D. G. (1951). Some problems in the theory of queues. *J. Roy. Statist. Soc. Ser. B* **13**, 151–185.

Kendall, D. G. (1953). Stochastic processes occurring in the theory of queues and their analysis by the method of the imbedded Markov chain. *Ann. Math. Statist.* **24**, 338–354.

Kendall, M. G., and Stuart, A. (1963). "The Advanced Theory of Statistics," Vol. 1, Distribution Theory, 2nd ed. Griffin, London, and Hafner, New York, 1958.

Kent, J. (1978). Some probabilistic properties of Bessel functions. *Ann. Probab.* **6**, 760–770.

Khintchine, A. Y. (1960). "Mathematical Methods in the Theory of Queueing." Griffin, London.

Kiefer, J., and Wolfowitz, J. (1955). On the theory of queues with many servers. *Trans. Amer. Math. Soc.* **78**, 1–18.

Kimura, M. (1956). Random genetic drift in a tri-allelic locus: exact solution with a continuous model. *Biometrics* **12**, 57–66.

Kimura, M. (1964). Diffusion models in population genetics. *J. Appl. Probab.* **1**, 177–232.

Kingman, J. F. C. (1966). On the algebra of queues. *J. Appl. Probab.* **3**, 285–326.

Kleinrock, L. (1975). "Queueing Systems," Vol. I, Theory, and Vol. II, Computer Applications. Wiley, New York.

Koopman, B. O. (1956a). The theory of search. I. *Oper. Res.* **4**, 324–346.

Koopman, B. O. (1956b). The theory of search. II. *Oper. Res.* **4**, 503–531.

Krall, H. L., and Frink, O. (1949). A new class of orthogonal polynomials: the Bessel polynomials. *Trans. Amer. Math. Soc.* **65**, 100–115.

Kummer, E. E. (1836). Über die hypergeometrische Reihe $F(a;b;x)$. *J. Reine Angew. Math.* **15**, 39–83.

Laha, R. G. (1954). On some properties of the Bessel function distributions. *Bull. Calcutta Math. Soc.* **46**, 59–71.

Lahiri, M. (1971). On a generalisation of Hermite polynomials. *Proc. Amer. Math. Soc.* **27**, 117–121.

Latouche, G. (1979). Queues with paired customers. Tech. Rep. No. 79, Laboratoire d'Informatique Théorique, Université Libre, Brussels, Belgium.

Latouche, G. (1981). On a Markovian queue with weakly correlated interarrival times. *J. Appl. Probab.* **18**, 190–203.

Lauricella, G. (1893). Sulle funzioni ipergeometriche a più variabili. *Rend. Circ. Mat. Palermo* **7**, 111–158.

Lebedev, N. N. (1965). "Special Functions and Their Applications" (Translated from the Russian and edited by R. A. Silverman). Prentice-Hall, Englewood Cliffs, New Jersey.

Ledermann, W., and Reuter, G. E. H. (1954). Spectral theory for the differential equations of simple birth and death processes. *Philos. Trans. Roy. Soc. London Ser. A* **246**, 321–369.

Lee, A. M. (1966). "Applied Queueing Theory." Macmillan, London.

Leslie, R. T. (1969). Recurrence times of clusters of Poisson points. *J. Appl. Probab.* **6**, 372–388.

Lévy, P. (1925). "Calcul des Probabilités." Gauthier-Villars, Paris.

Lindley, D. V. (1952). The theory of queues with a single server. *Proc. Cambridge Philos. Soc.* **48**, 277–289.

Lloyd, E. H., and Saleem, S. D. (1979). A note on seasonal Markov chains with gamma or gamma-like distributions. *J. Appl. Probab.* **16**, 117–128.

Luchak, G. (1956). The solution of the single-channel queuing equations characterized by a time-dependent Poisson-distributed arrival rate and a general class of holding times. *Oper. Res.* **4**, 711–732.

Luchak, G. (1957). The distribution of the time required to reduce to some pre-assigned level a single-channel queue characterized by a time-dependent Poisson-distributed arrival rate and a general class of holding times. *Oper. Res.* **5**, 205–209.

Luke, Y. L. (1962). "Integrals of Bessel Functions." McGraw-Hill, New York.

Luke, Y. L. (1969). "The Special Functions and Their Approximations," Vols. I and II. Academic Press, New York.

Luke, Y. L. (1975). "Mathematical Functions and Their Approximations." Academic Press, New York.

Luke, Y. L. (1977). "Algorithms for the Computation of Mathematical Functions." Academic Press, New York.

McBride, E. B. (1971). "Obtaining Generating Functions," Springer Tracts in Natural Philosophy. Springer-Verlag, Berlin and New York.

MacCrimmon, K. R., and Ryavec, C. A. (1964). An analytical study of the PERT assumptions. *Oper. Res.* **12**, 16–37.

McNolty, F. (1968). A note on radial distributions. *Oper. Res.* **16**, 211–215.

McNolty, F. (1968a). Expected coverage for targets of nonuniform density. *Oper. Res.* **16**, 1027–1040.

Madansky, A. (1976). Optimal initial conditions for a simulation problem. *Oper. Res.* **24**, 572–577.

Magnus, W., Oberhettinger, F., and Soni, R. P. (1966). "Formulas and Theorems for the Special Functions of Mathematical Physics." Springer-Verlag, Berlin and New York.

Manocha, H. L. (1968). On a generalized Laguerre distribution. *Univ. Nac. Tucumán Rev. Ser. A* **18**, 147–151.

Massy, W. F., Montgomery, D. B., and Morrison, D. G. (1970). "Stochastic Models of Buying Behavior." MIT Press, Cambridge, Massachusetts.

Mathai, A. M. (1972). Products and ratios of generalized gamma variates. *Skand. Aktuarietidskr.* **10**, 193–198.

Mathai, A. M., and Saxena, R. K. (1966). On a generalized hypergeometric distribution. *Metrika* **11**, 127–132.

Mathai, A. M., and Saxena, R. K. (1968). Distributions having the Gamma properties. *Metrika* **13**, 10–16.

Mathai, A. M., and Saxena, R. K. (1973). "Generalized Hypergeometric Functions with Applications in Statistics and Physical Sciences." Springer-Verlag, Berlin and New York.

Mathai, A. M., and Saxena, R. K. (1978). "The *H*-function with Applications in Statistics and Other Disciplines. Wiley (Halsted), New York.

Medhi, J. (1975). Waiting time distribution in a Poisson queue with a general bulk service rule. *Management Sci.* **21**, 777–782.

Meyer, K. H. F. (1971). "Wartesysteme mit variabler Bearbeitungsrate," Lecture Notes in Economics and Mathematical Systems, Vol. 61. Springer-Verlag, Berlin and New York.

Miller, A. J. (1961). A queuing model for road traffic flow. *J. Roy. Statist. Soc. Ser. B* **23**, 64–75.

Miller, R. G. (1959). A contribution to the theory of queues. *J. Roy. Statist. Soc. Ser. B* **21**, 320–337.

Mitrinović, D. S. (1972). "Specijalne Funkcije." Gradjevinska Knjiga, Beograd.
Molina, E. C. (1946). Some fundamental curves for the solution of sampling problems. *Ann. Math. Statist.* **17**, 325–335.
Montgomery, D. B. (1969). A stochastic response model with application to brand choice. *Management Sci.* **15**, 323–337.
Moran, P. A. P. (1961). Entropy, Markov process and Boltzmann's H-theorem. *Proc. Cambridge Philos. Soc.* **57**, 833–842.
Morgan, B. J. T. (1971). On the solution of differential equations arising in some attachment models of virology. *J. Appl. Probab.* **8**, 215–221.
Morse, P. M. (1958). "Queues, Inventories and Maintenance." Wiley, New York.
Morse, P. M., and Feshbach, H. (1953). "Methods of Theoretical Physics." McGraw-Hill, New York.
Naor, P. (1956). On machine interference. *J. Roy. Statist. Soc. Ser. B* **18**, 280–287.
Nath, P. (1951). Confluent hypergeometric function. *Sankhyā* **11**, 153–166.
Natvig, B. (1974). On the transient state probabilities for a queuing model where potential customers are discouraged by queue length. *J. Appl. Probab.* **11**, 345–354.
Natvig, B. (1975). On a queuing model where potential customers are discouraged by queue length. *Scand. J. Statist.* **2**, 34–42.
Neuts, M. F. (1964). The distribution of the maximum length of a Poisson queue during a busy period. *Oper. Res.* **12**, 281–285.
Neuts, M. F. (1967a). A general class of bulk queues with Poisson input. *Ann. Math. Statist.* **38**, 759–770.
Neuts, M. F. (1967b). Waiting times between record observations. *J. Appl. Probab.* **4**, 206–208.
Neuts, M. F. (1974). The Markov renewal branching process. "Mathematical Methods in Queueing Theory" (*Proc. Conf. Western Michigan Univ.*), pp. 1–21. Springer-Verlag, Berlin and New York.
Neuts, M. F. (1975a). Probability distributions of phase type. *In* "Liber Amicorum Professor Emeritus H. Florin," pp. 173–206. Dept. of Mathematics, Univ. of Louvain, Louvain, Belgium.
Neuts, M. F. (1975b). Computational uses of the method of phases in the theory of queues. *Comput. Math. Appl.* **1**, 151–166.
Neuts, M. F. (1977). An annotated bibliography on computational probability, Part 1. *J. Comput. Appl. Math.* **3**, 53–74.
Neuts, M. F. (1977a). Algorithms for the waiting time distributions under various queue disciplines in the $M/G/1$ queue with service time distributions of phase type. *TIMS Stud. in Management Sci.* **7**, 177–197.
Neuts, M. F. (1978). Renewal processes of phase type. *Naval Res. Logist. Quart.* **25**, 445–454.
Nishimura, S. (1974a). Random collision processes and their limiting distribution using the discrimination information. *J. Appl. Probab.* **11**, 266–280.
Nishimura, S. (1974b). Random collision processes with transition probabilities belonging to the same type of distribution. *J. Appl. Probab.* **11**, 703–714.
Novomestky, F. (1974). Asymptotic expressions for the unit-step and Dirac delta functions. *SIAM J. Appl. Math.* **27**, 521–525.
Oakes, D. (1979). Matched pairs in independent Poisson processes. *J. Appl. Probab.* **16**, 780–793.
Oliver, R. M. (1961). A traffic counting distribution. *Oper. Res.* **9**, 802–810.
Ott, T. J. (1977). The stable $M/G/1$ queue in heavy traffic and its covariance function. *Adv. in Appl. Probab.* **9**, 169–186.
Page, E. (1972). "Queueing Theory in OR." Crane Russak, New York.
Panda, R. (1974). Some multiple series transformations. *Jñānabha Sect. A* **4**, 165–168.
Panda, R. (1977). The reducible cases of certain hypergeometric functions of several variables. *Nederl. Akad. Wetensch. Proc. Ser. A* **80** (*Nederl. Akad. Wetensch. Indag. Math.* **39**), 469–476.

Panda, R. (1977a). Certain integrals involving the *H*-function of several variables. *Publ. Inst. Math. (Beograd) (N.S.)* **22 (36)**, 207–210.

Panda, R. (1977b). Integration of certain products associated with the *H*-function of several variables. *Comment. Math. Univ. St. Paul.* **26**, Fasc. 2, 115–123.

Panda, R. (1977c). On a multiple integral involving the *H*-function of several variables. *Indian J. Math.* **19**, 157–162.

Panda, R. (1977d). On a new class of polynomials. *Glasgow Math. J.* **18**, 105–108.

Pearson, K. (1924a). On the moments of the hypergeometrical series. *Biometrika* **16**, 157–162.

Pearson, K. (1924b). On a certain double hypergeometrical series and its representation by continuous frequency surfaces. *Biometrika* **16**, 172–188.

Pearson, K. (1924c). Note on the relationship of the incomplete *B*-function to the sum of the first *p* terms of the binomial $(a + b)^n$. *Biometrika* **16**, 202–203.

Pearson, K. (1957). "Tables of the Incomplete Gamma-Function." Cambridge Univ. Press, London and New York.

Pearson, K. (1968). "Tables of the Incomplete Beta-Function," 2nd ed. Cambridge Univ. Press, London and New York.

Pedler, P. J. (1971). Occupation times for two state Markov chains. *J. Appl. Probab.* **8**, 381–390.

Picard, Ph. (1965). Sur les modèles stochastiques logistiques en démographie. *Ann. Inst. H. Poincaré Sect. B* **2**, 151–172.

Prabhu, N. U. (1965a). "Queues and Inventories." Wiley, New York.

Prabhu, N. U. (1965b). "Stochastic Processes." Macmillan, New York.

Prajneshu (1980). Time-dependent solution of the logistic model for population growth in random environment. *J. Appl. Probab.* **17**, 1083–1086.

Prasad, Y. N., and Singh, S. N. (1977). An application of the *H*-function of several complex variables in the production of heat in a cylinder. *Pure Appl. Math. Sci.* **6**, 57–64.

Pugh, E. L. (1960). The Bayesian approach to the reliability-confidence relation for exponential failure. *Oper. Res.* **8**, 721–724.

Pugh, E. L. (1963). The best estimate of reliability in the exponential case. *Oper. Res.* **11**, 57–61.

Puri, P. S. (1969). Some new results in the mathematical theory of phage-reproduction. *J. Appl. Probab.* **6**, 493–504.

Raaijmakers, F. J. M. (1980). The lifetime of a standby system of units having the Birnbaum and Saunders distribution. *J. Appl. Probab.* **17**, 490–497.

Rainville, E. D. (1960). "Special Functions." Macmillan, New York.

Ramasubban, T. A. (1958). The mean difference and the mean deviation of some discontinuous distributions. *Biometrika* **45**, 549–556.

Rao, S. S. (1965). Queuing with balking, reneging and interruptions. *Oper. Res.* **13**, 596–608.

Rao, S. S. (1966). Balking and Reneging in Waiting Line Systems. Ph.D. Thesis, Univ. of Delhi, India.

Rao, S. S., and Jaiswal, N. K. (1969). On a class of queuing problems and discrete transforms. *Oper. Res.* **17**, 1062–1076.

Rehder, W. (1980). The asymptotic distribution of random molecules. *Adv. in Appl. Probab.* **12**, 640–654.

Reich, E. (1958). On the integrodifferential equation of Takács. I. *Ann. Math. Statist.* **29**, 563–570.

Reich, E. (1959). On the integrodifferential equation of Takács. II. *Ann. Math. Statist.* **30**, 143–148.

Renshaw, E. (1974). Stepping stone models for population growth. *J. Appl. Probab.* **11**, 16–31.

Renshaw, E. (1979). Waveforms and velocities for non-nearest-neighbour contact distributions. *J. Appl. Probab.* **16**, 1–11.

Reuter, G. E. H. (1957). Denumerable Markov processes and the associated contraction semigroups on *l*. *Acta Math.* **97**, 1–46.

Reuter, G. E. H., and Ledermann, W. (1953). On the differential equations for the transition probabilities of Markov processes with enumerably many states. *Proc. Cambridge Philos. Soc.* **49**, 247–262.

Reynolds, J. F. (1968). The stationary solution of a multiserver queuing model with discouragement. *Oper. Res.* **16**, 64–71.

Riordan, J. (1962). "Stochastic Service Systems." Wiley, New York.

Rivlin, T. J. (1974). "The Chebyshev Polynomials." Wiley, New York.

Robbins, H., and Whitehead, J. (1979). Sequential rank and the Pólya urn. *J. Appl. Probab.* **16**, 213–219.

Rothkopf, M. H. (1969). A model of rational competitive bidding. *Management Sci.* **15**, 362–371.

Rothkopf, M. H. (1971). An addendum to "A model of rational competitive bidding". *Management Sci.* **17**, 774–777.

Roy, A. D. (1960a). A note on prediction from an autoregressive process using pistimetric probability. *J. Roy. Statist. Soc. Ser. B* **22**, 97–103.

Roy, A. D. (1960b). Some notes on pistimetric inference. *J. Roy. Statist. Soc. Ser. B* **22**, 338–347.

Ruben, H. (1960). Probability content of regions under spherical normal distributions. I. *Ann. Math. Statist.* **31**, 598–618.

Ruben, H. (1962). Probability content of regions under spherical normal distributions. IV: the distribution of homogeneous and non-homogeneous quadratic functions of normal variables. *Ann. Math. Statist.* **33**, 542–570.

Runnenberg, J. T. (1965). On the use of the method of collective marks in queuing theory. *Proc. Symp. Congestion Theory*, Univ. of North Carolina, Monograph Series in Probability and Statist. No. 2, 399–438.

Saaty, T. L. (1957a). Résumé of useful formulas in queuing theory. *Oper. Res.* **5**, 161–200.

Saaty, T. L. (1957b). A. K. Erlang. *Oper. Res.* **5**, 293–294.

Saaty, T. L. (1960). Time-dependent solution of the many-server Poisson queue. *Oper. Res.* **8**, 755–771.

Saaty, T. L. (1961a). "Elements of Queueing Theory." McGraw-Hill, New York.

Saaty, T. L. (1961b). Some stochastic processes with absorbing barriers. *J. Roy. Statist. Soc. Ser. B* **23**, 319–334.

Sasai, T. (1978). The monodromy group and the reducibility conditions of the one dimensional section of Appell's hypergeometric equation for $F_3(\alpha, \alpha', \beta, \beta'; \gamma; x, y)$. *Proc. Japan. Acad. Ser. A* **54**, 59–61.

Sasieni, M. W. (1961). Double queues and impatient customers with an application to inventory theory. *Oper. Res.* **9**, 771–781.

Saunders, R., Lefèvre, C., and Kryscio, R. J. (1979). A note on the expected number of survivors in supercritical carrier-borne epidemics. *J. Appl. Probab.* **16**, 646–650.

Saxena, R. K., and Dash, S. P. (1979). The distribution of the linear combination and of the ratio of products of independent random variables associated with an *H*-function. *Vijnana Parishad Anusandhan Patrika* **22**, 57–65.

Scheuer, E. M. (1973). Maximum and minimum service times for batches of jobs with an unknown number of servers. *Oper. Res.* **21**, 1163–1167.

Schmeiser, B. W. (1980). Generation of variates from distribution tails. *Oper. Res.* **28**, 1012–1017.

Schweitzer, P. J. (1967). Initial provisioning with spare deterioration. *Oper. Res.* **15**, 513–529.

Sengupta, J. K. (1972). Chance-constrained linear programming with chi-square type deviates. *Management Sci.* **19**, 337–349.

Seth, A. (1963). The correlated unrestricted random walk. *J. Roy. Statist. Soc. Ser. B* **25**, 394–400.

Shortley, G. (1965). A stochastic model for distributions of biological response times. *Biometrics* **21**, 562–582.

Siddiqui, A. (1979). Integral involving the *H*-function of several variables and an integral function of two complex variables. *Bull. Inst. Math. Acad. Sinica* **7**, 329–332.

Siegel, M. J. (1976). The asymptotic behavior of a divergent linear birth and death process. *Adv. in Appl. Probab.* **8**, 315–338.

Singh, N. (1970). Spherical probable error and its estimation. *Metrika* **15**, 149–163.

Singh, N., and Bhashyam, N. (1972). Effect of the variable offset distance on the kill probability. *Metrika* **19**, 39–47.

Singhal, B. M. (1974). On the reducibility of Lauricella's function F_D. *Jñānabha Sect. A* **4**, 163–164.

Singhal, J. P., and Srivastava, H. M. (1972). A class of bilateral generating functions for certain classical polynomials. *Pacific J. Math.* **42**, 755–762.

Skellam, J. G., and Shenton, L. R. (1957). Distributions associated with random walk and recurrent events. *J. Roy. Statist. Soc. Ser. B* **19**, 64–111.

Skinner, C. E. (1968). A numerical inversion of the Pollaczek–Khintchine formula. Manuscript. IBM Corp., Poughkeepsie, New York.

Slater, L. J. (1960). "Confluent Hypergeometric Functions." Cambridge Univ. Press, London and New York.

Slater, L. J. (1966). "Generalized Hypergeometric Functions." Cambridge Univ. Press, London and New York.

Sneddon, I. N. (1961). "Special Functions of Mathematical Physics and Chemistry," 2nd ed. Oliver and Boyd, London.

Springer, M. D. (1979). "The Algebra of Random Variables." Wiley, New York.

Srinivasan, S. K., and Mehata, K. M. (1972). A stochastic model for polymer degradation. *J. Appl. Probab.* **9**, 43–53.

Srivastava, H. M. (1968). On an extension of the Mittag–Leffler function. *Yokohama Math. J.* **16**, 77–88.

Srivastava, H. M. (1969). An infinite summation formula associated with Appell's function F_2. *Proc. Cambridge Philos. Soc.* **65**, 679–682.

Srivastava, H. M. (1969a). An extension of the Hille–Hardy formula. *Math. Comput.* **23**, 305–311.

Srivastava, H. M. (1969b). Some bilinear generating functions. *Proc. Nat. Acad. Sci. U.S.A.* **64**, 462–465.

Srivastava, H. M. (1970). Certain results involving generalized hypergeometric functions. *SIAM J. Math. Anal.* **1**, 75–81.

Srivastava, H. M. (1971). Certain double integrals involving hypergeometric functions. *Jñānabha Sect. A* **1**, 1–10.

Srivastava, H. M. (1971a). Some formulas of J. Meixner. *Arch. Rational Mech. Anal.* **43**, 363–366.

Srivastava, H. M. (1971b). The Weierstrass–Laguerre transform. *Proc. Nat. Acad. Sci. U.S.A.* **68**, 554–556.

Srivastava, H. M. (1972). A pair of dual series equations involving generalized Bateman k-functions. *Nederl. Akad. Wetensch. Proc. Ser. A* **75** (*Nederl. Akad. Wetensch. Indag. Math.* **34**), 53–61.

Srivastava, H. M. (1972a). A class of integral equations involving the H-function as kernel. *Nederl. Akad. Wetensch. Proc. Ser. A* **75** (*Nederl. Akad. Wetensch. Indag. Math.* **34**), 212–220.

Srivastava, H. M. (1972b). A contour integral involving Fox's H-function. *Indian J. Math.* **14**, 1–6.

Srivastava, H. M. (1972c). Some formulas of Hermite and Carlitz. *Rev. Roumaine Math. Pures Appl.* **17**, 1257–1263.

Srivastava, H. M. (1973). On the reducibility of Appell's function F_4. *Canad. Math. Bull.* **16**, 295–298.

Srivastava, H. M. (1974). Note on certain generating functions for Jacobi and Laguerre polynomials. *Publ. Inst. Math.* (*Beograd*) (*N.S.*) **17** (**31**), 149–154.

Srivastava, H. M. (1974a). Analytic continuation of Appell's function F_4. *Rev. Math. Hisp.-Amer.* (*4*) **34**, 151–156.

Srivastava, H. M. (1975). An equivalence theorem on generating functions. *Proc. Amer. Math. Soc.* **52**, 159–165.

Srivastava, H. M. (1976). An integral equation involving the confluent hypergeometric function of several complex variables. *Applicable Anal.* **5**, 251–256.

Srivastava, H. M. (1976a). A note on a generating function for the generalized Hermite polynomials. *Nederl. Akad. Wetensch. Proc. Ser. A* **79** (*Nederl. Akad. Wetensch. Indag. Math.* **38**), 457–461.

Srivastava, H. M. (1977). A note on certain generating functions for the classical polynomials. *Atti Accad. Naz. Lincei Rend. Cl. Sci. Fis. Mat. Natur.* (8) **63**, 328–333.

Srivastava, H. M. (1977a). A Watsonian theorem for multiple series. *Rend. Sem. Mat. Univ. Padova* **58**, 241–245.

Srivastava, H. M. (1978). An operational representation for the Laguerre polynomials. *Nordisk Mat. Tidskr.* **25**/**26**, 137–138.

Srivastava, H. M. (1978a). Certain theorems on bilateral generating functions. *Indian J. Math.* **20**, 9–13.

Srivastava, H. M. (1978b). A note on certain summation theorems for multiple hypergeometric series. *Simon Stevin* **52**, 97–109.

Srivastava, H. M. (1979). A note on certain identities involving generalized hypergeometric series. *Nederl. Akad. Wetensch. Proc. Ser. A* **82** (*Nederl. Akad. Wetensch. Indag. Math.* **41**), 191–201.

Srivastava, H. M. (1979a). The Laplace transform of the modified Bessel function of the second kind. *Publ. Inst. Math.* (*Beograd*) (*N.S.*) **26** (**40**), 273–282.

Srivastava, H. M. (1979b). Some generalizations of Carlitz's theorem. *Pacific J. Math.* **85**, 471–477.

Srivastava, H. M. (1980). Remarks on some expansions in series of G and H functions. *Nederl. Akad. Wetensch. Proc. Ser. A* **83** (*Nederl. Akad. Wetensch. Indag. Math.* **42**), 83–91.

Srivastava, H. M. (1980a). Some bilateral generating functions for a certain class of special functions. I and II. *Nederl. Akad. Wetensch. Proc. Ser. A* **83** (*Nederl. Akad. Wetensch. Indag. Math.* **42**), 221–246.

Srivastava, H. M. (1980b). Quelques fonctions génératrices pour une classe de polynômes définis par une formule générale du type de Rodrigues. *C. R. Acad. Sci. Paris. Sér. A-B* **290**, A993–A996.

Srivastava, H. M. (1980c). Some generating functions for Laguerre and Bessel polynomials. *Bull. Inst. Math. Acad. Sinica* **8**, 571–579.

Srivastava, H. M. (1980d). Une généralisation de certains théorèmes de Brown, Calvez et Génin concernant les fonctionnes génératrices. *Boll Un. Mat. Ital.* (5) **17A**, 498–502.

Srivastava, H. M. (1981). Some generalizations of Carlson's identity. *Boll. Un. Mat. Ital.* (5) **18A**, 138–143.

Srivastava, H. M., and Brenner, J. L. (1974). Bounds for Jacobi and related polynomials derivable by matrix methods. *J. Approx. Theory* **12**, 372–377.

Srivastava, H. M., and Buschman, R. G. (1974). Some convolution integral equations. *Nederl. Akad. Wetensch. Proc. Ser. A* **77** (*Nederl. Akad. Wetensch. Indag. Math.* **36**), 211–216.

Srivastava, H. M., and Buschman, R. G. (1975). Some polynomials defined by generating relations. *Trans. Amer. Math. Soc.* **205**, 360–370.

Srivastava, H. M., and Buschman, R. G. (1976). Mellin convolutions and H-function transformations. *Rocky Mountain J. Math.* **6**, 331–343.

Srivastava, H. M., and Buschman, R. G. (1977). "Convolution Integral Equations with Special Function Kernels." Wiley (Halsted), New York.

Srivastava, H. M., and Buschman, R. G. (1977a). Adendum to "Some polynomials defined by generating relations." *Trans. Amer. Math. Soc.* **226**, 393–394.

Srivastava, H. M., and Daoust, M. C. (1969). On Eulerian integrals associated with Kampé de Fériet's function. *Publ. Inst. Math.* (*Beograd*) (*N.S.*) **9**(**23**), 199–202.

Srivastava, H. M., and Daoust, M. C. (1969a). Certain generalized Neumann expansions associated with the Kampé de Fériet function. *Nederl. Akad. Wetensch. Proc. Ser. A* **72** (*Nederl. Akad. Wetensch. Indag. Math.* **31**), 449–457.

Srivastava, H. M., and Daoust, M. C. (1972). A note on the convergence of Kampé de Fériet's double hypergeometric series. *Math. Nachr.* **53**, 151–157.

Srivastava, H. M., and Exton, H. (1973). On Laplace's linear differential equation of general order. *Nederl. Akad. Wetensch. Proc. Ser. A* **76** (*Nederl. Akad. Wetensch. Indag. Math.* **35**), 371–374.

Srivastava, H. M., and Exton, H. (1974). A transformation formula relating two Lauricella functions. *Atti Accad. Naz. Lincei Rend. Cl. Sci. Fis. Mat. Natur.* (8) **56**, 38–42.

Srivastava, H. M., and Exton, H. (1979). A generalization of the Weber–Schafheitlin integral. *J. Reine Angew. Math.* **309**, 1–6.

Srivastava, H. M., and Joshi, C. M. (1974). Integral representation for the product of a class of generalized hypergeometric polynomials. *Acad. Roy. Belg. Bull. Cl. Sci.* (5) **60**, 919–926.

Srivastava, H. M., and Lavoie, J.-L. (1975). A certain method of obtaining bilateral generating functions. *Nederl. Akad. Wetensch. Proc. Ser. A* **78** (*Nederl. Akad. Wetensch. Indag. Math.* **37**), 304–320.

Srivastava, H. M., and Panda, R. (1973). Some operational techniques in the theory of special functions. *Nederl. Akad. Wetensch. Proc. Ser. A* **76** (*Nederl. Akad. Wetensch. Indag. Math.* **35**), 308–319.

Srivastava, H. M., and Panda, R. (1973a). Some hypergeometric transformations involving Horn's functions H_3. *Rend. Mat.* (6) **6**, 859–864.

Srivastava, H. M., and Panda, R. (1975). Some analytic or asymptotic confluent expansions for functions of several variables. *Math. Comput.* **29**, 1115–1128.

Srivastava, H. M., and Panda, R. (1975a). Some expansion theorems and generating relations for the *H* function of several complex variables. *Comment. Math. Univ. St. Paul.* **24**, Fasc. 2, 119–137.

Srivastava, H. M., and Panda, R. (1975b). On the unified presentation of certain classical polynomials. *Boll. Un. Mat. Ital.* (4) **12**, 306–314.

Srivastava, H. M., and Panda, R. (1976). An integral representation for the product of two Jacobi polynomials. *J. London Math. Soc.* (2) **12**, 419–425.

Srivastava, H. M., and Panda, R. (1976a). Some bilateral generating functions for a class of generalized hypergeometric polynomials. *J. Reine Angew. Math.* **283/284**, 265–274.

Srivastava, H. M., and Panda, R. (1976b). Expansion theorems for the *H* function of several complex variables. *J. Reine Angew. Math.* **288**, 129–145.

Srivastava, H. M., and Panda, R. (1976c). Some expansion theorems and generating relations for the *H* function of several complex variables. II. *Comment. Math. Univ. St. Paul.* **25**, Fasc. 2, 167–197.

Srivastava, H. M., and Panda, R. (1978). Certain multidimensional integral transformations. I and II. *Nederl. Akad. Wetensch. Proc. Ser. A* **81** (*Nederl. Akad. Wetensch. Indag. Math.* **39**), 118–144.

Srivastava, H. M., and Panda, R. (1979). Some multiple integral transformations involving the *H*-function of several variables. *Nederl. Akad. Wetensch. Proc. Ser. A* **82** (*Nederl. Akad. Wetensch. Indag. Math.* **41**), 353–362.

Srivastava, H. M., and Panda, R. (1979a). A note on certain results involving a general class of polynomials. *Boll. Un. Mat. Ital.* (5) **16A**, 467–474.

Srivastava, H. M., and Panda, R. (1980). New generating functions involving several complex variables. *Rend. Circ. Mat. Palermo* (2) **29**, 23–41.

Srivastava, H. M., and Pathan, M. A. (1979). Some bilateral generating functions for the extended Jacobi polynomials. *Comment. Math. Univ. St. Paul.* **28**, Fasc. 1, 23–30.

Srivastava, H. M., and Pathan, M. A. (1980). Some bilateral generating functions for the extended Jacobi polynomials. II. *Comment. Math. Univ. St. Paul.* **29**, Fasc. 2, 105–114.

Srivastava, H. M., and Raina, R. K. (1981). New generating functions for certain polynomial systems associated with the H-functions. *Hokkaido Math. J.* **10**, 34–45.

Srivastava, H. M., and Singh, N. P. (1981). The integration of certain products of the multivariable H-function with a general class of polynomials. (Abst.). *Amer. Math. Soc. Abstracts* **2**, 404–405.

Srivastava, H. M., and Singhal, J. P. (1971). A class of polynomials defined by generalized Rodrigues' formula. *Ann. Mat. Pura Appl.* (*4*) **90**, 75–85.

Srivastava, H. M., and Singhal, J. P. (1972a). On a class of generalized hypergeometric distributions. *Jñānabha Sect. A* **2**, 1–9.

Srivastava, H. M., and Singhal, J. P. (1972b). A unified presentation of certain classical polynomials. *Math. Comput.* **26**, 969–975.

Srivastava, H. M., and Singhal, J. P. (1973). New generating functions for Jacobi and related polynomials. *J. Math. Anal. Appl.* **41**, 748–752.

Srivastava, H. M., and Srivastava, A. (1978). A new class of orthogonal expansions for the H-function of several complex variables. *Comment. Math. Univ. St. Paul.* **27**, Fasc. 1, 59–69.

Srivastava, H. M., Gupta, K. C., and Handa, S. (1975). A certain double integral transformation. *Nederl. Akad. Wetensch. Proc. Ser. A* **78** (*Nederl. Akad. Wetensch. Indag. Math.* **37**), 402–406.

Srivastava, H. M., Lavoie, J.-L., and Tremblay, R. (1979). The Rodrigues type representations for a certain class of special functions. *Ann. Mat. Pura Appl.* (4) **119**, 9–24.

Srivastava, H. M., Goyal, S. P., and Agrawal, R. K. (1981). Some multiple integral relations for the H-function of several variables. *Bull. Inst. Math. Acad. Sinica* **9**, 261–277.

Srivastava, H. M., Koul, C. L., and Raina, R. K. (1981a). A class of convolution integral equations. Report No. DM-243-IR, Univ. of Victoria.

Srivastava, R. (1980). An expansion formula for the generalized Lauricella functions. *Bull. Inst. Math. Acad. Sinica* **8**, 95–99.

Srivastava, R. (1980a). New generating functions for a class of polynomials. *Boll. Un. Mat. Ital.* (5) **17A**, 183–186.

Srivastava, R. (1981). Definite integrals associated with the H-function of several complex variables. *Comment. Math. Univ. St. Paul.* **30**, Fasc. 2, 125–129.

Stacy, E. W. (1962). A generalization of the Gamma distribution. *Ann. Math. Statist.* **33**, 1187–1192.

Stephan, F. F. (1958). Two queues under preemptive priority with Poisson arrival and service rates. *Oper. Res.* **6**, 399–418.

Stirzaker, D. (1980). Inheritance. *Adv. in Appl. Probab.* **12**, 574–590.

Stone, L. D. (1969). On the distribution of the supremum functional for semi-Markov processes with continuous state spaces. *Ann. Math. Statist.* **40**, 844–853.

Stone, L. D. (1970). Distribution of time above a threshold for semi-Markov jump processes. *J. Math. Anal. Appl.* **30**, 576–591.

Stone, L. D., Belkin, B., and Snyder, M. A. (1970). Distribution of time above a threshold for Markov processes. *J. Math. Anal. Appl.* **30**, 448–470.

Sud, A. R., and Sud, K. (1978). Analytic continuations of the Lauricella function. *J. Math. Phys.* **19**, 2485–2490.

Sud, K., and Wright, L. E. (1976). A new analytic continuation of Appell's hypergeometric series F_2. *J. Math. Phys.* **17**, 1719–1721.

Sudbury, A. (1981). The expected population size in a cell-size-dependent branching process. *J. Appl. Probab.* **18**, 65–75.

Sweet, A. L., and Hardin, J. C. (1970). Solutions for some diffusion processes with two barriers. *J. Appl. Probab.* **7**, 423–431.

Swensson, O. (1958). An approach to a class of queuing problems. *Oper. Res.* **6**, 276–292.

Szegö, G. (1975). "Orthogonal Polynomials," 4th ed., Amer. Math. Soc. Colloq. Publ. 23. Amer. Math. Soc., Providence, Rhode Island.

Takács, L. (1957a). On certain sojourn time problems in the theory of stochastic processes. *Acta Math. Acad. Sci. Hungar.* **8**, 169–191.

Takács, L. (1957b). On limiting distributions concerning a sojourn time problem. *Acta Math. Acad. Sci. Hungar.* **8**, 279–294.

Takács, L. (1961). The transient behavior of a single server queuing process with a Poisson input. *Proc. Berkeley Symp. Math. Statist. Probab., 4th*, Vol. II, pp. 535–567, Univ. of California Press, Berkeley, California.

Takács, L. (1962). "Introduction to the Theory of Queues." Oxford Univ. Press, London and New York.

Takács, L. (1964). Occupation time problems in the theory of queues. *Oper. Res.* **12**, 753–767.

Takács, L. (1966). The distribution of the content of a dam when the input process has stationary independent increments. *J. Math. Mech.* **15**, 101–112.

Takács, L. (1974). Occupation time problems in the theory of queues. "Mathematical Methods in Queueing Theory" (*Proc. Conf. Western Michigan Univ.*), pp. 91–131. Springer-Verlag, Berlin and New York.

Takács, L. (1976). On the busy periods of single-server queues with Poisson input and general service times. *Oper. Res.* **24**, 564–571.

Tandon, O. P. (1980a). Finite summation formulas for the H-function of n variables. *Pure Appl. Math. Sci.* **11**, 23–30.

Tandon, O. P. (1980b). Contiguous relations for the H-function of n variables. *Indian J. Pure Appl. Math.* **11**, 321–325.

Tata, M. N. (1969). On outstanding values in a sequence of random variables. *Z. Wahrscheinlichkeitstheorie Verw. Gebiete* **12**, 9–20.

Tomko, J. (1972). The rate of convergence in limit theorems for service systems with finite queue capacity. *J. Appl. Probab.* **9**, 87–102.

Toscano, L. (1972). Sui polinomi ipergeometrici a più variabili del tipo F_D di Lauricella. *Matematiche (Catania)* **27**, 219–250.

Tukey, J. W., and Wilks, S. S. (1946). Approximation of the distribution of the product of Beta variables by a single Beta variable. *Ann. Math. Statist.* **17**, 318–324.

Uppuluri, V. R. R., and Blot, W. J. (1974). Asymptotic properties of the number of replications of a paired comparison. *J. Appl. Probab.* **11**, 43–52.

van Danzig, D. (1947). Sur la méthode des fonctions génératrices. *Colloq. Internat. du C.N.R.S.* **13**, 29–45.

van Doorn, E. A. (1981). The transient state probabilities for a queuing model where potential customers are discouraged by queue length. *J. Appl. Probab.* **18**, 499–506.

Wall, H. S. (1948). "Analytic Theory of Continued Fractions." Van Nostrand, New York.

Watson, G. N. (1944). "A Treatise on the Theory of Bessel Functions," 2nd ed. Cambridge Univ. Press, London and New York.

Weiss, G. H. (1965). On the spread of epidemics by carriers. *Biometrics* **21**, 481–490.

White, H., and Christie, L. S. (1958). Queuing with preemptive priorities or with breakdown. *Oper. Res.* **6**, 79–95.

White, J. A., Schmidt, J. W., and Bennett, G. K. (1975). "Analysis of Queueing Systems." Academic Press, New York.

Whittaker, E. T., and Watson, G. N. (1952). "A Course of Modern Analysis." Cambridge Univ. Press, London and New York.

Widder, D. V. (1941). "The Laplace Transform." Princeton Univ. Press, Princeton, New Jersey.

Wilks, S. S. (1932). Certain generalizations in the analysis of variance. *Biometrika* **24**, 471–494.

Williams, T., and Ancker, C. J., Jr. (1963). Stochastic duels. *Oper. Res.* **11**, 803–817.

Winkler, R. L., Roodman, G. M., and Britney, R. R. (1973). The determination of partial moments. *Management Sci.* **19**, 290–296.

Woodside, C. M., Pagurek, B., and Newell, G. F. (1980). A diffusion approximation for correlation in queues. *J. Appl. Probab.* **17**, 1033–1047.

Wright, E. M. (1933). On the coefficients of power series having exponential singularities. *J. London Math. Soc.* **8**, 71–79.

Wright, E. M. (1934). The asymptotic expansion of the generalized Bessel function. *Proc. London Math. Soc.* (2) **38**, 257–270.

Wright, E. M. (1935). The asymptotic expansion of the generalized hypergeometric function. *J. London Math. Soc.* **10**, 286–293.

Wright, E. M. (1940). The asymptotic expansion of the generalized hypergeometric function. *Proc. London Math. Soc.* (2) **46**, 389–408.

Wright, S. (1937). The distribution of gene frequencies in populations. *Proc. Nat. Acad. Sci. U.S.A.* **23**, 307–320.

Yadin, M. (1965). On a Random Walk in the Positive Orthant of the Plane and a Study of Queueing System with Alternating Priorities. D.Sc. Thesis, Technion – Israel Institute of Technology, Haifa, Israel.

Yadin, M. (1970). Queuing with alternating priorities, treated as random walk on the lattice in the plane. *J. Appl. Probab.* **7**, 196–218.

Yeo, G. F. (1974). A finite dam with exponential release. *J. Appl. Probab.* **11**, 122–133.

Yeo, G. F. (1975). A finite dam with variable release rate. *J. Appl. Probab.* **12**, 205–211.

Index

Numbers in italics refer to the pages on which the complete references are listed.

C

M

W

Y

Z